Jordan Howard Sobel has long been recognized as an important figure in philosophical discussions of rational decision. He has done much to help formulate the concept of causal decision theory.

In this volume of essays, Sobel explores the Bayesian idea that rational actions maximize expected values, where an action's expected value is a weighted average of its agent's values for its possible total outcomes. Newcomb problems and the Prisoners' Dilemma are discussed, and Allais-type puzzles are viewed from the perspective of causal world Bayesianism. The author establishes principles for distinguishing options in decision problems, and studies ways in which perfectly rational causal maximizers can be capable of resolute choices. Several of the essays concern games, with interacting ideally rational and well-informed maximizing rationality. Sobel also views critically David Gauthier's revisionist ideas about maximizing rationality.

This collection will be a desideratum for anyone working in the field of rational choice theory, whether in philosophy, economics, political science, psychology, or statistics.

"Howard Sobel's work in decision theory is certainly among the most important, interesting, and challenging that is being done by philosophers."
– David Gauthier

Taking chances

**Cambridge Studies in Probability,
Induction, and Decision Theory**

General editor: Brian Skyrms

Advisory editors: Ernest W. Adams, Ken Binmore, Jeremy Butterfield, Persi Diaconis, William L. Harper, John Harsanyi, Richard C. Jeffrey, Wolfgang Spohn, Patrick Suppes, Amos Tversky, Sandy Zabell

This series is a forum for the most innovative and challenging work in the theory of rational decision. It focuses on contemporary developments at the interface between philosophy, psychology, economics, and statistics. The series addresses foundational theoretical issues, often quite technical ones, and therefore assumes a distinctly philosophical character.

Other titles in the series
Ellery Eells, *Probabilistic Causality*
Richard Jeffrey, *Probability and the Art of Judgment*
Robert Koons, *Paradoxes of Strategic Rationality*
Cristina Bicchieri and Maria Luisa Dalla Chiara (eds.), *Knowledge, Belief, and Strategic Interaction*
Patrick Maher, *Betting on Theories*
Cristina Bicchieri, *Rationality and Coordination*

Forthcoming
Patrick Suppes and Mario Zanotti, *Foundations of Probability with Applications*
Clark Glymour and Kevin Kelly (eds.), *Logic, Confirmation, and Discovery*

Taking chances
Essays on rational choice

Jordan Howard Sobel
University of Toronto

CAMBRIDGE UNIVERSITY PRESS
Cambridge, New York, Melbourne, Madrid, Cape Town, Singapore, São Paulo

Cambridge University Press
The Edinburgh Building, Cambridge CB2 8RU, UK

Published in the United States of America by Cambridge University Press, New York

www.cambridge.org
Information on this title: www.cambridge.org/9780521416351

© Cambridge University Press 1994

This publication is in copyright. Subject to statutory exception
and to the provisions of relevant collective licensing agreements,
no reproduction of any part may take place without the written
permission of Cambridge University Press.

First published 1994
This digitally printed version 2007

A catalogue record for this publication is available from the British Library

Library of Congress Cataloguing in Publication data
Sobel, Jordan Howard.
Taking chances: essays on rational choice / Jordan Howard Sobel.
p. cm. – (Cambridge studies in probability, induction, and
decision theory)
Includes bibliographical references and index.
ISBN 0-521-41635-3 (hc)
1. Bayesian statistical decision theory. 2. Statistical decision.
I. Title. II. Series.
QA279.5.S635   1994
003′.56 – dc20   93-4800
                          CIP

ISBN 978-0-521-41635-1 hardback
ISBN 978-0-521-03898-0 paperback

Contents

Preface   *page* ix

Part I   World Bayesianism

1 Utility theory and the Bayesian paradigm   3
   0. Introduction   3
   1. Utility theory   4
   2. The problem   5
   3. A reaction: Insist on complete basic alternatives   9
   4. Utility theory within a more general theory   13
   5. Summing up   26

Part II   Problems for evidential decision theory

2 Newcomblike problems   31
   0. Introduction   31
A. *Analysis and variety*   32
   1. An analysis of Newcomblike problems   32
   2. The variety of Newcomblike problems   33
B. *Objections*   36
   1. These are problems for space cadets   36
   2. Deliberators cannot view their choices as signs of prior conditions   38
   3. Agents who believe in causes for their actions   41
   4. Agents who believe in causes for their choices   45
   5. Newcomblike problems are not possible for ideally rational agents   48
   6. Conclusions   58
   Appendix I: Statements of some Newcomblike problems   60
   Appendix II: Further structure for "Brian Skyrms's Uncle Albert" and "How do you spell relief?"   63
   Appendix III: Dominance problems   65
   Appendix IV: Further structures for Newcomb's Problem   66
   Appendix V: Ratificationism   68

3	**Not every prisoners' dilemma is a Newcomb problem**	77
	0. Introduction	77
	1. Near-certainty Newcomb problems	78
	2. Near-certainty prisoners' dilemmas	80
	3. Near-certainty Newcomb problems and prisoners' dilemmas compared	80
	4. Conclusions	86
4	**Some versions of Newcomb's Problem are prisoners' dilemmas**	89
	0. Introduction	89
	1. Preliminaries	90
	2. Some, but not all, versions of Newcomb's Problem are prisoners' dilemmas	92
	3. Further questions	94
	4. Concluding remarks	96
5	**Infallible predictors**	100
	0. Introduction	100
	1. Predictors who are certainly incapable of error	101
	2. Predictors who are certainly unerring	105
	3. Conclusions	111
6	**Kent Bach on good arguments**	119
7	**Maximizing and prospering**	126
	0. Introduction	126
	1. Gauthier's pragmatism	127
	2. Implications for Newcomb problems	128
	3. Possibly disturbing aspects of these implications	130
	4. Judgmental conclusions	132
	Appendix: Socializing sentiments	136

PART III CAUSAL DECISION THEORY

8	**Notes on decision theory: Old wine in new bottles**	141
	0. Introduction: Bayesian decision theory	141
	1. Jeffrey's logic of decision	143
	2. The popcorn problem	145
	3. A causal decision theory	152
	4. Conclusions	164
	Appendix: Demonstrations of partition theorems	165
9	**Partition theorems for causal decision theories**	174
	0. Introduction	174
	1. A causal decision theory	175

	2. Causal decision theories need partition theorems	178
	3. Two partition theorems	181
	4. Fishburn and conditional acts	185
	5. Armendt on conditional preferences	187
	6. *Partitions for U* and *Exclusive Partitions* compared	191
	7. Uses of partition theorems	192
	Appendix: A theorem for sufficiently fine partitions	195
10	**Expected utilities and rational actions and choices**	197
	0. Introduction	197
	1. Definitions, assumptions, and restrictions	198
	2. The ideal stability of rational decisions	200
	3. Principles that apply tests of expected utility to actions	202
	4. A principle that confines tests of expected utility to choices	212
	5. A principle for agents who are sure they can make mixed choices	215
	6. Conclusion	215
11	**Maximization, stability of decision, and actions in accordance with reason**	218
	0. Introduction	218
	1. Positions	218
	2. Arguments and perspectives	227
	3. The maximization and stability theory: Restatement and elaboration	233
12	**Useful intentions**	237
	0. Introduction	237
	1. Senses in which the rational cannot intend irrational actions	238
	2. Forming and adopting intentions	240
	3. Magical bootstrapping, and rational intentions and preferences	243
	4. Rational adoption of intentions to do things that would otherwise be irrational	246
	5. Conclusion	251
	6. Postscript	252

PART IV   INTERACTING CAUSAL MAXIMIZERS

13	**The need for coercion**	257
	0. Introduction	257
	1. The hyperrational community	258

	2. The community's need for coercion	262
	3. The individual's need for coercion	270
	Appendix: The Farmer's Dilemma and mutual trust	276
14	**Hyperrational games**	283
	0. Introduction	283
	1. The concept of a hyperrational normal-form game	283
	2. Resolutions of hyperrational games	294
	Necessary conditions for, or limitations on, resolutions	294
	Sufficient conditions for, and ways to, resolutions	303
	3. Problems for members of hyperrational communities	313
	Games in which they would not do well	313
	Games in which they could not do anything at all	315
	The significance of these problems	324
	Appendix: Lewis domination	325
15	**Utility maximizers in iterated prisoners' dilemmas**	330
	0. Introduction	330
	1. The defeat of utility maximizers in iterated prisoners' dilemmas	331
	2. The significance of this defeat	338
16	**Backward induction arguments: A paradox regained**	345
	0. Introduction	345
	1. Indicative rationality and belief premises	346
	2. A backward argument for defection	348
	3. Subjunctive rationality and belief premises	351
	4. Interlude: Defense of common-knowledge premises	358
	5. The paradox and its solution	363
References		367
Index of names		375

Preface

The Bayesian idea with which the essays of this volume are concerned is that rational actions maximize agents' expected values. I understand an agent's expected value for an action to be a weighted average of his values for its possible outcome, where the weight for the value of a particular possible outcome is his probability for this outcome given the action.

The essay in Part I features an interpretation of "outcomes" according to which an action's possible outcomes detail all choice-relevant aspects of its possible worlds: "Outcomes" in expected values are "worlds for practical purposes." I explain how interpreting outcomes in this comprehensive manner leads to a theory that accommodates patterns of preference such as those framed by Maurice Allais that can embarrass less commodious Bayesianisms such as those of Leonard Savage and of R. Duncan Luce and Howard Raiffa.

The essays in Part II study problems that challenge so-called evidential decision theories, such as Richard Jeffrey's, that interpret "the probability for an outcome given an action" as a conditional probability that reflects the possible evidential bearing for the agent of this action on that outcome, a probability that would make an action's expected value a measure of the welcomeness to the agent of information that it will take place. I suggest in Chapter 2 that such theories – whether or not "metatickle enhanced," as Ellery Eells would have them be – yield wrong answers in some Newcomblike problems, and give wrong reasons even when they get right answers. Challenges to the coherence of Newcomblike problems are examined in that chapter; relations of Newcomb problems and prisoners' dilemmas are discussed in Chapters 3 and 4, and in Chapter 5 the two-box prescription of causal decision theories is defended even for infallible-predictor Newcomb problems. Causal maximizing arguments are held to be good in a sense identified in Chapter 6, even when the agent is sure that the predictor has correctly predicted his choice; for even then, causal maximizing – which still prescribes the two-box choice – is certain to lead to a payoff as great as would any other kind of practical thinking. Chapter 7 responds to challenges to the intelligence of maximizers, especially causal maximizers, that can be found in the writings of David Gauthier: I oppose the charge that maximizers, since they would in some circumstances not do as well as, are thereby less rational than, certain other thinkers.

Part III explores so-called causal decision theories already in play in Part II, theories that would interpret "the probability for an outcome given an action" as something like the unconditional probability of this action's causing that outcome. Such theories would make an action's expected value a measure of the welcomeness, in view of the differences it figures to make, of the fact that it will take place. A theory is described in Chapter 8 that features subjective probabilities for "chancy objective conditionals"; in calculating expected values, this theory uses not probabilities of causal action–outcome "would" conditionals but, rather, probability-weighted averages of chances in action–outcome "chance" conditionals. An account is proposed in Chapter 10 of partitioning agents' options, and accounts are proposed in Chapters 8 and 9 of partitions of circumstances for applications of causal decision theories. Chapters 10 and 11 contain the suggestion that a rational action must not only maximize causal expected value; additionally, a decision for that action must, in a certain sense, be "ratifiable." These conditions are said to be individually necessary and jointly sufficient for an action's being, in a certain thin sense, *rational*. Chapter 12 explores the ways in which agents who are in this sense rational may be able to adopt useful intentions and make resolute choices, including intentions and choices to perform antecedently non-maximizing irrational actions.

Part IV deals with interactions of causal maximizers. I maintain in Chapters 13 and 14 that not even ideally rational and well-informed causal maximizers would invariably do well when interacting: There are situations in which they would do less well than other agents. I contend, further, that situations exist in which they, in their perfect and pure causal maximizing transparency, could no nothing at all. I maintain in Chapters 15 and 16 that perfectly rational and well-informed causal maximizers not only do less well than other agents in prisoners' dilemmas but do so even when they realize they are in ongoing relationships and will meet in a sequence of such dilemmas.

Debts are acknowledged with appreciation in notes to the essays. Special thanks are due to Włodek Rabinowicz and Willa Freeman-Sobel for their comments and criticisms. Essays were revised and a manuscript for this volume assembled at St. Andrews during Whitsunday term in 1992, when I was a fellow at the Centre for Philosophy and Public Affairs, to which I am grateful. This book is dedicated to my teachers and friends Richard Cartwright and David Falk.

I

World Bayesianism

1

Utility theory and the Bayesian paradigm

Abstract. A problem for utility theory – that it would have an agent who was compelled to play Russian roulette with one revolver or another, to pay as much to have a six-shooter with four bullets relieved of one bullet before playing with it as he would be willing to pay to have a six-shooter with two bullets emptied – is explained. A less demanding theory that does not have this problem, causal world Bayesianism, is described. This theory would have an agent maximize expected values of worlds in which his actions, complete with their risk dimensions, might take place. Utility theory is located within that theory as valid for agents who satisfy certain formal conditions: It is valid for agents that are in terms of that more general theory indifferent to certain dimensions of risks.

0. INTRODUCTION

Utility theory is characterized provisionally in Section 1, after which a problem for this theory – the Zeckhauser-Gibbard problem – is set out. A reaction to this problem that would have one insist on "complete basic alternatives" is then considered, and it is observed that it would make the theory inapplicable to preferences of some reasonable people if certain attitudes to risk are allowed to be reasonable. Next, a general theory based on complete alternatives – "practical worlds" – is formulated, and utility theory is established within it as valid for agents who satisfy certain conditions of risk neutrality. Consideration of Savage- and Raiffa-style arguments against common preferences in our problem is then resisted. Addenda take up implications for game theory and relations between utilities and values. A final "Summing Up" comments on the recent history of the Bayesian idea that would have rational choices and preferences be determined by weighted averages of values of possible outcomes.

© 1989 Kluwer Academic Publishers. Reprinted with revisions by permission of Kluwer Academic Publishers from *Theory and Decision* 26 (1989), pp. 263-93.

1. UTILITY THEORY

Let a *lottery* be a distribution of chances to states or lotteries; let a *lottery set* include precisely the members of some set of incompatible basic alternatives: every simple lottery with only basic prizes; every first-order compound lottery with only basic outcomes and simple lotteries as prizes; and, for every n, every nth-order compound lottery with only basic outcomes, simple lotteries, and less than nth-order compound lotteries as prizes. Let a *utility function* be an assignment of numbers to states and lotteries in a lottery set such that the number assigned to a lottery L is its expected value, that is, the weighted average of the numbers assigned to its possible prizes (whether these be lotteries or states) in which the weights employed are the chances L accords directly to these prizes.

Provisionally, we state utility theory, as a theory of ideally rational preferences, thus:

> An agent's preferences for mutually incompatible states are ideally rational only if (i) the agent has pairwise preferences for all members of a lottery set based on these states, and
> (ii) a utility function represents these preferences.

A somewhat more demanding principle would add: (iii) the agent prefers X from subset S (supposing a free choice among precisely the members of S) only if, for each X' in S other than X, he pairwise prefers X to X' or is pairwise indifferent regarding them. Applications of utility theory generally take for granted the satisfaction of this more demanding principle, but in what follows we are concerned only with the less demanding one.

Luce and Raiffa (1957) explain how relevant pairwise preferences between members of a lottery set might be made explicit, although not all relevant pairwise preferences can be made explicit because, when there is more than one basic alternative state, there are infinitely many lotteries to compare (Luce and Raiffa 1957, p. 15). The idea is that utility functions would represent certain conditional pairwise preferences. An agent who would make one of these explicit supposes that his choice is just between members of a certain pair (i.e., that he can and must choose one of them), and then chooses in imagination. He makes epistemic suppositions – suppositions that operate to shift his epistemic perspective by conditionalization.

For future reference, I note that if an agent has pairwise preferences for all members of a lottery set, and these preferences are represented by a utility function, then they satisfy the following familiar principles addressed to members of this set.

Transitivity of indifference. For any states or lotteries X, Y, and Z, if $X \approx Y$ and $Y \approx Z$ then $X \approx Z$.

Substitution. For any lotteries X and X' and states or lotteries Y and Y', let $Y \approx Y'$. Let X' be like X except that X' includes, instead of the chances X has for Y, like chances for Y'. Then $X \approx X'$.

The intended sense of this substitution principle is clear enough. One route to a more articulated statement would be by way of a formal language for utility theory with reference to which one could speak literally of substitutions of one term (state term or lottery term) for another within a lottery term.

Reduction. For any lotteries X and X', if ultimate chances (perhaps as revealed by computations) for basic alternatives are the same in X and X', then $X \approx X'$.

Here again things could be cleaner in a theory in which terms were distinguished from states and lotteries designated by them. Lottery sets could then be redefined as containing only basic alternatives and all lotteries with only these alternatives as prizes, and the simple/compound distinction would pertain not to lotteries themselves but only to lottery terms for them. And one could say that when one of the lottery terms X' comes from another X by syntactical operations corresponding to the computations alluded to in Reduction, then X is identical with X' and so of course $X \approx X'$. That arrangement would make plainer that Reduction, in contrast with Substitution, is not a substantive condition (cf. Jarrow 1987, p. 100).

2. The problem

What follows elaborates on a problem sketched by Allan Gibbard at a workshop in 1984. The numbers are his, as well as the main lines of the deduction. The problem is a "variant of a problem of Richard Zeckhauser's that Kahneman and Tversky (1979, p. 283) relate" (Jeffrey 1987, p. 227).

2.1. The case for the problem

Suppose that an agent will play Russian roulette either with a revolver whose six chambers contain two bullets, or with one whose six chambers contain four bullets. Suppose further that, whichever gun the agent plays with, he will have a choice whether to play with it as it stands, or to pay certain amounts to have a bullet or bullets removed before playing.

Suppose there is a maximum payment that he would be willing to make in order to have both bullets removed from the first gun – a payment sufficient to move the agent to say, "That much, but no more; if more I would rather take my chances." To make plausible the reasonableness of this assumption, let possible "payments" include not only transfers of assets but also liabilities to penalties, including liabilities to episodes of torture after he has played and if he survives. The payment here consists in the acceptance of a liability to torture. This payment is made whether or not the agent survives and the torture takes place.

Suppose the agent is rich (R) and that, after paying the most he is willing to pay to have both bullets removed from the first gun before playing with it, he would be poor (P). Suppose that he will be dead (D) if and only if he plays and loses. Let him be sure that his chances of dying equal the number of bullets in the gun, divided by 6. And let the agent be exactly indifferent between (i) paying the most he would be willing to pay to have two bullets removed from the first gun, thereby ensuring that he does not die but is rendered poor, and (ii) not paying anything and taking his chances with that gun as loaded:

(1) $\qquad (\sim D \& P) \approx [\tfrac{2}{6}(D \& R), \tfrac{4}{6}(\sim D \& R)]$.

To make plausible the idea that the most he would be willing to pay should strike this balance, we include as possible payments all *chances* of payments. Suppose also that this agent does not care whether he dies rich or poor:

(2) $\qquad (D \& R) \approx (D \& P)$.

Such indifference, while rare (since most people take an interest in their posthumous estates), is not unknown; even if it be ungenerous and inconsiderate, it is not necessarily unreasonable. Suppose now that this agent has pairwise preferences for the four compound states $(\sim D \& R)$, $(\sim D \& P)$, $(D \& R)$, and $(D \& P)$, and for all lotteries based on these states.

Our conditions are so far certainly consistent with the agent's being reasonable. To test the general idea that preferences are reasonable only if they are represented by a utility function, we suppose finally that this agent's preferences are represented by a utility function.

2.2. *The problem of this case*

The problem lies in what these conditions imply. By Substitution we have, given (1), the indifference

(3) $\quad [\tfrac{1}{2}(\sim D \& P), \tfrac{1}{2}(D \& P)] \approx [\tfrac{1}{2}[\tfrac{2}{6}(D \& R), \tfrac{4}{6}(\sim D \& R)], \tfrac{1}{2}(D \& P)]$;

by Substitution, given (2), we have the indifference

(4) $$[\tfrac{1}{2}[\tfrac{2}{6}(D \& R), \tfrac{4}{6}(\sim D \& R)], \tfrac{1}{2}(D \& P)] \approx [\tfrac{1}{2}[\tfrac{2}{6}(D \& R), \tfrac{4}{6}(\sim D \& R)], \tfrac{1}{2}(D \& R)]$$

From (3) and (4), by Transitivity we have the indifference

(5) $$[\tfrac{1}{2}(\sim D \& P), \tfrac{1}{2}(D \& P)] \approx [\tfrac{1}{2}[\tfrac{2}{6}(D \& R), \tfrac{4}{6}(\sim D \& R)], \tfrac{1}{2}(D \& R)].$$

And from (5), two applications of Reduction and Transitivity yield

(6) $$[\tfrac{3}{6}(\sim D \& P), \tfrac{3}{6}(D \& P)] \approx [\tfrac{4}{6}(D \& R), \tfrac{2}{6}(\sim D \& R)].$$

This last indifference (6) is the problem, as the following considerations show. The lottery to the left – $[\tfrac{3}{6}(\sim D \& P), \tfrac{3}{6}(D \& P)]$ – is the situation faced by the agent who pays to have *one* bullet removed from the *second* gun and who pays precisely his maximum amount to have *both* bullets removed from the *first* gun. Paying that much would make him poor – either not dead and poor, or dead and poor – after playing with the partially unloaded second gun. He would, on paying this amount, have a $\tfrac{3}{6}$ chance of ending up poor and alive and a $\tfrac{3}{6}$ chance of ending up poor and dead: having paid this amount, the agent would be poor, and one of the four bullets would be removed from the gun. Indifference (6) has on the right $[\tfrac{4}{6}(D \& R), \tfrac{2}{6}(\sim D \& R)]$, which is the situation faced by the agent who chooses not to pay and to accept a $\tfrac{4}{6}$ chance of dying rich. Indifference (6) says that the agent is indifferent between these situations. In other words,

(6*) The most this agent would be willing to pay to have both bullets removed from the first gun, supposing he were required to play with it, is exactly equal to the most he would be willing to pay to have one bullet removed from the second gun, supposing he were required to play with *it*.

This exact equality follows no matter how much, or how little, is the greatest amount the agent would be willing to pay to have both bullets removed from the first gun, provided only that

(i) his highest price is such that he would as soon pay it as not,
(ii) he is indifferent to the effects on his estate were he to die,
(iii) he has pairwise preferences for all relevant lotteries, and
(iv) *these preferences are represented by a utility function.*

Indifference (6) follows even if the agent is suicidal, so that the most he would be willing to pay is some "negative payment" that would be better described as "the least he would be willing to accept." For clarity in a suicidal case, one might interchange *R* and *P*; accepting a payment to have bullets removed would make a suicidal agent richer, not poorer.

2.3

Should (6*) be true of every agent who satisfies the conditions of the case other than the one being tested? *Must* it be true of any such agent, if he is rational in his preferences? Suppose it is settled that a person must play with the first gun. Couldn't it be reasonable for him to pay more to have both bullets removed from this first gun than he would be willing to pay to have only one bullet removed from the second gun if *it* were the gun with which he was required to play? Surely this variance in acceptable prices could be reasonable for someone who: was not eager to pay his highest price in the case of the first gun (condition (1)); was indifferent to his estate were he to die (condition (2)), so that he does not care whether he dies rich or poor; and had preferences for all relevant lotteries by pairs.

It can indeed seem that a variance in acceptable prices here would be not only possible but *mandatory* for most persons. It can seem that for anyone who had an interest in the outcome, it could not be reasonable to pay exactly as much to improve his chances for life from $2/6$ to $3/6$ as to improve his chances from $5/6$ to a certainty. But this strong conclusion cannot be maintained. Consider a person who wants to live. He might be willing without error or unreason to pay as much for the smaller improvement from $2/6$ to $3/6$ in chances for life, just because the smaller improvement in his chances for life would take place in circumstances where his chances of dying were in any event substantial, and he believes that "You can't take it with you." (Recall that our agent is of a can't-take-it-with-you turn of mind: he does not care whether he dies rich or dies poor.) It may be asked, "How *could* he be willing to pay as much to have just one bullet removed from the four-bullet gun as he would be willing to pay to have both removed from the two-bullet gun?"; one possible answer is that the money paid (or the risk incurred of torture) in the first case could be worth less to him (cf. Kahneman and Tversky 1979, p. 283). It evidently can be reasonable to pay as much to have just one bullet removed from the second gun as one would pay to have both removed from the first gun. (I owe recognition of this point to Włodek Rabinowicz.)

Although we can't say that the price for having two bullets removed from the first gun would, for *any* reasonable agent who wanted to live, be too high a price to pay for having just one bullet removed from the second gun, this price would surely be too high for *some* reasonable agents.

I believe that many reasonable agents who wanted to live would, contrary to indifference (6), prefer not to pay this price, and would persist in this preference even after thoughtful reflection, and that the reverse would prove true of many reasonable agents who wanted to die.

An argument for indifference (6) goes through, we may note, even for an agent who does not care whether he lives or dies. The most such an agent would be willing to pay to have the first gun emptied would be nothing, so that $P = R$ and thus (i) $(\sim D \& R) = (\sim D \& P)$. Furthermore, since he does not care whether he lives or dies, presumably (ii) $(\sim D \& P) \approx (D \& P)$. Indifference (6) is an easy consequence of (i) and (ii) and principles of utility theory. However, in this odd case in which agents satisfy our initial conditions, (6) says what one expects, and is not a problem for utility theory. The problem posed by indifference (6) is that it seems *not* to say what must be true of every reasonable agent, including of course every thoughtful and reflective agent who does care whether he lives or dies.

3. A REACTION: INSIST ON COMPLETE BASIC ALTERNATIVES

3.1

The problem, one may feel, is not with utility theory as a theory of ideally rational preferences, but only with that theory as provisionally stated; or, equivalently, with the over-hasty application of the theory in our case. Utility theory is perhaps properly applicable only to preferences for members of lottery sets founded on alternatives that are, relative to all relevant pairwise comparisons, *complete* with respect to things of interest to the agent. The suggestion with regard to our problem could be that its threat is only a prima facie one, since the four compound states $(\sim D \& R)$, $(\sim D \& P)$, $(D \& R)$, and $(D \& P)$ can be expected to be not relevantly complete nor the "real alternatives" in the situation (Luce and Raiffa 1957, p. 28). For example, it certainly seems that for many agents it would *not* be all the same whether they lived poor, $(\sim D \& P)$, when they had a choice between this for sure and taking the chance $[\frac{2}{6}(D \& R), \frac{4}{6}(\sim D \& R)]$, or lived poor after taking the chance $[\frac{3}{6}(\sim D \& P), \frac{3}{6}(D \& P)]$ when they could instead have taken the chance $[\frac{4}{6}(D \& R), \frac{2}{6}(\sim D \& R)]$. It seems that reasonable agents' attitudes toward the four basic outcomes of the problem can be sensitive to shapes of lotteries from which these alternatives might issue, and to shapes of alternative lotteries. But then, according to the present idea, utility theory cannot be tested by applications to preferences of such agents for lotteries based on our four relatively simple compound states; by hypothesis, these states would not be complete for such agents.

3.2

Perhaps utility theory should be addressed only to lottery sets founded on complete basic alternatives that are (in the way just now indicated) complete, and its provisional statement amended thus:

> An agent's preferences for any mutually incompatible states *that are complete with respect to all pairs of members of a lottery set based on them* are ideally rational only if he has pairwise preferences for all members of a lottery set based on these states, and these pairwise preferences are represented by a utility function.

If this is right then the problem in our case is with its over-hasty application of utility theory, though by this I mean not merely that the application discussed would be improper for some persons but that, for at least some reasonable persons, utility theory would be *quite inapplicable.* The theory as amended is applicable only if all lotteries founded on certain basic alternatives are *internally coherent,* and are by pairs *coherently comparable.* But it seems that for some rational agents and situations there may be no set of basic alternatives satisfying these logical conditions and satisfying also the condition that its members are complete (or fully specific) with respect to all things of interest to the agent in pairwise preferences for lotteries based on them.

Suppose, for example, that it matters to an agent whether or not A obtains as an outcome of a lottery in which A and B have chances of $\frac{1}{2}$. Let A' be the version of A in which it is an outcome of such a lottery. Then neither $[\frac{1}{3}(A'), \frac{2}{3}(B)]$ nor (assuming that C does not entail B) $[\frac{1}{2}(A'), \frac{1}{2}(C)]$ is internally coherent. In contrast, both $[\frac{1}{2}(A'), \frac{1}{2}(B)]$ and (understanding B' similarly to A') $[\frac{1}{2}(A'), \frac{1}{2}(B')]$ are internally coherent. Suppose next that it matters to an agent whether or not basic alternative A results in a situation in which he can choose either A or a lottery L in which A is not a possible outcome. Let A'' be A in such a situation – that is, A when lottery L could have been chosen (i.e., "A instead of L"). Then, for $L' \neq L$, A'' and L' are for this agent not coherently comparable, for it is not logically possible for him to have a choice between just A'' and L' (as A'' is open to choice only when L is), and so he cannot (for purposes of a shift of his epistemic perspective by conditionalization) suppose that his choice is between just A' and L' and then choose in imagination. Objection: "But these are not reasonable attitudes; it should not matter by what means one obtains A, or what one's alternatives are." Response: "Oh?"

Conditions for lottery sets of comparable members can be expected to clash with the requirement that basic alternatives be complete, when an agent's preferences "reflect intrinsically comparative views of payoffs" (Jeffrey 1987, p. 225). This clash will occur when preferences do that in a certain manner: It will happen when an agent's preferences are informed by certain attitudes toward the shapes of lotteries and alternatives – certain attitudes that are not derivative of how he thinks he would feel afterwards (e.g., regretful, chagrined, delighted) if he were to win or lose. (Recall that, in the problem, the agent can be supposed to know that he will not feel regret if he loses – that he will not feel anything.)

It may be useful to recall a distinction. In one case, an agent realizes that if he were to run a risk and lose then he would regret running the risk and be bothered, but *not* because he would think it had been a mistake (bothersome feelings of regret aside). Here we have a "minimalist" interpretation of the idea of a preference's "reflecting an intrinsically comparative view of payoffs" (Jeffrey 1987, pp. 225-6). In a second case, an agent realizes that if he were to run some risk and lose then he would regret having run it and also be bothered because he would realize that it *had* been a mistake to run it (bothersome feelings of regret again aside). Suppose an agent thought that taking two aspirins would ensure that he would in no case experience regret. He might reasonably take them and run the risk in the first case (depending on his view of the merits of the risk, threats of regret aside), but perhaps not in the second. His aversion in the first case is based partly on matters extrinsic to the risk – on a probable, and conceivably blockable, consequence of running it. In the second case his aversion can, for all that has been said, be entirely intrinsic and based on the risk's nature, all merely possible consequences of running it quite aside (cf. Sobel 1988c, p. 542, n. 6). Whether or not it is irrational to be prone to regrets of the first kind, they do not make theoretical problems; extrinsic psychological effects can enter into payoffs of all possible lotteries, and are consistent with all possible alternative pairings of lotteries. In contrast, regrets of the second kind reflect intrinsically comparative views of payoffs interpreted in nonminimalist ways, views that can create problems for the applicability of utility theory.

3.3

It has been said that when applying utility theory to a case "it may be *necessary* to use a richer set of basic alternatives in order for [the theory] to be ... valid" (Luce and Raiffa 1957, p. 29, emphasis added). The problem I am pressing is that it may not always be *possible* to enrich an initial set of basic alternatives in all relevant ways. Making basic alternatives complete with respect to things of practical interest to a given rational agent, can, depending on what things interest him, be inconsistent with making all lotteries that would be founded on refined alternatives both possible and pairwise comparable.

Raiffa has implied that for certain individuals in Allais's experimental situation it may be necessary to replace the money consequences in the lotteries by more fully specified ones, but that this is something we can always do: "we can introduce new consequences" (Raiffa 1968, p. 86). The issue on which I am expressing a negative opinion is whether we *can* do this always for rational agents, and do it in a way consistent with the framework requirement that all lotteries with properly enriched consequences be possible and comparable (pp. 369–70 in Krantz et al. 1971).[1] This inconsistency would, I think, obtain in our problem for some who cared whether they lived or died, and would obtain in the end even on ideally thorough thought and reflection.[2]

Addendum: Implications for game theory

Let the *strong principle* be that. An agent's pairwise preferences for any mutually incompatible states and all lotteries based on them are ideally rational only if they are represented by a utility function. Parts of game theory that feature randomizations designed to enlarge strategy and outcome spaces, if these parts claim to be relevant to all rational agents, suppose this strong and possibly false principle.

Pure strategies in a game determine pure-strategy interactions that, viewed as basic alternatives, may not be complete for all agents with respect to all lotteries based on them. Even so, it is standard in game theory to compute values for interactions involving individual mixed strategies by the rule of expected values applied to the lotteries on pure-strategy interactions that they determine. The minimax theorem for mixed strategy extensions of two-person zero-sum finite pure-strategy games, and Nash's theorem for the existence of an equilibrium in the mixed strategy extension of any n-person finite pure-strategy game, both depend on values of mixed strategy interactions that are not only functions of values of pure-strategy interactions but also linear functions according to the rule of expected utility.

A *bargaining problem* consists (in part) of a convex set of n-tuples representing the values to participants of possible bargains. (See Luce and Raiffa 1957, pp. 116–17; Roth 1979.) *Convexity* is secured if it is assumed that possible bargains include all randomizations over possible bargains, and if it is assumed that participants' pairwise preferences between possible bargains and all lotteries based upon them are represented by utility functions. Convexity is also secured for a bargaining problem if it is based on a cooperative game in the sense that possible bargains correspond to all possible "joint randomized strategies" (or lotteries with pure interactions as their outcomes), where these joint strategies are evaluated according to the rule of expected values (cf. Roth 1979, p. 4). Convexity is important to the uniqueness of Nash solutions to bargaining problems (Luce and Raiffa 1957, p. 126, n. 4) and is also important to the optimality of Kalai–Smorodinsky solutions (Roth 1979, pp. 99ff).

David Gauthier writes:

> The general theory of rational bargaining is underdeveloped territory. Whether there are principles of rational bargaining with the same context-free universality of application as the principle of expected utility-maximization has been questioned, notably by Alvin Roth. . . . Undaunted by Roth's scepticism . . . we shall outline our own theory. (Gauthier 1986, p. 129)

My claim is that even if there are principles of rational bargaining that apply as widely as the aforementioned strong principle, they may not (for that principle may not) apply to all rational agents. Insofar as principles for rational bargains are dependent for their applications on the assumption of classical utility functions for possible bargains – as, of course, all solution concepts for Nash bargaining problems are – these principles can claim no greater range of application than that of the principle of expected utility maximization, and so may fall far short of "context-free universality of application."

4. Utility theory within a more general theory

Even if not always applicable, utility theory, when restricted to lottery sets based on complete alternatives, is I think true of ideally rational preferences. But one wonders under what conditions, otherwise expressed, alternatives will be relevantly complete or at least sufficiently so. I offer an answer to this question that locates utility theory within a general and quite unrestricted theory that is "simple and roomy" (Jeffrey 1987, p. 223), and that in its fundamental principle takes the idea of complete basic alternatives to its limit.

4.1

The general theory – *causal world Bayesianism* – measures the choice-worthiness of an action an agent is sure he can choose, as well as the welcomeness or desirability of any state he is sure is causally possible, by a probability-weighted average EV of values V of possible ways in which it might work out, where each way or "world" w is completely specific with respect to all things of practical interest and relevance:

$$\mathrm{EV}(a) = \sum_{w:\, w \in W \,\&\, \Pr(w \text{ given } a) > 0} [\Pr(w \text{ given } a) \cdot V(w)].$$

For an agent for whom aspects of some possible actions and states are themselves of interest and practically relevant, worlds will include these and thus not be detached and general results that can attend just any action or state (cf. Weirich 1986a).

According to the fundamental principle of the theory, rational choices and preferences are determined by these probability-weighted averages.[3] The idea of the theory is that choices and preferences of an ideally rational agent for actions and states would be determined by (i) the attractions and repulsions exerted by the various comprehensive ways they might take place, (ii) the agent's credences for their taking place in these various ways, and (iii) "the proportion [viewed geometrically] that all these things have together" (*The Port-Royal Logic,* last chapter; see note 1 of Chapter 8 in this volume). The idea of the theory is that each possible comprehensive way would eventually, in an ideal reflective and open-minded review, contribute independently and stably to the inclination for its action or state; pushing toward or pulling away proportionally to the contributing way's attraction or repulsion for the agent, and proportionally to his credence for its being the way the action or state would take place; and all this in the even-handed manner explicated by the idea of a probability-weighted average of values. Weights of values of *comprehensive* ways would be credences, on pain otherwise of double-counting attitudes toward risks.

4.2

The theory, we note, is for ideal rational agents: those whose credences and potential responses to evidence are quite definite and confident everywhere and are represented by unique probability measures Pr, and for whom relative directed differences in preferences for worlds are also definite everywhere and are represented by world-value measures V that are bounded and unique to positive linear transformations.

To be more specific regarding Pr: for propositions p and q, an ideal rational agent is to consider p at least as likely as q if and only if $\Pr(p) \geq \Pr(q)$; for propositions p, q, r, and s, he is to consider the potential evidential bearing of q on p to be at least as great as that of r on s if and only if $\Pr(p/q) \geq \Pr(r/s)$. The probability measure Pr is to take as arguments all propositions and all ordered pairs of propositions; axioms for such a combined probability and conditional probability function can be gathered from Sobel (1987a, pp. 60-1).

Turning to world-value measures V (see Krantz et al. 1971, pp. 150-2), for any V and for worlds w, x, y, and z, the agent is to find the directed difference in degrees of welcomeness of w and x, or in his intensities of preference or desire for them (supposing it is to be one of these two worlds) at least as great as that between y and z (supposing it is one of these two) if and only if $V(w) - V(x) \geq V(y) - V(z)$.

For credences, potential responses to evidence, and preferences of a less-than-ideal agent I would suppose that, instead of representations by unique Pr-functions and V-functions that are unique to positive linear transformations, we would have representations by members of *sets* of Pr- and V-functions that are consistent with his credences, etc., insofar as his mind is made up about these things. But I defer consideration of such matters to another time.

4.3

It turns out that in this theory the choiceworthiness of a made-up lottery L that an agent is sure he can choose need not be measured, as utility theory would have it be, by

$$\Sigma_i [c_i \cdot \text{EV}(O_i)],$$

where outcome O_i has chance c_i in L; though the choiceworthiness of L is necessarily measured by

$$\Sigma_i [c_i \cdot \text{EV}(L \& O_i)].$$

This formula for the expected value of a lottery that an agent is sure he can choose comes from a theorem for expected values of options under adequate partitions S of states or possible circumstances. According to this theorem,

$$\text{EV}(a) = \Sigma_{s: \, s \in S \, \& \, \text{Pr}(s \text{ given } a) > 0} [\text{Pr}(s \text{ given } a) \cdot \text{EV}(a \& s)].$$

Partitions that are "adequate" for this rule in causal decision theories are studied in Sobel 1985a, 1986a [Chapter 8 in this volume], and 1989a [Chapter 9].

For a made-up lottery L in which chances are given, $\text{Pr}(O_i \text{ given } L) = c_i$. Furthermore, the partition constituted of a lottery's possible outcomes will be adequate for an analysis of this lottery's expected value. Thus we have the formula displayed previously for the expected value of lottery L: $\Sigma_i [c_i \cdot \text{EV}(L \& O_i)]$ (cf. Sobel 1978, pp. 140b, 140c; for a similar idea see Jeffrey 1982, p. 729). A similar formula holds for the expected value of a lottery on an epistemic condition, and, in particular, for its expected value conditional epistemically on the agent's having to choose between it and some one other lottery:

$$EV[L/(L \text{ or } L')] = \Sigma_i \, (Pr_{L \text{ or } L'}(O_i \text{ given } L)) \cdot EV[(L \& O_i)/(L \text{ or } L')]).$$

Since the chance c_i for O_i is given in the specification for an artificial lottery L, we have $Pr_{(L \text{ or } L')}(O_i \text{ given } L) = c_i$ and

$$EV[L/(L \text{ or } L')] = \Sigma_i \, (c_i \cdot EV[(L \& O_i)/(L \text{ or } L')]).$$

Here and in the sequel, the formula "$(L$ or $L')$" abbreviates "one can choose either L or L', and must choose one or the other of these." Expected values from epistemic perspectives reached by conditionalizing on a proposition are defined as follows:

$$EV(a/q) = \Sigma_{w:\, w \in W \,\&\, Pr(m \text{ given } a) > 0} \, [Pr_q(w \text{ given } a) \cdot V(w)].$$

Here Pr_q comes from Pr by conditionalization on q. The theorem for conditional expected values under adequate partitions is that

$$EV(a/q) = \Sigma_{c:\, c \in C \,\&\, Pr_q(c \text{ given } a) > 0} \, ([Pr_q(w \text{ given } a)] \cdot EV[(a \& c)/q]).$$

4.4

These principles lead to a sufficient condition for utility theory's way with lotteries. The necessary condition stressed in Section 3 for the applicability of utility theory – that all lotteries founded on basic alternatives be possible and pairwise comparable – is presupposed by and is a necessary part of this sufficient condition.

According to causal world Bayesianism, rational pairwise preferences are determined by conditional expected values. Hence, for comparable members of a lottery set L and L',

$$L \geqslant L' \text{ if and only if } EV[L/(L \text{ or } L')] \geq EV[L'/L \text{ or } L')].$$

By the partition theorem for conditional expected values,

$$EV[L/(L \text{ or } L')] = \Sigma_i \, (c_i \cdot EV[(L \& O_i)/(L \text{ or } L')]).$$

So a sufficient two-part condition for the correctness of utility theory for a rational Bayesian agent and a lottery set whose members are all comparable may be expressed as follows:

(i) *Expected values of outcomes must be independent of both lotteries and alternatives;* that is,

$$EV(O) = EV[(L \& O)/(L \text{ or } L')]$$

for any lottery L, member L' of the lottery set, and outcome O that has a positive chance in L.

(ii) *Expected values of basic alternatives must be independent of alternatives;* that is,

$$EV(A) = EV[A/(A \text{ or } L)]$$

for any basic alternative A and member L' of the lottery set.

(This two-part condition could be simplified if lottery sets were confined to lotteries, with sure-thing lotteries $1(A)$ standing in for basic alternatives A; in that case, (ii) could be omitted.) Given this condition, for any lottery L in which outcome O_i has chance c_i, and any member L' of the lottery set,

$$EV(L/L \text{ or } L') = \sum_i [c_i \cdot EV(O_i)].$$

Pairwise preferences for members of this lottery set are measured on a single scale and, as utility theory would have them, are represented by a utility function. They are represented by the function U where, for any basic alternative A,

$$U(A) = EV(A),$$

and where, for any lottery L in which outcome O_i has chance c_i,

$$U(L) = \sum_i [c_i \cdot EV(O_i)].$$

When this two-part condition is satisfied, a rational Bayesian agent is, with respect to a lottery set, indifferent to certain dimensions of risk: his pairwise preferences are determined by weighted average expected values of outcomes without regard to other characteristics of lotteries by pairs. For an agent of this mind regarding members of a lottery set founded on basic alternatives $0, \$M$, and $\$5M$, "a zero is a zero"; it is all the same whether he wins nothing after taking a chance when he can choose $\$M$ for sure rather than $[^{10}\!/_{100}(\$5M), {}^{89}\!/_{100}(\$M), {}^{1}\!/_{100}(\$0)]$, or wins nothing after "going for the big one" in a gamble between $[^{10}\!/_{100}(\$5M), {}^{90}\!/_{100}(\$0)]$ and $[^{11}\!/_{100}(\$M), {}^{89}\!/_{100}(\$0)]$.[4] If our condition is satisfied for a lottery set then its basic alternatives are, for purposes of utility theory, complete enough.

4.5

Causal world Bayesianism (CWB) leaves open the possibility, advanced in Section 3, that utility theory when restricted to lottery sets founded on complete basic alternatives is sometimes for some agents quite inapplicable. Left open in particular is that utility theory may not be applicable to the set based on the agent's worlds. This point – that CWB does not presume that all world lotteries are both possible and comparable – is, I note, consonant with a feature of the Bolker–Jeffrey theory in which the preference relation need not extend to all probability distributions over "holistic states of nature": terms of the relation can "form a much thinner set than that" (Jeffrey 1974, p. 78).

It has been said that modern expected utility theory – "a theory for coherent and consistent choice among alternative courses of action with uncertain outcomes" (Fishburn 1988, p. 267) – has been challenged "by Maurice Allais's research in the early 1950s . . . and . . . a host of subsequent experiments that demonstrate persistent and systematic violations of expected utility" (p. 267). But this is misleading if, as Peter Fishburn intends, by "expected utility theory" is meant the body of diverse articulations of Daniel Bernoulli's idea. For included among them are world Bayesianisms as first exemplified in the logic of decision of Jeffrey (1965b). And these "simple and roomy" theories (Jeffrey 1987, p. 223) can accommodate just the experimental results that violate expected utility theory as narrowly conceived along lines proposed by "Oskar Morgenstern, John von Neumann, Frank P. Ramsey, and Leonard J. Savage" (Fishburn 1988, p. 267).[5] In this connection there is truth to the observation that "not . . . much room [is provided] for the construction of gambles" (Krantz et al. 1971, p. 411) within world Bayesian frameworks: care and work are required, and theories for gambles afforded by that parsimonious "mono-set" framework may not be as simple and bold as classical theories erected on mixed ontologies. But these are virtues, not deficiencies, of that curiously neglected amongst specialists "framework for preferences."

It has been written that "nearly everyone seems to agree that there are chance events to which probabilities adhere, consequences which exhibit utilities, and decisions that are more or less arbitrary associations of consequences to events" (Krantz et al. 1971, p. 411). Regarding their own ontology, Luce and Krantz stipulate in passing that "the consequences . . . are [to be] in no way inherently related to the chance events" (Krantz et al. 1971, p. 370). It is, as we have seen, a virtue of world Bayesianism that its basic principles imply the need for such restrictive conditions, if classical formulas for values of gambles are to hold. And it is a virtue of the mono-set framework of sets of worlds (propositions) that it allows for space within the formal theory for such conditions. Not surprisingly, therefore, it is no longer true that nearly everyone agrees that decision theory is best erected on mixed foundations. Many who come to decision theory from philosophy favor mono-set frameworks, and consequently find that they must go out of their own ways to discover in what sense Allais's problem, for example, is a problem.

4.6

We have seen that CWB affords a theoretically sufficient condition for the correct applicability of utility theory: it offers another way of saying when basic alternatives are, if not quite complete, then at least complete enough. But I do not think that this condition is a practical one, or that it should be easy to tell, in advance of a consideration of an agent's preferences for a range of lotteries, whether or not his attitudes toward these lotteries actually satisfy this condition and are thus such as to make available modes of analysis that utility theory underwrites. I think that advance assurance on this point is not possible. If this is right, then the moral (assuming that thought-out and reflected-upon attitudes toward risks are to be respected) is that decision analyses underwritten specifically by utility theory need always be viewed not as means to definite solutions of complex problems given accepted solutions of simple ones, but only as possibly illuminating – though also possibly distorting – perspectives on complex problems. Decision analyses underwritten by utility theory but not also by unrestricted Bayesianism, especially ones backed up by tricky arguments for adjustments in initial relatively unreflective or uncertain preferences, should always be tentatively deployed as merely so many things cautiously to think about when reaching for fully informed reflective preferences.

4.7

As illustrations of tricky arguments, I offer adaptations to the Zeckhauser-Gibbard problem of reactions to Allais's problem.

4.7.1. First "tricky argument." Imagine that at 9:00 A.M. you see there is a .5 chance that you will be killed without further ado, as well as a .5 chance that instead you will have a choice at 9:05 A.M. between beggaring yourself or playing with our first gun as loaded. More dramatically, suppose this:

There are two revolvers, one half-loaded, the other ⅓ loaded [as is our first gun]. The cylinders are spun, and the revolvers held at your head by two of your captors. The trigger of the half-loaded one is to be pulled [at 9:00 A.M. – you have no choice at this time], and then [at 9:05 A.M.] – [if you survive that first trial, and] if you don't pay – the trigger of the ⅓-loaded one [is to be pulled]. (Jeffrey 1987, p. 233; cf. Jeffrey 1988a, p. 121)

(See Figure 1.) If you reach the 9:05 A.M. decision node then you will be indifferent between A and B: You will be confronted with our first-gun choice, regarding which you are (by condition (1) of our case) indifferent.[6]

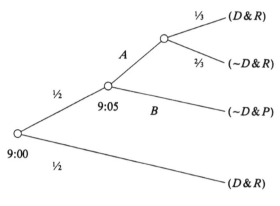

Figure 1

But then you should even now, at 9:00 A.M., be indifferent regarding a choice between A and B. You should be "prepared to make yourself penniless at the beginning to escape" a second risk of death supposing you survive the first one (Jeffrey 1987, p. 234).

How much would you pay *after* the first play, if you survive it, to avoid the second? How much would you pay *at the very beginning* to avoid the second play if you survive the first? ... Perhaps you'd pay different prices ... depending on whether [you'd be] out of your mind with dread or with relief [if you had just survived the first play], but if you're amazingly cool you'll see [the issues] as on a par. (Jeffrey 1988a, p. 121)

On the one hand, you should be indifferent between making yourself penniless to escape that second risk, should you survive the first one,

$$[\tfrac{1}{2}(D\&R), \tfrac{1}{2}(\sim D\&P)],$$

or equivalently (by condition (2) and Substitution),

$$[\tfrac{3}{6}(D\&P), \tfrac{3}{6}(\sim D\&P)];$$

and, on the other hand, running that second risk while remaining rich, should you survive the first risk,

$$(\tfrac{1}{2}(D\&R), \tfrac{1}{2}[\tfrac{1}{3}(D\&R), \tfrac{2}{3}(\sim D\&R)]),$$

or equivalently (by Reduction),

$$[\tfrac{4}{6}(D\&R), \tfrac{2}{6}(\sim D\&R)].$$

This, however (see the underlined), is indifference (6), and says that you should be indifferent between on the one hand paying to have a bullet removed from the *second* gun, and on the other hand not paying and playing with it fully loaded.

Perhaps, in thinking that it might be reasonable to pay more to have the first gun unloaded than you are prepared to pay to have the second relieved of one of its four bullets, "you were dazzled by some of the numbers.... Well, what do you think?" (cf. Raiffa 1968, p. 83). Richard Jeffrey suggests that the moral of the story is that "rationality is a form of steadfastness" (Jeffrey 1987, p. 234; Jeffrey 1988a, p. 121).

Important to an assessment of this counterargument is the distinction between a conditional preference addressed to a possible future choice and a conditional preference addressed to a choice between possible precommitments with regard to a future possible choice supposing no precommitments are made. For the former conditional preference you suppose that the future time is now and that the choice has arisen. This supposition operates to shift your epistemic perspective to one in which you are sure that the choice is immediately upon you. To elicit this conditional preference you then (consequent to that epistemic shift) choose in imagination. Certainly at 9:00 A.M. your conditional preference in this sense for a choice at 9:05 A.M. between A and B should be the *same* as you predict, at 9:00 A.M., that your choice at 9:05 A.M. will be if you survive to make one. The conditional preference should be the same as the conditionally predicted preference, unless you think that, if you survive to 9:05 A.M., untoward influences will have changed your mind in ways of which you cannot presently approve.

However, that there should be this qualified sameness at 9:00 A.M. – between your conditional preference for a choice at 9:05 A.M. and the preference you are prepared to conditionally predict for that possible choice at 9:05 A.M. – implies nothing for your proper conditional preference for precommitments (i.e., choices) at 9:00 A.M. for what (but for precommitments) may be options open to you at 9:05 A.M., supposing such precommitments (i.e., choices in advance) are possible and that you are confronting a decision between them. Your realization that you will be indifferent between A and B *then,* at 9:05 A.M. (if you live that long) does not mean that, if you are steadfast and anticipate no untoward influences, you must in imagination be indifferent between precommitments now, at 9:00 A.M., to A or to B at 9:05 A.M. should you live that long. It is obvious that, without changing your mind or being unsteadfast or spinelessly inconsistent, you can see in one light a choice at 9:00 A.M. between precommitments either to A or to B at 9:05 A.M., should you "luck out" and A and B become possible for you; and that you can see in a very different light a choice between A and B at 9:05 A.M., should this choice arise – that is, should you both survive and not be precommitted. These two imagined choices would be addressed to very different alternative arrays of chances. The first of these imagined choices (the one regarding possible precommitments) is, we have observed, tantamount to a choice between beggaring yourself and playing with the *second* gun; while the second of these imagined choices is between beggaring yourself and playing with the *first* gun.

These two hypothetical or conditional choices would have you address and choose in imagination between what, for most persons, are palpably and relevantly different options that involve very different risks! The question, put at 9:00 A.M.,

> Supposing you have a choice at 9:05 A.M. between A and B; what do you choose, or are you indifferent?

is ambiguous between these very two very different hypothetical issues. And only one of these hypothetical issues is, for the steadfast (and barring anticipated untoward influences), tied to the predictive question,

> Supposing you have a choice at 9:05 A.M. between A and B; what *will* you choose, or will you be indifferent?

(cf. McClennen 1983, sec. 4, pp. 121–6; Weirich 1986a, pp. 431–2; and the last section of Sobel 1989b.) Of course, "*if* after due consideration you see the ultimate... payoffs as the only significant considerations [and are in this sense *risk neutral*], then" if rational you answer these questions alike (Jeffrey 1987, p. 234). *Of course.*

Jeffrey reports that it was by essentially this Raiffa-style argument that "some years ago, David Lewis persuaded me that my initial inclination toward the analysis of Zeckhauser's problem... was right – thus diagnosing a specific sort of irrationality in my unease with the Bayesian consequences of that analysis" (Jeffrey 1987, p. 233). I think that Jeffrey may have been tricked by that argument, and that after further reflection he may feel that his unease was due to certain attitudes toward risk about which a person need not be embarrassed or apt to be ashamed (Jeffrey 1988a, p. 120).

4.7.2. Second "tricky argument." To adapt Savage's words (1972, p. 103) to our case, we use the following abbreviations: $D^* = [(D \& R) \vee (D \& P)]$, $R^* = (\sim D \& R)$, and $P^* = (\sim D \& P)$.

One way in which lotteries related to the second gun could be realized is by a draw from six tickets, with prizes according to the schedule:

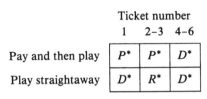

If one of the tickets numbered 4 through 6 is drawn then it will not matter which gamble I choose. I therefore focus on the possibility that one of the tickets numbered 1 through 3 will be drawn,

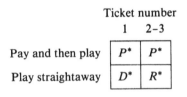

in which case the situation is exactly parallel to the one I confront if forced to play with the first gun, which could be realized thus:

	Ticket number		
	1	2-3	4-6
Pay and then play	P^*	P^*	P^*
Play straightaway	R^*	D^*	R^*

The subsidiary decision in the second-gun case, just as the decision in the first-gun case, depends on whether I would prefer some chance or other p of surviving poor, or prefer instead a $\frac{2}{3}(p)$ chance of surviving rich together with a $\frac{1}{3}(p)$ of dying.

Against this argument I say that my attitude toward a chance p of surviving poor, as compared with a $\frac{2}{3}(p)$ chance of surviving rich joined with a $\frac{1}{3}(p)$ chance of dying, can reasonably depend on p and on the character of outcomes for which I have remaining chances, if any (there are of course no remaining chances if $p=1$). However, if this is so than focusing (as Savage would have me do) can be a way of ensuring that I ignore things that matter to me and so reach decisions that are not fully informed by credences and preferences, including most prominently preferences that display fully reflected upon nonneutral attitudes toward risk.

4.7.3. Two notes. First, relating to the Zeckhauser–Gibbard problem, tricky arguments produced in response to Allais's problem consist in part of making adjustments for the fact that our problem does not directly challenge sure-thing principles but rather a related common-ratio principle (Jeffrey 1987, p. 231). Second, it is significant that neither of these arguments supposes that outcomes of lotteries are complete with respect to things that matter to the decision maker, or even with regard to things that are not specific to particular lotteries or pairs of lotteries. Each argument would defend utility theory as first stated in Section 1.

Addendum: Utilities and values

Suppose that our two-part condition for the existence of a utility function is satisfied. Suppose, indeed, that a somewhat stronger condition is satisfied – replacing (ii), which is confined to basic alternatives, by the condition that $EV(L) = EV[L/(L \text{ or } L')]$ for any members L and L' of the lottery set. Then this agent's pairwise preferences are determined by unconditional expected values of members (both basic alternatives and lotteries) of the lottery set; they are represented by utility functions. Furthermore, it seems that in this case pairwise preferences must be determined by values simpliciter of members of the lottery set, since expected values are based on values of worlds that are completely specific regarding things of interest to the agent, including all risk factors; such expected values should be values simpliciter.

Even so, because our two-part condition is merely sufficient for the existence of a representing utility function, it does not follow that whenever representing utility functions exist they are value functions. Whether or not they are so remains a question, albeit one that has recently received a well-argued negative answer (reprised in what follows). I conjecture a negative answer to the narrower question of whether or not utility functions are value functions whenever the preferences they represent are reasonable.

Let V be a value function that represents comparisons of directed differences between intensities of preferences for members of a lottery set L if and only if: (i) for any p, q, r, and s in L, the directed difference between the agent's preference for p and his preference for q taken in that order is at least as great as the directed difference between his preference for r and his preference for s taken in that order – that is (using an asterisk to distinguish comparisons of directed preference differences from ordinary numerical comparisons), $(p, q) \geq^* (r, s)$ if and only if $V(p) - V(q) \geq V(r) - V(s)$; and (ii) every function V' that satisfies (i) is a positive linear transformation of V.

Bouyssou and Vansnick (1988) show that even when utility functions exist for pairwise preferences between members of a lottery set, and value functions that are unique to positive linear transformations for comparisons of directed differences between preferences for these things also exist, they may be "scaled differently" (Bouyssou and Vansnick 1988, p. 107). These authors are of the view that the crucial assumption for representing value functions for comparisons of differences in preferences among members of a lottery set, coinciding with representing utility functions for pairwise preferences among this set's members, is simply that such a utility function exist (p. 110). But they stress that this condition is not strictly sufficient. The main additional requirement, they hold, is satisfaction of the condition

A2: $(p, [½(p), ½(q)]) =^* ([½(p), ½(q)], q)$,

or, in words: that the directed difference between a subject's preference for p and his preference for the even-chance lottery in which it is mixed with q is equal to the directed difference between his preference for this even-chance lottery and his preference for q. The only condition employed by Bouyssou and Vansnick that actually connects "$\geq$" and "$\geq^*$" is

A1: $p \geq q$ if and only if $(p, q) \geq^* (q, q)$,

which says that lottery p is weakly preferred by an agent to q if and only if the directed difference between his preference for p and his preference for q equals or exceeds the null difference between his preference for q and itself (p. 106). This, as Bouyssou and Vansnick maintain, seems an obvious compatibility condition for preferences and directed differences between preferences. Theorem 1 of their paper says that if pairwise preferences for members of a lottery set are represented by a utility function U, and if comparisons of directed differences in preferences for members are represented by a value function V, then V comes from U by a positive linear transformation and "utilities are values" if and only if A1 and A2 hold.

These authors view both A2 and representability of pairwise preferences by utility functions as rationality conditions (Bouyssou and Vansnick 1988, p. 110). I think, however, that representability of pairwise preferences by utility functions is not a condition of rationality, and I suspect that A2 is not such a condition either – not even when restricted to persons whose preferences are represented by utility functions.[7] On the first point, certainly, "likings for prospects should be based on likings for outcomes and their probabilities" (Ellsberg 1954, p. 536), but it does not follow that likings for prospects must, if rational, be determined by mathematical expectations or weighted means of values of outcomes. It seems that a rational person's pairwise preferences for lotteries can be based not only upon *means* of lotteries, but also upon "mode[s] ... median[s] ... range[s], variance or other properties" (Ellsberg 1954, pp. 536-7; contrary to Savage 1972, p. 97).

5. Summing up

The Bayesian idea is that rational actions maximize weighted average values of possible outcomes, the weight for an outcome's value being the agent's probability for this outcome given the action. Actions are viewed as natural lotteries in which subjective probabilities stand in for stipulated chances. Hence it is tempting to think that a theory of rational actions might be imbedded in an extension of utility theory, where this is a theory of rational preferences for artificial lotteries in which chances and prizes or outcomes are stipulated. This approach has special attraction for theorists who are interested mainly in game theory and who concern themselves with decision theory per se only in passing. Utility theory provides nice foundations for mixed and jointly randomized strategies.

But notwithstanding its prima facie plausibility and certain powerful ulterior recommendations, the approach to a Bayesian theory of rational action via utility theory is I think suspect and relatively unpromising. For one thing, the basic Bayesian idea – that rational actions would maximize expected values – owes its considerable plausibility in part to its contact with bona fide values. And yet, even when rational preferences are represented by utility functions, it seems that these functions may not need to be measures of proper values. Another (and weightier) point is that problems such as those of Allais and Zeckhauser-Gibbard raise the possibility that pairwise preferences of at least some rational agents for members of certain lottery sets are not represented by utility functions, and cannot be reconciled with utility theory by the device of enriching outcomes and addressing the theory to other lottery sets.

To approach the theory of rational actions through utility theory once seemed a good idea, but now should be viewed as getting things the wrong way 'round. Rather than try to locate a theory for rational actions within an extension of utility theory, in order to put utility theory into proper perspective one should, I think, start with a Bayesian theory of maximizing expected value, a theory suited with little question to choices by all rational agents regardless of their attitudes toward risk, and then locate utility theory within that theory as valid for pairwise preferences of rational agents with relatively neutral attitudes toward risks. This approach leaves open the question of whether or not utility theory is valid for all rational agents and for all lottery sets – which seems to be the right stance for the present. Briefs on the question, both for and against utility theory's general validity, are still coming in (cf. Wakker 1988 and Sobel 1989c).

NOTES

1. Paul Weirich makes similar claims. He thinks that ideally "results" would include "*everything that would be attributable to the option*" if the option were realized" (Weirich 1986a, p. 423); and he observes that classical axioms impose restrictions: "For example ... that any probability mixture of consequences is an option" (p. 423). "Consequences" in utility theory can, Weirich suggests, be no more than general results that "can be produced with arbitrary probabilities by [many] options" (p. 424). I note that Weirich does not attend explicitly to the troublesome requirement of utility theory that all lotteries be pairwise comparable; noting this requirement strengthens his argument.
2. Peter Hammond tells us that consequences are to be complete with respect to things of proper relevance to choice: "Consequentialism requires everything which should be allowed to affect decisions to count as [part of] a relevant consequence. If regrets, sunk costs, *even the structure of the decision tree itself are relevant to normative behaviour,* they are therefore already in the consequence domain" (Hammond 1988, p. 26; emphasis added). But his "consequences" must not be "logically specific" to particular decision problems: "[Consequentialism] applies to all logically possible finite decisions whose terminal nodes have consequences in the given domain. This is the assumption that there is an *unrestricted domain of consequential decision trees*" (p. 27). But if structures of some trees *are* parts of their consequences then, contrary to the assumption of unrestricted domain, these consequences will be specific to these trees (a point made in other words in Sobel 1988c, p. 539, n. 3). Hammond suspects that the assumption of unrestricted domain "may be restrictive," since it would rule out as a "partial consequence" some forms of regret (Hammond 1988, p. 76, n. 4), but does not mention the radical inconsistency of structures of trees as "partial consequences" with this assumption. His approach quite excludes "the structure of the decision tree itself [being] relevant to normative behaviour" (p. 26).
3. This theory serves present issues, though I favor a theory that (i) uses weights more complicated than any suggested by the form "Pr(w given a)" (see Sobel 1986a [Chapter 8 in this volume]), and that (ii) makes stability on ideal reflection a condition coordinate with maximization for rational choices (see Sobel 1983 [Chapter 10] and Sobel 1990a [Chapter 11]).
4. For most agents in Allais's problem: (i) these things are *not* all the same; (ii) $EV[(L1\,\&\,\$0)/(L0\,\&\,L1)] \neq EV[(L2\,\&\,\$0)/(L2\,\&\,L3)]$; and so (iii) not both of these conditional expected values equal his unconditional expected value $EV(\$0)$; hence our condition of risk neutrality is not satisfied.
5. Jeffrey's theory is not mentioned in this article, though it is the subject of a section in Fishburn's (1981) review of expected utility theories.
6. Actually, if you reach the 9:05 A.M. node then you will confront a choice with this difference: You will have "just survived a harrowing 50-50 chance of dying" (Jeffrey 1987, p. 233). Jeffrey holds that this difference could give you "no reason to lower the ransom you'd pay to empty the gun" (p. 234). I disagree, but nothing here turns on whether I am right about this.

7. A2 is implied by the following operational definition for "$=^*$": for members x, y, w, and z of the lottery set, $(x, y) =^* (z, w)$ if and only if $[½(x), ½(w)] \approx [½(y), ½(z)]$. Those who use this operational definition may consider A2 to be secure – that is, in a way, "analytic." However, this definition imports attitudes toward chances into comparisons of differences between strengths of preference, attitudes that, I think (following Fishburn), are not integral to these comparisons (cf. Fishburn 1970, pp. 81–2).

II

Problems for evidential decision theory

2

Newcomblike problems

Abstract. Newcomblike problems are cataloged and defended as coherent challenges to evidential decision theories, even elaborate theories that are enhanced with "metatickles." Criticism of arguments to the contrary feature not only nondominance and third-person Newcomblike decision problems concerning what to do but also nonaction Newcombesque desire problems concerning what to want and hope.

0. INTRODUCTION

Richard Jeffrey's logic of decision is the best-known evidential decision theory. In it the "Desirability" of an action is a weighted average of the Desirabilities of its possible outcomes, where average weights are conditional probabilities of outcomes on the action. Conditional probabilities are assumed to measure possible evidential bearings of propositions. Connectedly, the Desirability of a proposition is said to measure its relative welcomeness as a possible item of news.

To say that A is ranked higher than B [i.e., Des $A >$ Des B] means that the agent would welcome the news that A is true more than he would the news that B is true. (Jeffrey 1965b, p. 72)

(For reservations concerning this interpretation of Desirability see Sobel 1986a, p. 410, and Eells 1987.)[1]

"The Bayesian principle [of this theory, which] is to *perform an act which has maximum desirability*" (Jeffrey 1965b, p. 1), relates Desirability to choiceworthiness. This principle can be given the aspect of a truism.

If the agent is deliberating about performing act A or act B . . . , [then there is no effective difference between asking whether he prefers A to B as a news item or as an act, for he makes the news. (Jeffrey 1965b, pp. 73-4, corrected as in Jeffrey 1983, p. 84)

This selection first appeared in *Philosophy of the Human Sciences,* Midwest Studies in Philosophy, vol. XV (French, Uehling, & Wettstein, eds.). © 1990 by University of Notre Dame Press. Reprinted with revisions by permission.

Since in choosing *A* over *B* the agent will make for himself the news that *A*, it can seem obvious that he should do this only if he would prefer the news that *A*. "What else?" Jeffrey tempts us to ask.

For present purposes the important feature of Jeffrey's theory is, put positively, its purely epistemic and unmetaphysical character. Put negatively, this feature is its acausal character, a feature that has always been first among the theory's Humean virtues in Jeffrey's view: "The theory is *non-causal* in the sense that [no] causal notion is taken as primitive" (Jeffrey 1965a, p. 292). "It [is] the principal virtue of the theory, that it makes no use ... of any ... causal notion" (Jeffrey 1965b, p. 409). Newcomblike problems challenge this assessment of the evidential, acausal character of Jeffrey's theory. As Jeffrey himself reports, that "causal imputations play no explicit role [in his theory is seen by Newcomb critics and] partisans of causal decision theory as a defect" (added in December 1981b to Jeffrey 1981, p. 492).

In part A, an analysis of Newcomblike problems is proposed and their considerable variety is canvassed. In part B, I defend Newcomblike problems from important objections and maintain that such problems make coherent challenges to evidential decision theories, even refined ones. I am convinced that they make more than this, that Newcomblike problems refute evidential decision theories. But I am not concerned in this chapter to press this final substantive point.

A. Analysis and variety

1. An analysis of Newcomblike problems

Newcomblike problems for Jeffrey's theory would have the following elements. (Problems cited by name are set out in Appendix I.)

> *Element A.* In a Newcomblike problem the agent would be sure that certain features relevant to the values of outcomes of his actions – for example, whether or not there is $M in the second box (Newcomb's problem), or whether or not the other prisoner is going to confess (Prisoners' Dilemma) – are *causally independent* of his possible actions, and are things he can in no way influence.

> *Element B.* These features would be for him *epistemically* or *evidentially dependent* on his actions, so that news of these actions would provide him with signs, that is, with evidence for and against these features.

Element C. It is maintained, largely on the basis of element A, that a certain action would be uniquely choiceworthy and rational.

Element D. It is maintained, largely on the basis of element B, that the evidential Desirability of this action is exceeded by that of some other action that is open to the agent. Given element C, this entails that the choiceworthy action is *not* the action of which news would be most welcome.

2. The variety of Newcomblike problems

Problems vary in the grounds provided within them for these elements. Problems vary in how they purport to secure the dependencies and independencies in elements A and B, and in the character of explicit or implicit arguments for elements C and D. Important differences among Newcomblike problems relate mainly to elements B and C, but there is variety in the other elements as well.

2.1. Grounds for causal independence

In many cases the agent is sure that some feature is causally independent of his actions, because he is sure that this feature is already settled one way or the other. This may be because the feature relates to the past (e.g., a past prediction), or because he views it as relating to the already determined future (e.g., the development of a disease whose causes are already in place). But a belief that the feature is "already settled" is not a part of every problem. For example, it is not a part of every Prisoners' Dilemma. Whether or not the other prisoner will confess can be viewed as not yet decided and as not already settled. That a feature is already settled is only a particularly simple way of making reasonable an agent's belief that his actions can have no causal bearing on it.

2.2. Grounds for epistemic dependence

Most important differences between problems have to do with how, notwithstanding the absence of perceived causal bearings, *evidential* bearings of actions on certain features are made plausible.

2.2.1. Predictor cases. The agent may think that a feature depends on some good predictor's prediction, so that news of an action would provide a good sign that it had been predicted, and that the feature (for example, $M in the second box) was as a consequence in place or not. The agent may be supposed to think that the predictor works with a theory, perhaps a causal theory. But predictor cases need not be in any explicit way causal. For example, an agent may be supposed to think that the predictor is an intimate who understands him well and knows what he is going to do better and sooner than he does himself, but who knows the agent's future actions only in something like the way in which the agent himself often knows and can predict what he is going to do. Such a predictor will not be supposed, while reflecting from the agent's deliberating point of view, to think of the agent's actions as things that will be caused, any more than the agent himself does from that point of view.

2.2.2. Causal cases. (a) The agent may think that there is a possible cause for the feature, and a cause for his action, and that these causes go together so that either both are present or both are absent. For example, he may think that it is possible that he has inherited a gene for a disease from a person who also carries a gene for tendencies to intellectualism (Nozick 1969, pp. 125-6).

(b) He may think that there is a possible *common cause* for the feature and his action; for example, that there is a single gene that both inclines possessors to smoke and makes them independently prone to cancer ("Sir Ronald lights up").

(c) He may think that the feature would cause his action – for example, that a signal on the screen flashing when and only when there is popcorn to be had would register subliminally and cause him to go for popcorn (the popcorn problem).

Causal Newcomblike problems admit of further divisions. Suppose for example that the agent believes in a cause for his action. Then he may think that this cause will operate directly, or only indirectly and by way of causing his decision and choice for it. And, if indirectly by way of his decision, he may think that this will be caused (i) by way of conscious factors (e.g., his credences and preferences) of his decision making's being caused; or (ii) by way of unconscious factors being caused (e.g., his implicit rule for translating credences and preferences into a decision – he might think that the common cause would cause him to be a Desirability maximizer – cf. Nozick 1969, pp. 127-9); or (iii) by way of his being caused to be, in his choice, carried away or confused; or by a combination of these ways.

2.2.3. Similar decision processes. If the feature is an action, he may think that in whatever way he decides what to do, similar processes will lead to the other action. For example, he may believe that the other prisoner will think as he does, and he may believe this while being in considerable doubt concerning how he himself is going to think and decide (cf. Sobel 1985a, p. 271).

2.2.4. Signs of character. If the feature is a possible aspect of the agent's character or personality – for example, ruthlessness (the "rising young executive") or charisma (Solomon's problem) – an agent's action could in yet another way be a good sign of something of which it was not considered to be a possible cause. He could think he was a person who learns things about himself by listening to what he says, and seeing what he does. Most people are somewhat like that. "'How can I tell what I think till I see what I say?' – E. M. Forster" (Lewis 1981a, pp. 9–10). How can I know, really know, what I am like, for example, what I am and am not capable of, until I see what I do when the chips are down? These, for most persons, are sometimes apt rhetorical questions.

2.3. Arguments for the choiceworthiness of bad news actions

Choiceworthinesses of bad news actions are made plausible in best-known Newcomblike problems by dominance arguments, but this device is not a feature of every Newcomblike problem. Some use the more widely applicable common-sense notion that an action is choiceworthy if it would probably have best consequences. For example, not going for popcorn recommends itself in the popcorn problem even though the agent would prefer to go for popcorn if he thought there were some. Not going for popcorn is held to be his rational course, because he is nearly sure that there isn't any popcorn and that he would thus be wasting his time in going for some.

2.4. Arguments for the desirability of another action

There is little variation in how relative Desirabilities of actions are established. Most problems use assumptions concerning cardinalities of Desirabilities of possible outcomes. One class of exceptions are probability-of-1 problems. In these, ordinal relations of Desirabilities for outcomes suffice. There are other exceptions, including a version of the popcorn problem (Sobel 1986a, pp. 414–15 [Chapter 8 in this volume, Section 2]).

Summing up, three bifurcations of Newcomblike problems divide these problems into predictor and nonpredictor problems, into problems that involve causal hypotheses and problems that do not, and into problems that feature dominance arguments and problems that do not. Crisscrossing these bifurcations produces an eightfold partition, each part of which contains a problem that is either already in the literature or readily constructable. It will be evident that very few objections to Newcomblike problems are, either as they stand or by easy extensions, objections to *all* Newcomblike problems.

B. Objections

Objections to be opposed in this part are of several sorts. First (1) come complaints that Newcomblike problems are too fantastic or unrealistic to be taken seriously and counted as refutations of otherwise satisfactory, ensconced theories. Next (2) come objections that these would-be problems are impossible, in that they would have clearheaded deliberating agents view their possible actions as potential evidence for prior conditions. Then (3) comes the objection that causal problems that would be test cases for theories of rational action are impossible, in that they would have agents view their actions as – though caused – avoidable by choice. Following this (4) are objections for cases in which agents believe their actions will be caused indirectly by way of their *choices* being caused. And finally (5) come objections that Newcomblike problems do not threaten properly qualified and formulated theories of rational choice that are addressed at least in the first place to choices of ideally rational and sophisticated agents, for such agents would be conscious of their credences and preferences and, by the time for making a decision, could not learn anything from news of what action they were about to do. For them, it has been argued, choiceworthiness must in the end be determined exactly by both evidential Desirability *and* causal utility, so that the kinds of theories that Newcomblike problems would set at odds, when properly qualified and formulated, will in every case agree. (Whereas evidential Desirabilities are computed using measures of evidential bearings of actions on likely outcomes, measures of probable causal bearings would be used in computations of causal utilities. The simplest causal decision theory uses unconditional probabilities of causal conditionals, $Pr(a > o)$, instead of the conditional probabilities $Pr(o/a)$, i.e. $Pr(a \& o)/Pr(a)$, used in Jeffrey's logic of decision.)

1. These are problems for space cadets

Against predictors it is sometimes suggested that their powers in problem cases need to be viewed as extraordinary, hardly human, and that cases based on them are always bizarre and far-fetched. However, when the stakes are high, an agent can in fact have a very low opinion of the predictor's power.

Define *average estimated reliability* as the average of (A) the agent's conditional degree of belief that the predictive process will predict correctly, given that he takes his thousand, and (B) his conditional degree of belief that the process will predict correctly, given that he does not take his thousand.... Let r be the ratio of the value of the thousand to the value of the million: .001 if value is proportional to money.... We have a disagreement between two conceptions of rationality if and only if the expected value [i.e., the Desirability] of taking the thousand is less than that of declining it, which is so if and only if the average estimated reliability exceeds $(1+r)/2$. (That is, .5005 if value is proportional to money.) This is not a very high standard of reliability. (Lewis 1979a, pp. 238-9)

For a proof, I employ certain simplifying numerical stipulations and make one weak assumption. Simplifying stipulations: Des $0 = 0$; Des $M = 1$. Recall that $r = $ Des $T/$Des M. By the second stipulation it follows that $r = $ Des T. Substantive assumption: Des($M + T) = $ Des $M + $ Des T. (This is weaker than the assumption that Desirabilities are linear with money.) Letting C abbreviate "the predictive process will predict correctly" and T "the agent takes his thousand," here is a Desirability matrix for the agent's decision problem:

	C	$\sim C$
T	r	$1+r$
$\sim T$	1	0

The agent's Desirabilities for $\sim T$ and T are:

$$\text{Des} \sim T = \Pr(C/\sim T);$$

$$\text{Des } T = \Pr(C/T)r + \Pr(\sim C/T)(1+r).$$

We are to prove that

Des $\sim T >$ Des T if and only if the average estimated reliability is greater than $(1+r)/2$.

Here is a proof of the "only if" part; turned upside down, it is a proof of the "if" part.

> Des $\sim T >$ Des T
> $\Pr(C/\sim T) > \Pr(C/T)r + \Pr(\sim C/T)(1+r)$
> $\Pr(C/\sim T) > \Pr(C/T)r + [1 - \Pr(C/T)](1+r)$
> $\Pr(C/\sim T) > \Pr(C/T)r + 1 + r - \Pr(C/T) - \Pr(C/T)r$
> $\Pr(C/\sim T) + \Pr(C/T) > 1 + r$
> $[\Pr(C/\sim T) + \Pr(C/T)]/2 > (1+r)/2$
> average estimated reliability $> (1+r)/2$.

Assuming that Desirabilities are linear with money, as Lewis reports, we have $r = .001$ and $[(1+r)/2] = .5005$. And, if .5005 is not already near enough to .5 to make no difference, then by reducing the monetary amount (to $1, to 1¢, etc.) it is possible to reduce the average estimated reliability required for the problem to a point where it is near enough to .5 to make no difference. It is not essential to a predictor case that the agent believe in a predictor of preternatural powers. The agent can think that the predictor is barely better than a random one. Nor need the agent consider the stakes to be extraordinarily high. Stakes for the agent, and his confidence in the predictor, can both be moderate as far as the mathematics of predictor problems is concerned.

Newcomblike problems need not be far-fetched, which is not to say that they are commonplace. But then decision theory as practiced by philosophers is not an empirical science, and as long as a problem is possible it should not matter whether it arises often or indeed ever. Even if Newcomblike problems were never encountered in practice and were all problems only "for space cadets," they would still not be problems "to which [in our theoretical exercises we could wisely] ... apply the [practical wisdom] of Esther Marcovitz (1866–1944): 'If cows had wings, we'd carry big umbrellas'" (Jeffrey 1983, p. 25).

2. Deliberators cannot view their choices as signs of prior conditions

2.1

Some critics object to the idea that choices of deliberating agents in Newcomblike problems would be for these agents evidence of causally independent states. Solomon, for example, when deliberating whether or not to send for the woman, should know "that he *is* deliberating"; he "*believes* that he has freedom of choice*" (Kyburg 1980, p. 162). But then he must believe that "there is no connection between how he decides and whether or not he has charisma" (p. 162). Since he "regards his choice as free, it cannot be taken [by him] as *evidence* of a prior state" (p. 164).

However, it seems that Solomon can believe his character is connected with his decisions without thinking that it is causally connected. And even if Solomon is a philosopher who thinks that connections between characters and actions can only be causal, still – as Ellery Eells has maintained – he needn't think of his actions as inexorable consequences of his character:

It seems perfectly consistent for Solomon to believe both that he has freedom of choice and that ... whether or not he has charisma is to some extent causally relevant to [and connected with] whether or not he sends for [the woman]. (Eells 1984a, p. 93)

These beliefs are consistent, at least according to ordinary understandings. (More is said about this in the next section.) And if Solomon believes his character is causally relevant to his actions, then I am sure Kyburg would agree there is no reason why he cannot find in his actions evidence of that character. Others could learn of Solomon's character through his actions and, assuming that he believes his character is causally relevant to these actions, so can he.

Furthermore, even if there is a problem with deliberating agents viewing their possible actions as evidence of their causes (due to a problem with deliberating agents thinking their actions have causes), this difficulty does not extend to all Newcomblike problems. Shifting to an acausal problem, it is quite clear that I can consistently take my free choice to be evidence of what a wise friend and mentor has predicted I shall do of my own free will. That I think my predicting friend regards my choice as free could hardly be a reason for my thinking otherwise; nor would the fact that I regarded my choice as free rule out my having confidence in my friend's predictive powers. Kyburg, if he means his objection (to deliberating agents viewing their possible actions as evidence) to apply against all Newcomblike problems, implicitly denies these facts. Perhaps Kyburg thinks that, in order for actions to be evidence for states, the actions must be viewed as connected with these states either as causes or effects. But this is not true, and it would not make Kyburg's point if it were.

2.2

Huw Price (1986, p. 195) promises to show that "a free agent cannot take the contemplated action to be probabilistically relevant to its causes" in certain kinds of Newcomblike problems. However, freedom plays no role in his argument, which (if successful) would show only that agents for whom conditional credences are relevant to choice, and whose relevant conditional credences for causes of actions on these actions would be formed by direct inferences for statistical correlations made with attendance to all available evidence, cannot take contemplated actions as relevant to their causes. I will consider, and reject, his argument for this limited result without considering the usefulness for decision theory of the result, supposing it could be established.

Suppose, for example, that Fred is an evidential desirability maximizer or "something like" one, and that in his conditional credences he abides by "the *principle of total evidence*" (Price 1986, p. 199). Assume that he is a smoker who believes that "the cancer gene occurs in 20% of smokers, and in 2% of nonsmokers" (p. 196). Can he have conditional probabilities based by direct inference on these statistics? (These conditional probabilities would be $\Pr(G/S) = .2$ and $\Pr(G/\sim S) = .02$.) Price says that he cannot, for conditional credences based on these statistics would not, once he had formed them, be based on *all* of the available evidence that needed then to be taken into account. These conditional probabilities would themselves, Price avers, be new evidence that conflicted with the old and so made a difference to his view of the pertinent correlation between smoking and the gene among "Fred-alikes." These conditional credences would, Price tells us, "be negatively correlated with smoking; they [would for Fred-alikes as evidential Desirability maximizers] increase the attractiveness of not smoking" (p. 200).

Price contends (at least implicitly) that any conditional probabilities that would make Fred's smoking evidence for or against the gene can be shown to be for him untenable. His argument, however, is not sound even for the particular conditional probabilities he considers. For though the evidence provided by these conditional probabilities themselves might lead to revisions in the pertinent statistic, contrary to Price it is not true that it would necessarily do this. Let us grant that these conditional probabilities would be "negatively correlated with smoking" (Price 1986, p. 200). Grant also that Fred would expect there to be relatively fewer smokers within the class of Fred-alikes* (i.e., Fred-alikes all of whom, for one resemblance to Fred, have those conditional probabilities) than in the population at large, or in the larger class of Fred-alikes that includes some who (perhaps through inattention and lack of reflection) do not have those conditional probabilities. Suppose indeed that Fred would think that almost everyone in the class of Fred-alikes* would decide not to smoke, and that only a very few of this vast majority would then backslide, change their minds, and actually smoke. Even so, he could think that in this now refined sample space the cancer gene was still possessed by 20% of (the relatively few Fred-alike*) smokers, and 2% of (the relatively many Fred-alike*) nonsmokers.

Fred could consistently think that the evidence of those conditional probabilities was to the 20%/2% correlations in the populations of smokers and nonsmokers at large (and in the classes of Fred-alike* smokers and nonsmokers) as evidence that a fire in an empty building was started in a rubbish bin could be to correlations between fires in empty buildings started by arsonists and ones started in other ways (cf. Price 1986, pp. 198–9). He could think that the new evidence provided by those .2 and .02 conditional probabilities of his made no difference to pertinent correlations of the gene with smoking and not smoking by his alikes. And if he did think this, then the new evidence provided by those conditional probabilities need not, according to Price, be considered, since it would not "[conflict] with the old" (p. 200).

Alternatively, it could be allowed that, even so, such evidence would need to be considered: If the evidence provided by these conditional probabilities themselves makes no correlational difference then the probabilities are not "self-defeating" or "inherently unstable" (Price 1986, p. 201). Contrary to Price, notwithstanding the relevance of conditional probabilities to the actions of an evidential Desirability maximizer, and notwithstanding a commitment to the principle of total evidence, these conditional probabilities *can* be supposed without contradiction to belong to this maximizer. So, in this case, he will "take the contemplated action[s] to be probabilistically relevant" (p. 195) and will get what Price (p. 199) considers to be the wrong answer; he will not smoke.

Summing up, it seems (Section 2.1) that deliberating agents *can* take their well-reasoned choices and actions to be, though not causes, still evidence for prior conditions that would be relevant to the values of outcomes. And (Section 2.2) Price has not demonstrated (for what this result would be worth) that at least some agents for whom such evidence would be choice-relevant – for example, evidential Desirability maximizers committed to the principle of total evidence – cannot, in what would be common-cause Newcomblike problems, take their reasoned actions to be evidence for their causes. Other efforts (due mainly to Eells) to show that ideally rational and sophisticated agents – or at least ideally rational and sophisticated evidential Desirability maximizers – cannot take their actions to be evidence for certain things are considered in Section 4 and especially in Section 5.

3. AGENTS WHO BELIEVE IN CAUSES FOR THEIR ACTIONS

3.1

Problem cases for Jeffrey's theory would be decision problems in which choiceworthiness and Desirability seem to diverge. But for an agent to have a decision problem it is necessary that he be sure that there is a decision he can make; it is necessary that he be sure that there are actions he can choose to do, each of which he could do were he to so choose. However, it may be suggested – not against all Newcomblike decision problems, but against all of the causal ones – that an agent cannot consistently be sure that he *has* a choice, and that several actions are really open to him, if he also thinks that he may be *caused* to do what he does.

Whether or not an agent can have a choice, if what he will do will be caused, is an old issue, one stand on which underlies the present objection to some Newcomblike problems. I do not take that stand. I think that an agent can have a choice even if he will be caused to do what he will do. I hold that a person can believe that, under certain circumstances that he thinks may already obtain (e.g., POPCORN! being flashed on the screen), he would be moved to do a certain thing; and, consistent with this thought, be sure that even if he is going to be so moved and though he will be caused to do the thing and not another, he will not *have* to do it and that he still has a choice. He can be sure, even when he is certain that causes for his action are already in place, that he can still choose to do otherwise than he will in fact do, otherwise than he will in fact be caused to do. This is I think all consistent given ordinary notions of agency and causation, and these (as distinct from various possible philosophic explications of them) are the notions of present relevance. The issue as I take it is whether agents can, in their own terms, consistently think that they may be caused to do certain things and be sure that, even if they are being so caused or moved to do them, they may still choose not to do them.

There is, I think, no general difficulty here for ordinary thinking about causes and choices. Of course, there are some beliefs in causes for actions that are not consistent with viewing these actions as avoidable by choice. If, for example, an agent believes that some predictor "hypnotically controls" his actions, this being what makes the predictor so reliable (Mackie 1977, p. 218), or that some predictor will "interview [if necessary] via telekinesis . . . to *make* [him] choose the alternative she has 'predicted'" (Talbott 1987, p. 421), then he cannot think that he has a choice in the matter or that what he does is up to him. (For another example see Sobel 1988b, p. 9 [Chapter 5, Section 1].) But not all ordinary beliefs in causes for actions are beliefs in such causes, or in causes that work in ways that quite remove control from the agent. Ordinary beliefs in subliminal stimuli and in character- and disposition-shaping genes can be of these stimuli and genes as causes that an agent can, even if he rarely or never does, override.

Some causes of actions are ordinarily viewed as contravenable. Furthermore, though not all philosophers approve of such views and provide explications for them, some do and thus can (as far as present objections go), consistently with their technical philosophies, countenance such causes in causal Newcomblike decision problems. This can be so even for a philosophic determinist. Suppose, for example, that a philosopher believes that he will do A and that this is related to an already-in-place determining cause C by a law of nature L such that $(C \& L)$ entails A. Suppose furthermore that he believes that even if he were to choose to act otherwise, $ch(\sim A)$, this would make no difference to C, and that whatever he were to do would have a determining cause. He can, even so, think that he can choose to do otherwise, ch($\sim A$), and that if he were to choose he would actually do otherwise, $\sim A$. For he can think that: (i) up to the time of ch($\sim A$) there would be no change (C would thus remain in place); (ii) at that time there would be a small miracle (he could think that the "choice otherwise" would be this miracle, which – he might want to caution – though his and made by him would not itself be an "action" of his or something he "did"); and (iii) from that time the world would unfold through $\sim A$ in accordance with the laws of the actual world modified minimally (presumably in ways peculiarly relevant to the agent's psychology and personal history) to allow for, and to make lawful, that small miracle (which while contrary to actual laws, given that the past to its time would remain fixed, would not be contrary to what, given it, would be the laws – indeed, this small miracle might itself have a determining cause under the laws that would prevail were it to obtain). See Gibbard and Harper (1978, in Campbell and Sowden 1985, pp. 136, 157-8, n. 2) and Lewis (1981c, p. 294). Free will and forms of determinism are discussed in Sobel (in press).

But enough of metaphysical fancies. What is important to our subject is not whether beliefs in effective choices are consistent with various speculative technical opinions concerning causes and the past or with the semantics of "non-backtracking" act-consequence counterfactuals. What is important is that persons (including I suspect most philosophers when they are not doing philosophy but simply getting on with life) can, in their own natural and everyday terms, think that even when they are under the influence of causes to do things they still have choices and capacities *not* to do these things, and so to override those causes. What is important to the cogency of causal Newcomblike problems is that this be true of natural and everyday thought, and not whether – in the opinion of this or that metaphysician (opinions are of course divided) – good philosophic sense can be made of the possibility of choosing to act otherwise, and then acting according to that choice.

Decision theories are not specific to the enlightened any more than they are specific to the humane:

> The logic of decision is concerned neither with the agent's belief function nor with his underlying value function. . . . [T]he Bayesian framework . . . admits belief functions that would be entertained only by a fool, and value assignments that would be entertained only by a monster. (Jeffrey 1965b, p. 199)

Decision theories are certainly not specific to philosophic determinists. And so, though it is interesting that causal Newcomb problems are possible even for some philosophic determinists, it is not important.

3.2

An agent can have a decision problem even if he thinks that his action will be caused. However, though the belief that his action will be caused is consistent with his having a decision problem, the belief that it will be caused quite regardless of his decision and choice is not. An agent has a "proper" decision problem only if he is sure that what he does can depend on his choice or decision: an agent in a proper decision problem must be sure that he has choices which if made would be efficacious. (I have maintained elsewhere that an agent's options in a proper decision problem *are* choices, i.e., "organizations of will" in which actions are made certain, or given specific objective chances. In the theory I endorse, the requirement that choices be viewed as efficacious for chosen options in a proper decision problem is satisfied by trivializing identification. See Sobel 1983, p. 159 [Chapter 10, Section 0].)

Consider now these words of Jeffrey's:

> Fisher's problem belongs to the Newcomb species only if it is the *choice* of smoking that the agent takes the bad gene to promote directly.... If he takes the performance itself to be directly promoted by presence of the bad gene, there is no question of preferential choice: the performance is compulsive. (Jeffrey 1983, p. 25)

Jeffrey's words suggest, and it can seem true, that an agent in a proper decision problem cannot think that his action will be caused *directly* by some already-in-place condition, and not merely indirectly and only by way of this condition causing his choice or decision. I think, however, that what are excluded are not beliefs in direct causation but only (as has already been said) beliefs in *quite regardless* and *uncontrovertible* causation. And, I maintain, to say that an action is caused, caused this way or that, directly or whatever, is *not* to say that it will take place quite regardless and that the agent has no choice in its connection. Consider that even if an action will be caused directly by some already-in-place condition (e.g. a subliminal stimulus), it is possible that it will also be chosen, and possible for both the in-place direct cause and the choice to be necessary causes as well as sufficient in each other's presence. It is possible for an action to be counterfactually dependent on a directly operating cause, and also counterfactually dependent on a choice: it is possible that if that cause were absent then so would be the action, notwithstanding the continued presence when its time comes of this choice; it is also possible that, if this choice were not merely absent but replaced by an alternative choice, then the action would be absent and indeed replaced by that chosen action, with the action's not taking place this time in spite of the continued presence of that already-in-place direct cause.

But is it possible for an action caused directly, and not by way of its choice being caused, to be rational? Yes it is, just as it is possible for such an action to be irrational, and for things the agent could have done by choice instead to be rational. Howsoever rational actions are related to credences and preferences, this caused action can likewise be related to credences and preferences, or not so related. And this is true whether the required relation is purely structural (e.g., having maximum evidential Desirability or maximum causal utility) or not purely structural but rather, as Talbott (1987, p. 423) would have it, somehow causal. A proper causal condition for rational actions must, I think, tolerate the causal pattern described in the previous paragraph.

4. Agents who believe in causes for their choices

4.1

Causal Newcomblike problems do not need to be indirect causation problems, for direct causation of actions is not necessarily causation quite regardless. Still, cases in which agents believe their actions are subject only to indirect causation, and to causation by way of their choices, are important. It may be objected that such cases cannot be Newcomblike.

There are no special difficulties with an agent's believing that his action will be caused by way of his choices being caused, where these in turn will be caused by way of his credences and preferences. For he can still think that his choice will be up to him, and that he will not have to make the choice that he does make. He can think that he need not choose as he will, and that were he to choose otherwise he would act otherwise, and think that these things are so whether or not his act will be rational relative to his credences and preferences, and even though his choice will be caused by way of his credences and preferences being caused.

But it can seem that an agent who believes that his actions will be caused only indirectly by way of his credences and preferences and thence his choice being caused cannot be in a Newcomblike decision problem. For it can seem that, *if* an agent believes (just before the time for choice) that just before the time for his choice his credences and preferences and thence his choice will, depending on certain crucial circumstances (e.g., whether or not the popcorn message is flashing), be caused to take on a particular character, *then* he will at least at this last minute have views about those circumstances, views that will bring into alignment the choiceworthiness and Desirabilities of his possible actions. The objection is not that an agent in a decision problem cannot have the beliefs in question. It is rather that if he does have them, then at least in the end his case cannot constitute a Newcomblike decision problem and so be a case in which, for Newcomblike reasons, choiceworthiness and Desirability diverge.

Now come several responses to this objection. The objection supposes that an agent will, at least just before the time for his choice, be aware of the character of his beliefs and preferences, and that he will not then be able to learn from news of his impending action anything regarding its causes that he has not already learned. But a competent agent need not be supposed to have perfect and complete self-awareness; he need not be supposed to be at the time of choice fully and accurately aware of his perhaps just-then-attained beliefs and preferences. Another difficulty with the objection is that, even if we concentrate on agents who are always aware of their credences and preferences, we need not suppose that they always know what time it is; agents can, in important ways, sometimes not know what time it is. In particular, an agent can fail to know, just before the last time for a choice, that the time then *is* just before the last time for that choice. All this is true of actual agents, including very competent ones. Verging toward science fiction, I note that a self-styled superagent could think that there never *were* times just before the times of his last times for choice. He could think that until an action was actually done there was always time left for him to reconsider and change his mind.

Finally, in addition to these difficulties with the objection now under discussion, we should note its limitations. This objection is only to problems in which it is believed that choices may be caused by various features of situations affecting the credences and preferences on which these choices will be based. But circumstances can affect choices in other ways, ways that need not involve intermediate conditions in place before the decision and discoverable then by introspection. Circumstances can be supposed to determine an agent's choices by affecting the very way he will translate his ultimate credences and preferences into action: Circumstances can be supposed to affect choices by affecting how, when making them, the agent will take into account his credences and preferences – something which, even just before the decision is made, may still remain to be seen. The present form of objection admits of no easy extension to cases in which the agent believes that his choice may, depending on circumstances, be caused in this fashion.

The fact seems undeniable that a competent and clearheaded agent need not think that he knows what he is going to do, not even just before the time for choice, and not even if he thinks that his decision will be caused by his final credences and preferences. Thus he can learn, from news of his impending action, things he has not already learned. He can learn not only what he is going to do, but also things of which such news would be evidence. A competent and clearheaded agent can be in causal Newcomblike problems, both direct and indirect ones.

4.2

Although such problems are possible for ordinary agents, perhaps they are not possible for appropriately ideal agents. Reasons due mainly to Eells for this view will be taken up at some length in the next section. I conclude here by noting, without discussion, a reason that has affinities to ideas of Talbott's (1987, esp. sec. 6 and sec. 9). It can be argued with some plausibility that – even though competent and clearheaded agents can be in indirectly causal Newcomblike problems – *ideally* competent, clearheaded and self-confident agents cannot be. However, it could be contended that such agents have views concerning possible causes of aspects of their credences, preferences, ways of thinking, and actions that imply that they, the agents, are possibly *not,* in the circumstances of the case, ideally competent and clearheaded and responsive only to objective evidence and good reasons.

For a related Talbottlike point, I note that we might argue (i) that it is a condition of every properly qualified theory of rational action that an action be rational in the subjective circumstances of a case only if the agent's belief in that rationality would not be undermined by ideal reflection on these circumstances; (ii) that in any indirectly causal Newcomblike problem, ideal reflection conducted by a competent, clearheaded, and self-confident deliberator *would* undermine his confidence in his competence in a way that would undermine any conclusion he had tentatively reached concerning what was rational to do; and therefore (iii) that even though indirectly causal Newcomblike problems in which causal utilities and evidential Desirabilities differ are possible for competent, clearheaded, and self-confident agents, no indirectly causal Newcomblike problems are possible for them for which *properly qualified* causal and evidential theories of rational action diverge. The suggestion is that properly qualified theories, when applied to such cases, will all say that nothing – that is, no action – would be rational, since no belief in the rationality of any action would be maintainable on ideal reflection. (I endorse a principle that makes it a condition of a rational action that a decision for that action would be ideally stable; see Sobel 1983 [Chapter 10], Sobel 1985c, and Sobel 1990a [Chapter 11]. The condition presently before us is that a certain *belief,* specifically the belief that the action would be rational, would be ideally maintainable. While these conditions are different – the maintainable belief condition, in contrast with the stable decision condition, builds into practical rationality a modicum of theoretical rationality – they are formally similar.)

5. NEWCOMBLIKE PROBLEMS ARE NOT POSSIBLE FOR IDEALLY RATIONAL AGENTS

Eells has two arguments to show that Newcomblike problems are not possible for ideally rational agents. He tells us that a difficulty with the first one makes welcome the second. If he is right – if, in particular, causal utility and evidential Desirability maximizing agree for ideally rational and sophisticated agents in the manner maintained in his second argument – then a fully general evidential theory that, without in any way "going causal," prescribes correctly for all agents can seem to be in the offing.

After considering Eells's original defense of evidential decision theory and the objections to it, I take up his second one and give reasons why, as a defense of evidential decision theory, it can be only a partial success. One reason to be stressed is that ideally rational and sophisticated evidential and causal maximizers would not agree in their actions in all Newcomblike problems in which ordinary agents can find themselves. These paragons would, I claim, end up deciding for different actions in some nondominance problems. In these problems the decisions that would be made by ideally rational and sophisticated evidential maximizers would be wrong – if not for themselves, then at least for the ordinary agents in these problems that would be their clients. Another reason for considering Eells's second defense as at best only a partial success is that all-of-a-piece maximizers of these two sorts would end up hoping for different acts in some third-person Newcomblike problems and for different states in some nonaction Newcombesque problems. This is relevant if, as I think, decision theory, or the theory of the choiceworthiness of acts, should fit smoothly into a theory of the desirabilities of possible facts as distinct from the desirabilities of possible items of news.

5.1. Eells's first static argument – his "original defense"

We may paraphrase Eells (1984a, pp. 73–6) as follows:

> In a common-cause problem an ideally rational agent would be sure that the feared cause C could cause his act S only by affecting the credences and preferences R on which his decision will be based. Such an agent, in his sophistication, would be fully conversant with these credences and preferences. This

information would "screen off" possible further information concerning that common cause, and make that cause probabilistically irrelevant to his act. But then, by symmetry of irrelevance, his act would be made probabilistically irrelevant to that cause. There could not be, for an ideally rational agent, the kind of evidential bearing of acts on causes that is essential to a common-cause Newcomblike problem. For such an agent, there would be probabilistic independence as well as causal independence of circumstances from acts. The two paradigms, causal utility maximization and evidential Desirability maximization, would thus agree in their prescriptions for any ideally rational agent, and they would both be right.[2]

5.2. Horwich's objection

The argument is, as Eells has come to realize, at least incomplete. To say that an agent is sure that certain of his credences and preferences convey all of some possible common cause's *causal* relevance to his act is not to say that his knowledge of these encompasses all relevant information that could be provided by knowledge of that common cause, or that these credences and preferences already have for him all of the *evidential* relevance to his decision and eventual act that news of that common cause could provide. This is so because, though an ideal agent knows for sure what his credences and preferences are, Eells takes care not to assume for this argument that his ideal agent knows what rule he will use to translate his credences and preferences into a decision:

> It is worth noting that the argument does not assume that DM [the decision maker] actually will apply PMCEU [principle of maximizing conditional expected utility], or even that DM knows what decision rule he will use. The argument is intended to show that PMCEU would, *if applied,* give the right answer. (Eells 1985a, p. 212, n. 19)

(cf. Eells 1981, p. 324: "I do not assume that the agent knows what rule he will use.") Indeed, Eells does not insist that his rational agent, the decision maker of his argument, have a fixed decision rule. Eells supposes that "the assumption that an agent is rational should be enough to ensure that the presence or absence of a common cause will not affect which decision rule is used," but he writes that this latter condition "is not essential to [the] argument" (Eells 1985a, p. 203; I have transposed his text). It is thus clear that information concerning the presence or absence of that common cause could help the agent see what he will probably make of his preferences and beliefs (cf. Eells 1984a, p. 78), and so could be further evidence for what he was going to do, which conversely would then be evidence for it.

Eells (1984a, pp. 76ff) calls this objection to his original static defense "Horwich's objection." He thinks that it may be possible to meet the objection by insisting on very strong conditions of rationality, but rather than developing such a response he undertakes, "in the interest of defending evidential decision theory as applicable to agents who do not satisfy such strong conditions of rationality," a second dynamic answer to Horwich's objection: Eells tries "to show that the required independence ... should in fact, *eventually* ... hold of the agent's probabilities" (Eells 1984a, pp. 78-9, emphasis added). Eells leaves open the exact relationship between original defense and this new one:

> By taking a closer look at the "dynamics" of the process of deliberation, I shall try to put ... objections to rest, not trying to decide here whether the ideas involved ... constitute a revision, rather than an elabloration of the original defenses. (Eells 1984a, p. 72)

5.3. Eells's second dynamic argument – his theory of "continual CEU maximization"

Here is a paraphrase of this second defense, made with license but I trust without prejudice (cf. Eells 1984a, pp. 83-92). Let a problem be *ripe for decision* if and only if the agent in it is sure that he will receive "no new [possibly practically relevant] information 'from outside' [and that] all he will learn [of possible practical relevance] during the course of his deliberation is information about how his deliberation is going" (p. 87). According to Eells's second argument, an ideally rational Desirability maximizer can begin deliberation with credences and preferences that make a Newcomblike problem that is ripe for decision. Strong idealizing conditions are avoided that would bar him from, and thus compromise his relevance to, the problems of ordinary variously limited agents (including, perhaps, even "the dithery and the self-deceptive" ones; Lewis 1981a, p. 10).

However, though this ideal Desirability maximizer can *begin* deliberations in such a problem, he will proceed (very quickly if need be) to size up his credences and preferences according to his way of thinking and will not, Eells assures us, ever *end up* in such a problem. His initial credences and preferences in such a problem will, to his way of thinking, favor the good news act (e.g. not smoking). Reflecting on this, of which he has introspective knowledge, provides him with evidence that he will choose that favored act, and that a decision for it is what he would make of those initial credences and preferences were they his final credences and preferences. After conditionalizing on this evidence, he thinks again. Indeed, ideally rational agent that he is, he conditionalizes anew and rethinks – time permitting and as required in order to reach stable credences, preferences, and inclinations – again and again. As "the agent [deliberating continuously in this way] carries out sufficiently many calculations . . . before the time of action, accommodating [before recalculations] the information learned from previous calculations" (Eells 1984a, pp. 91–2), he becomes more and more confident both of what his final credences and preferences will be and of what he will make of them. Eventually, at some point before the moment of decision and from that point on, he "knows" (i.e., he is, at least for all practical purposes of the case, as near to certain as makes no difference) what his final credences and preferences will be, and knows what decision he will make of them, and perhaps what decision he is just about to make of them (cf. Eells and Sober 1986, p. 228, n. 5, and p. 235). That is, as an ideally rational and sophisticated Desirability maximizer deliberates, he gets "*precisely* the kind of information whose earlier absence was *precisely* the reason for the nonindependence between [common causes] and acts" that the "original defense" failed clearly to exclude (Eells 1984a, p. 88).

"Horwich's objection [is thus] answered by supposing that the agent continually . . . calculates [Desirabilities] with appropriate alterations in his subjective probabilities in the light of the results of previous calculations" (Eells 1984a, p. 92). An ideally rational evidential Desirability maximizer can find himself initially undecided in a ripe-for-decision Newcomblike problem in which feared common causes are probabilistically dependent on his acts; however, in the end (when he makes his decision), his knowledge of his credences and preferences and of what he is about to make of them screens off those causes from his acts, and makes them probabilistically independent of one another (either completely, or nearly enough for all practical purposes).

And so "the principle of maximizing [Desirability] will, in the end, even if only after a false start, prescribe the correct act," which is the same act that is then prescribed by the principle of maximizing causal utility (Eells 1984a, pp. 91–2). And it seems that this result opens the possibility of a completely general evidential decision theory that addresses itself to the problems of all agents, and not just to those of ideally rational evidential Desirability maximizers. Ignoring ties (in order to simplify), it can seem – as suggested by Sobel (1988a, p. 128, n. 5) – that this result provides a basis for the general rule that an act in a decision problem that is ripe for its agent is rational if and only if this act is the one an ideally rational evidential Desirability maximizer would decide to perform were he in precisely the same ripe-for-decision problem.

5.4

It may be that, for an ideally rational evidential Desirability maximizer, in the end there always will be probabilistic or evidential independence from acts of conditions that are seen to be causally independent of these acts. It may be that, even if such an agent begins deliberation in a Newcomblike problem, he will end up in a problem that is not Newcomblike. It may be (as Eells maintains) that plausible assumptions concerning the dynamics of ideally rational deliberation entail that no terminal Newcomblike problems are possible for suitably characterized ideally rational agents, even while allowing these agents *entry* into all kinds of decision problems. This would be a marvelous result, but it would not be sufficient for a defense of Desirability maximizing as a fully general theory of practical rationality.[3]

One problem is that, in order for the general theory envisioned to give right answers to all decision problems, it is necessary for ideally rational evidential maximizers to reach decisions in their ripe-for-decision problems that are always right not only for themselves, but right for all ordinary agents in identical problems. It would be necessary, one might say, for the decisions of these paragons of evidential reasoning to in every case also be right for their clients. It will be argued that this condition is not met and that, even giving "metatickles" every possible due, there remain decisions that ideally rational desirability maximizers would get wrong for their clients.

A second problem is that, even if this second defense of Eells's yielded a satisfactory fully general theory of rational *decisions,* it could not be made into a defense of the larger evidential theory of rational *desires* of which an evidential theory of rational decisions would be only a part. To get the larger umbrella theory right there seems to be no recourse but to "go causal" in at least parts, and to keep it simple it seems best to go causal throughout.

5.5

Eells's argument does not work as a fully general evidential Desirability maximizing theory of rational action. For even if (as is plausible) continual Desirability maximizing, by converting crucial probabilistic dependencies into independencies, gives correct answers for all dominance Newcomblike problems, it gives wrong answers for some nondominance problems staffed by ordinary agents who are not ideally rational evidential Desirability maximizers. I have used the popcorn problem to make this point (Sobel 1988a). Here, for a change, I consider Brian Skyrms's Uncle Albert.

Uncle Albert is not feeling well. He realizes that for him going to the doctor would be a further symptom that he is genuinely ill. But he knows that it would have no tendency to make him ill, and that staying home would certainly not help him to get well, if he really is ill. And so he goes to the doctor, notwithstanding the inferior subjective prospects of his doing so. "He's no fool" (Skyrms 1982, p. 700).[4] (Numbers are stipulated for this problem in Appendix II.)

Nor is Uncle Albert an ideally rational Desirability maximizer. But suppose that he were. Consider, that is, a ripe-for-decision case exactly like his – one in which, however, the agent, IDesM, is an ideally rational continual Desirability maximizer deliberator (thus his name). IDesM will calculate, discover that his calculation favors staying home, and take this good news as evidence that he is going to stay home and that he is not sick after all! Recalculations by this ideally rational Desirability maximizer will in this case only confirm first comparative results and first tentative dispositions to action. And so, even if (as is plausible) they lead to probabilistic independence of his health from his acts and to a non-Newcomblike problem, recalculations will not lead to the *same* non-Newcomblike problem to which the deliberations of an ideally rational causal utility maximizing agent would lead. Desirability and utility maximizers (whether or not they are ideally rational) would both decide to stay home in an ideally rational evidential Desirability maximizer's final problem, whereas they would both decide to go to the doctor in a ideally rational causal utility maximizer's final problem.

An ideally rational evidential Desirability maximizer in a ripe-for-decision problem exactly like Uncle Albert's would decide to stay home. And that seems the right decision for him because while deliberating he would acquire what for him would be good evidence that he was not sick after all. Ideally rational evidential Desirability maximizer that IDesM is,

[he believes] that, for him, the way in which [it could happen that his being ill would cause his going would be] by causing . . . beliefs and desires [on which] evidential decision theory [would] prescribe going. [We have seen that IDesM eventually] finds [that *not* going is] maximal [and prescribed by evidential decision theory]. This is evidence against [his illness]. (Eells 1989b)

Here I have turned Eells's references to Uncle Albert into references to our IDesM. For Uncle Albert himself is no kind of evidential Desirability maximizer: If *he* notices that staying home maximizes evidential Desirability, he quite sensibly makes nothing of this and is not at all inclined on that account to stay home.

Though IDesM would acquire good evidence that *he* was not ill, he need learn nothing of necessary relevance to Uncle Albert's condition. Indeed, if he knew (as we do) that Uncle Albert is not an evidential Desirability maximizer, let alone an ideally rational one, then IDesM would realize that he had learned nothing of relevance to Uncle Albert's condition, which IDesM would still (just as Uncle Albert does) fear was grave. But then IDesM's decision to stay home, while right for him, would be wrong for Uncle Albert (though IDesM, applying evidential decision theory to Uncle Albert's case, would deny this). The probability that Uncle Albert is really ill – both IDesM's probability for this and Uncle Albert's own probability – is high enough for us to agree (notwithstanding the benighted official positions of hypothetical evidential theorist kibitzers) that Uncle Albert is right: he had better go to the doctor.

The two paradigms diverge in Uncle Albert's ripe-for-decision problem, and would take ideally rational and sophisticated maximizers in different directions – the causal ones to, and the evidential ones away from, the doctor. And the way in which these theories can diverge in nondominance problems strongly suggests that, even if an evidential decision theory can (by metatickle elaborations) be made to give right answers in all dominance Newcomblike problems, it will not – in them, or in any problems, or in anyone's problems – give "right reasons" (Lewis 1981a, p. 10).

5.6

Eells's dynamic defense does not work. It is not true that the two paradigms converge at least for suitably characterized ideal agents, and that a theory of evidential Desirability maximizing can be made to deliver correct answers in all problems including nondominance ones, for all agents including most prominently agents who are not ideally rational evidential Desirability maximizers (as few if any real agents are), and deliver these answers "on the cheap" and without more or less explicit recourse to causal primitives. And even if it could do so, the gain in conceptual economy would be disappointingly local and of only limited Humean comfort. This is my second major response to Eells's dynamic defense of evidential decision theory.

A theory of rational decisions should fit smoothly into a general theory of rational hope (cf. Sobel 1985a, p. 199, n. 7), for it should be rational to choose to do what is rational to *hope* that one will do. I note that Jeffrey (1965b; 1983) situates his own evidential logic of decision, one that would have the choiceworthiness of acts go by their Desirabilities, in a theory of Desirabilities that extends beyond given agents' action propositions to propositions in general. However, even if (as may be so) the true choiceworthiness of acts of ideally rational and sophisticated evidential Desirabilities is in every case determined by their eventual evidential Desirabilities, the true hopeworthiness of facts in general, and not just facts concerning one's possible acts, need not be for *any* person invariably determined by this person's final evidential Desirabilities.

Consider a case in which I know that my friend is in a Newcomb problem, so that I am in what we can term a *third-person* Newcomb problem in which my question is: What, for his sake, should I hope that he will do? Every consideration that makes it seem that he should for his sake take both boxes makes it seem that I, as his friend, should hope that he will take both boxes. (See Appendix I for a description of these boxes.) For I know that – whatever he thinks, and whatever he has tentatively (or even finally) decided to do – if he takes both boxes then he will get more money than if he takes just the second box. I should, it seems, find his taking both boxes more hopeworthy, more desirable, and more welcome as a fact. Of course, if we suppose that I do not know what the predictor has predicted, am ignorant of what is in the second box, and have great confidence in that predictor's powers; then "news" (by which I mean a new subjective certainty that may be false, thus the scare quotes) that my friend is going to take just the second box, or that he already has, might be for me most welcome news. It might be, for me, far more welcome than would be news that he was going to take both boxes, even though I do hope that he will take both boxes. Regarding the complex issue of what I should hope to learn is a fact concerning his act, I can be of two minds without being in any way confused (cf. Sobel 1985a, p. 199, n. 7). The divergence between what it would be rational for me to hope is a *fact* in this case, and what might be for me best *news* and most welcome as a new subjective certainty, could survive any amount of reflection on my part about the case, including any amount of reflection on my part about the probable course of my friend's deliberations and efforts to decide what to do, and including any amount of reflection on my part about the course of my own efforts to make up my mind what to hope he will in fact do, and on my efforts (if any) to make up my mind what news to hope for concerning his act.

In this case, I should not hope for the fact that would be hoped for by an ideally rational person, IDesH, whose hopes are always determined by his final Desirabilities for propositions in the case. For I should hope for the best fact, which we have seen is that my friend takes both boxes, whereas IDesH would hope for the fact the news of which would be for him (and for me) best news – that my friend takes just the second box.[5]

5.7

It thus seems that even if a metatickle-enhanced evidential theory could be made to give right answers for all decision problems regarding what to make true, it cannot be made to fit smoothly into an evidential theory that gives right answers to all attitude problems regarding what to hope is true. But wait. My proposal – that the theory of rational acts should be a part of a general theory of rational desires for facts – is formally similar to the proposal on which Jackson and Pargetter (1983) build a response to Eellsian "tickle defenses." They hold that the first-person theory of rational decisions should be imbedded in a theory that yields not only agent assessments but also spectator assessments of the rightness of acts. And so it may seem that Eells's (1985b) reply to their argument can be revamped and addressed to mine. This has been tried. Regarding the case of my friend, Eells has said:

> I should take *my* subjective probablities and *his* desirabilities [with which, friend that he is, I presumably agree] and then *hypothetically* deliberate (evidentially), as if my beliefs and his desires are all my own. Then I should hope that he does the act that has maximal (evidential) expectation from my hypothetical perspective. . . . [D]oing this, with the evidential theory, will, with metatickles, give the appropriate prescription. (Eells 1989b)

This approach can be extended beyond friends to foes (whom I wish ill), and in general to all others, by having me deliberate from the hypothetical perspective of others with not only my own probabilities but also my own Desirabilities.

But there is a problem with this recycling of Eells's reply to Jackson and Pargetter, for my proposal is only somewhat similar to theirs. It differs most notably in being more general. They would imbed a theory for personal decisions in a general theory for prescriptions to all persons. I hold that a theory for what to do should be, as Jeffrey's theory is, imbedded in a general theory for what to hope. And so, even if a metatickle-enhanced evidential theory could be made to provide right answers for all decision problems of what to do as well as for all attitude problems of what to hope others will do, still more is required for a complete theory of what to hope. Even supposing that an evidential theory could be made to yield right decisions and hopes for all *acts,* a purely evidential theory cannot be made to yield right hopes for not only all possible acts but for absolutely all possible facts. Evidential theories are quite stymied, I think, by some nonaction Newcombesque problems. Here are two such problems.

5.7.1. How do you spell relief? Consider a young person who believes there is a gene that, in adults, causes cancer but tends to *prevent* heartburn.

What should he, in his youth, hope for when he grows up? Should he hope for heartburn? Given assumptions summarized in the matrices of Appendix II, heartburn should be for him most welcome as an item of news (or new subjective certainty) in view of its evidential potency. News that when he grows up he will be free from heartburn would be bad news. And (on these assumptions) the proposition that he will get heartburn when he grows up will excel in evidential Desirability, and could continue to excel throughout any course of ideal reflection on the matter that included reflection on the credences and hopes being firmed up. But even so, that he will be free from heartburn when he grows up should be most welcome as a fact in view of its probable causal potency and the differences it might make. For who wants heartburn, whether along with or without cancer?! (Numbers are stipulated for this problem in Appendix II.)

Three things may be stressed. First, while there is certain causal independence of value-relevant circumstances from possible facts, there is evidential dependence; that is, the case is Newcomblike. Second, this dependence could survive any amount of reflection on the youth's part to determine which proposition, $\sim H$ or H, to hope for as a fact or truth, or which to hope for as an item of news, regardless of what maximizing principles operated for him in these reflections. And third, there is no point of view to which his credences and preferences might be transported, from which ideal reflection on either of these issues would, by metatickle mechanisms, in the end necessarily resolve this dependence.

5.7.2. Pennies for heaven?

For a theological case to the same point, consider someone who believes "that a rich man shall hardly enter into the kingdom of heaven" (*Matthew* 19:23), that "it is easier for a camel to go through the eye of a needle, than for a rich man to enter into the kingdom of God" (19:24), and that "many that are first shall be last" (19:30). Let this person believe these things are so *not* because riches make it difficult for a person to be good and deserving, but rather because God, for His own reasons, rarely bestows both saving grace and worldly riches on the same person. Consider, that is, someone who believes that riches, while in no way causal obstacles to eternal bliss (which is already settled yea or nay before one is born), are reliable signs that it is not forthcoming. Should such a person want to be rich?

Nondominance Newcomblike problems such as Uncle Albert's argue that not even a correct limited theory designed specifically for rational decisions can be purely evidential.[6] Third-person Newcombesque problems argue, and nonaction Newcombesque problems argue more forcefully, that – even if a purely evidential theory of rational acts could be developed that yielded right answers for all agents in all decision problems –

no purely evidential theory for desires for facts will give right answers regarding even the desires of ideally rational evidential desirers, and that for right answers a general theory needs to be causal at least in part and so for simplicity is best made causal throughout. Third-person and nonaction problems provide further grounds for thinking that – though they can be made to give right answers to a range of questions concerning what acts to perform and what facts to hope for – evidential theories never give right reasons.

Summing up Section 5: Although Eells's dynamic metatickle defense of evidential theories is promising against dominance Newcomblike action problems, it fails against nondominance ones. And all metatickle defenses have difficulty with third-person Newcombesque problems including dominance ones, and cannot even get started against nonaction Newcombesque problems. Probably Eells would have given up on tickles, or never started on them, had he not concentrated exclusively on problems that are like Newcomb's problem itself – not only in that they are problems for which simple evidential and causal decision theories diverge, but also in that the action selected by causal theories is a dominant action – or if he had seen his task as that of defending a general evidential theory of the hopeworthiness of facts, rather than a special theory confined to the choiceworthiness of acts.

6. Conclusions

Problem cases for Jeffrey's (1965b) logic of decision come in many varieties, and, depending on their details, occasion objections of several sorts. The best cases for displaying the putative error of identifying an action's choiceworthiness with its Desirability are, I think, "predictor" cases, where the predictor is a friend of whose predictions the agent is reasonably respectful. Such cases avoid the science fiction-like aspects of Newcomb's problem, and need not include the idea that the agent thinks of his actions and/or choices as things that will be determined by causes. Prisoners' Dilemma cases are I think not as good for the purpose stated, owing to the relative difficulty of making plausible within them that a reasonable agent might still, even just before the time for action, view his actions (both the one he then presumably considers reasonable, and the alternative action) as sufficient evidence for like actions on the part of the other agent. (See Sobel 1985b, [Chapter 3].) But while I think that acausal predictor cases have dialectical advantages, in my view most Newcomblike cases that have been brought against Jeffrey's logic of decision are, when generously interpreted, coherent challenges that need to be taken seriously.

Critics of Jeffrey's theory hold that at least some Newcomblike problems are good challenges that dramatize the point that, when news of an act would provide evidence of states that the agent is sure it would not affect, although in his act "the agent indeed makes ... news, he does not make all the news his act bespeaks" (Gibbard and Harper 1978, p. 140). Critics hold that for such an agent, as Jeffrey (a sometime critic himself) puts it, preferring "A1 over A2 *as a news item* ... is compatible with [a] preference for A2 over A1 *as a course of action*" (Jeffrey 1981a, p. 381).

Partisans of causal decision theories think that Newcomblike problems should convince one that, in order to get things right, a theory needs to take more or less explicitly into account opinions about causes and objective chances. I note that Jeffrey himself at times agrees with this:

> Purists like de Finetti hold that the concept of objective chance is metaphysical hocus-pocus.... But I am persuaded that the notion of objective chance often plays a useful rôle in our practical deliberations – maybe, an essential rôle. Perhaps the clearest example is a ... Prisoner's Dilemma [in which the prisoners] are pretty sure they'll decide alike *but don't know how* [and in which I, a prisoner] see my decision as a rather good indicator of his.... Confession, the wiser choice, is ruled out by judgemental c.e.u. maximization if you see your choice as a strong but ineffectual symptom of his. (Jeffrey 1986, pp. 10–11)

Jeffrey also includes "a version of 'causal' decision theory" (1986, p. 16, n. 6).

I suspect, however, partly because of subtleties in the paper from which I have quoted, that Jeffrey is still inclined to think that the logic of decision (1965b), unadorned and unrevised, may yet take the high ground. He seems at times to think that one might gracefully withdraw from the fray and maintain the equation of rational preferences with Desirability maximization, not as a rule for making decisions, but as a fundamental principle of decision theory that connects consistent and rational decisions, credences, and preferences. He writes that, in a Prisoners' Dilemma in which the agents are

> pretty sure [they will] confess.... [T]he ... description, "They are pretty sure they'll decide alike *but don't know how*", is inaccurate. I do know how, not by c.e.u. maximization [that is, not by judgmental conditional expected utility or Desirability maximization] but by a dominance argument that refers to causal considerations. That's all right, or, anyway, not all bad. As Sandy Zabell urges, [the principle of] c.e.u. [maximization] enters decision theory as a way of connecting preference, probability, and utility: A is preferred to B iff $P(u/A) > P(u/B)$. It is a mistake to view this biconditional as showing that preference is determined by P and u, not vice versa. (Jeffrey 1986, p. 12)

Here $P(u/-)$ is "c.e.u. ... computed via judgmental prevision" or judgmental conditional probabilities (Jeffrey 1986, p. 11); $P(u/A)$, for example, is Des A.

I have by implication urged that, although the biconditional according to which final preferences for actions are determined extensionally by Desirabilities may be right for ideally sophisticated rational agents, third-person and nonaction problems suggest that it is wrong when extended from action propositions to all propositions. Whereas rational preferences for news may for the most part be determined by Desirabilities, rational preferences for facts sometimes are not. Reflection on Newcomb-like problems (especially nondominance ones) and on third-person and nonaction ones challenges the idea that Desirabilities enter decision theory, or the theory of rational preferences for action facts, in any important way at all.

APPENDIX I: STATEMENTS OF SOME NEWCOMBLIKE PROBLEMS

Newcomb's Problem

Suppose a being in whose power to predict your choices you have enormous confidence.... There are two boxes, (B1) and (B2). (B1) contains $1000. (B2) contains either $1,000,000 ($M), or nothing.... You have a choice between two actions: (1) taking what is in both boxes, (2) taking only what is in the second box. Furthermore, and you know this . . . : (I) If the being predicts you will take what is in both boxes, he does not put the $M in the second box. (II) If the being predicts you will take only what is in the second box, he does put the $M in the second box.... First, the being makes its prediction. Then it puts $M in the second box, or does not.... Then you make your choice." (Nozick 1969, pp. 114-15)

For a mechanically simpler version of Newcomb's Problem we have the following:

A preternaturally good predictor of human choices does or does not deposit a million dollars in your bank account, depending on whether he predicts you will reject or accept the extra thousand he will offer you just before the bank reopens after the weekend. Would it be wise on Monday morning to decline the bonus? (Jeffrey 1983, p. 15)

See Sobel (1985a, pp. 198-9, n. 6).

It seems the reasonable thing would be to take both boxes, or accept the offer of the thousand, and end a thousand ahead no matter what. Yet news that you were going to take just the second box, or reject the offer, would be good news; given that news, you would be confident that you were going to be rich.

The Prisoners' Dilemma

Two men are arrested for bank robbery. Convinced that both are guilty, but lacking enough evidence to convict either, the police put the following proposition to the men and then separate them. If one confesses but the other does not, the first will go free (amply protected from reprisal) while the other receives the maximum sentence of ten years; if both confess, both will receive lighter sentences of four years; and if neither confesses, both will be imprisoned for one year on trumped-up charges. . . . [T]he police are convinced that both will confess, even though both would be better off if neither confessed. (Jeffrey 1983, p. 15)

Reflection on the following matrix

	He confesses	He does not confess
I confess	−4	0
I do not confess	−10	−1

can explain why the police are convinced that each prisoner will confess. "But the Bayesian principle [of *The Logic of Decision,* Jeffrey 1965b] advises each prisoner *not* to confess, if each sensibly sees his own choice as a strong clue to the other's and therefore assigns high subjective probabilities . . . to the other prisoner's doing whatever it is . . . that he himself chooses" (Jeffrey 1983, p. 16).

Sir Ronald lights up

In 1959, R. A. Fisher argued that the undoubted correlation which had been demonstrated between cigarette smoking and lung cancer admitted of three explanations, among which [was] . . . (iii) Smoking and lung cancer are effects of a common cause, viz., a certain genetic makeup. Fisher urged that the evidence then available made (iii) a lively alternative. . . . Coupled with his reference to 'the mild and soothing weed,' this leads me to guess that Fisher was a smoker who thought that while abstinence might be of prognostic interest, it was probably useless either as prophylaxis or as therapy. (Jeffrey 1981b, p. 476)

Fisher's smoking, notwithstanding its predicting cancer, seems reasonable, if he would think that it had no chance of causing the disease.

The popcorn problem. I want very much to have some popcorn, but

I am nearly sure that the popcorn vendor has sold out. . . . And so . . . I have decided not to go for popcorn. . . . Still, I am also nearly sure that in *this* theatre, when and only when there *is* popcorn . . . the signal – POPCORN!!! – is flashed on the screen, though at a speed that permits only subliminal, unconscious awareness. . . . [And, since] I consider myself to be a highly *suggestible* person . . . I am nearly sure that I will change my mind and *go* for popcorn if and only if I am influenced by this subliminal signal to do so. . . . [And so] while I think it is very unlikely that I will go for popcorn . . . I think it is *much more* unlikely that I will go for popcorn though there is none . . . to be bought. . . . In short . . . my going for popcorn would provide me with a near certain sign of . . . there being popcorn. . . . And my *not* going would provide me with a near certain sign that there is *not* popcorn . . . . (Sobel 1986a [Chapter 8])

News that I was going for popcorn would be good news, but – since I am nearly sure that there is no popcorn – it would seem a mistake to go for some.

Evidential potencies "are . . . founded on beliefs concerning possible causes of my actions, but they could be provided with very different kinds of bases. For example, I could be supposed to think that the manager was a very reliable predictor who acted on his predictions in order to sell as much popcorn as possible, and waste as little as possible" (Sobel 1988a).

Uncle Albert's problem

Uncle Albert believes that, for him, going to the doctor is a symptom of being genuinely ill. Indeed, he believes that, whatever his other symptoms, if he finds himself in the doctor's office in addition, he is more likely to be ill than if not. This need not be an irrational belief. Now, in certain cases, conditional expected utility will recommend staying home because of diminished prospects associated with going in to be examined. But Uncle Albert goes anyway. He's no fool. He knows that staying home won't help him if he's really ill. (Skyrms 1982, p. 700)

The problem of the rising young executive

Robert Jones, rising young executive of International Energy Conglomerate Incorporated . . . and several other young executives have been competing for a very lucrative promotion. . . . Jones learns . . . that all the candidates have scored equally well on all factors except ruthlessness. . . . Jones [before the promotion decision is announced] must decide whether or not to fire poor old John Smith. . . . Jones knows that [his] ruthlessness factor . . . accurately predicts his behavior in just the sort of decision he now faces. (Gibbard and Harper 1978, p. 137)

If Jones wants not to fire old John then it seems he shouldn't, even though doing so would make him confident of the promotion he covets.

Solomon's problem. Solomon faces a situation like the one David faced. He covets a married woman and fears possible civic consequences of public adultery, but "he, unlike David, has studied works on psychology and political science which teach him" that "a king's degree of charisma . . . cannot be changed in adulthood," that "charismatic kings tend to act justly," that "successful revolts against charismatic kings are rare," and that it would not be the unjust act of public adultery that caused civic unrest and revolution, but rather the lack of charisma that it signaled and the sneakiness and ignoble bearing that go with lack of charisma (Gibbard and Harper 1978, pp. 135–6).

Problems for some pregnant women

Many physiological states produce quite specific symptomatic effects on choice behaviour. Pregnancy, for example, often affects what a woman chooses to eat. Any such physiological effect is a potential source of a medical Newcomb problem. All else that is needed is that the attitudes of the person concerned about the possession of the underlying physiological state and the performance of the symptomatic act should conflict.... If the agent is someone who knows that pregnancy tends to make her decline garlic, then the problem will arise if either she likes garlic (ceteris paribus) but wants to be (now) pregnant; or doesn't like garlic (ceteris paribus) and doesn't want to be pregnant. Clearly she shouldn't refuse the delectable garlic in order to be already pregnant, in the former case; or eat the distasteful garlic in order not to be, in the latter. (Price 1986, p. 196)

APPENDIX II: FURTHER STRUCTURE FOR "BRIAN SKYRMS'S UNCLE ALBERT" AND "HOW DO YOU SPELL RELIEF?"

Brian Skyrms's Uncle Albert

A possible consequence matrix for Uncle Albert's problem

	He is really ill	He is not really ill
Go to the doctor	genuine illness, bother, and medical assistance	bother
Do not go	genuine illness	—

Desirabilities and utilities of acts in circumstances

	Ill	~Ill
Go	−90	−10
~Go	−100	0

Conditional probabilities of circumstances on acts

	Ill	~Ill
Go	$14/15$	$1/15$
~Go	$4/15$	$11/15$

Probabilities of act–circumstance practical conditionals

	Ill	~Ill
Go	$9/15$	$6/15$
~Go	$9/15$	$6/15$

These probability matrices reflect the assumptions that Uncle Albert's probability for his being genuinely ill is $9/15$, and that acts Go and ~Go would be equally good further evidence, respectively, for his being genuinely ill and his not being genuinely ill - news of either would make a $\frac{5}{15}$ difference to the probability of the condition for which it would be evidence.

Given the assumptions summarized in these matrices, although $\sim$Go would be best news, Go would be the best act:

$$\text{Des(Go)} = -84\frac{2}{3} < \text{Des}(\sim\text{Go}) = -31\frac{2}{3},$$

whereas

$$\text{Util(Go)} = -58 > \text{Util}(\sim\text{Go}) = -60.$$

How do you spell relief?

A possible consequence matrix for this problem

	I have the gene	I do not have the gene
No heartburn	cancer	—
Heartburn	cancer and heartburn	heartburn

Evidential desirabilities and factual desirabilities for possible facts in circumstances

	Gene	$\sim$Gene
$\sim H$	-1000	0
H	-1001	-1

Conditional probabilities of circumstances on facts

	Gene	$\sim$Gene
$\sim H$	$20/100$	$80/100$
H	$2/100$	$98/100$

Probabilities of fact-circumstance practical conditionals

	Gene	$\sim$Gene
$\sim H$	$11/100$	$89/100$
H	$11/100$	$89/100$

Given the assumptions summarized in these matrices, although possible fact or proposition H would be best news, $\sim H$ would be the best fact:

$$\text{EviDes}(H) = \tfrac{2}{100}(-1001) + \tfrac{98}{100}(-1) = -21$$
$$> \text{EviDes}(\sim H) = \tfrac{20}{100}(-1000) + \tfrac{80}{100}(0) = -200,$$

whereas

$$\text{FacDes}(\sim H) = \tfrac{11}{100}(-1000) + \tfrac{89}{100}(0) = -110$$
$$> \text{FacDes}(H) = \tfrac{11}{100}(-1001) + \tfrac{89}{100}(-1) = -111.$$

That FacDes($\sim H$) exceeds FacDes(H) follows of course without calculation given that (i) $\sim H$ dominates H in the matrix for factual desirabilities in circumstances, and (ii) circumstances Gene and $\sim$Gene are independent of facts $\sim H$ and H in the probability matrix relevant to factual desirabilities.

APPENDIX III: DOMINANCE PROBLEMS[7]

A *(weak) dominance argument* has as premises (i) that circumstances of a partition for a problem are independent of options in this problem and (ii) that some option (weakly) dominates all others under this partition, and has as its conclusion that this option (weakly) maximizes - that is, that its expected value (weakly) excels in comparison with the expected values of all other options in the problem. I mean by a *dominance problem* one that, given the agent's credences and preferences, can be resolved by a useful dominance argument and can thus be resolved without even rough-and-ready calculations and comparisons of the expected values of options.

Let a partition C of circumstances be *irrelevant* to options in a problem if and only if, for every $c \in C$ and option x, the expected value of x in c equals the expected value of x simpliciter. And let a partition C of circumstances be *independent* of options if and only if, for every $c \in C$ and option x, the probability of c given x equals the probability that c. Then

> an option x maximizes in a problem if there is a partition
> of circumstances that is irrelevant to and independent of
> options in this problem, and under which partition x is
> (weakly) dominant.

Furthermore, all natural credences and preferences afford irrelevant and independent partitions for all problems. For example, the partition {Alpha Centauri is at least as large as the Sun, Alpha Centauri is smaller than the Sun} is both irrelevant to and independent of options in all problems that come readily to mind. But then, assuming natural credences and preferences,

> an option x maximizes if and only if there is a partition that
> is independent of options, under which partition x is
> (weakly) dominant - that is, if and only if there is a (weak)
> dominance argument for its maximizing.

For an option maximizes under a partition that is irrelevant to options if and only if this option is (weakly) dominant under this partition.

However, even though (assuming natural credences and preferences) an option maximizes if and only if there is a dominance argument for it, there need not be a *useful* dominance argument for its maximizing. It is clear that dominance arguments for an option's maximizing that are based on irrelevant partitions are never useful, since claiming dominance for an option under such a partition is equivalent to claiming that this option maximizes. A dominance argument, even a sound one, to the effect that a certain option maximizes in some decision problem would, if based on an irrelevant partition, plainly be of no use at all as a resolution of this problem, or as demonstration, *sans* calculations of the expected values, that that option maximizes.

APPENDIX IV: FURTHER STRUCTURES FOR NEWCOMB'S PROBLEM

Assume that the agent is interested only in the money, and that his confidence in the predictor is reflected in conditional probabilities of .9 for correct predictions whichever choice he makes. Let the agent's probability for their being $M in (B2) be $p > 0$; let p also be the agent's probability for the predictor's predicting that the agent takes only (B2). Two analyses of this problem are presented. Though they can seems to favor different results, they in fact agree in their conclusions.

Analysis featuring the dominance of "take both boxes"

A consequence matrix for Newcomb's Problem

	There is $M in (B2)	There is nothing in (B2)
Take both boxes	$M + $1000	$1000
Take only (B2)	$M	$0

Possible desirabilities and utilities of acts in circumstances

	$M	~$M
Both	1001	1
Box 2	1000	0

Conditional probabilities of circumstances on acts

	$M	~$M
Both	.1	.9
Box 2	.9	.1

Probabilities of act-circumstance practical conditionals

	$M	~$M
Both	p	$1-p$
Box 2	p	$1-p$

Probabilities in the last matrix reflect the agent's certainty that these states are causally independent of his possible acts. Given the assumptions summarized in these matrices, 'Box 2' would be best news but 'Both boxes' would be the best act, regardless of the value of p, because of the dominance of Both and the independence in the second probability matrix of circumstances from acts:

$$\text{Des(Box 2)} = .9(1000) + .1(0) = 900 > \text{Des(Both)} = .1(1001) + .9(1) = 101;$$

$$\text{Util(Both)} = p(1001) + (1-p)(1) = 1 + p(1000)$$
$$> \text{Util(Box 2)} = p(1000) + (1-p)0 = p(1000).$$

Analysis featuring the agent's confidence in the predictor

A consequence matrix for Newcomb's Problem

	Correct prediction	Incorrect prediction
Take both boxes	$1000	$M + $1000
Take only (B2)	$M	$0

Possible desirabilities and utilities of acts in circumstances

	Correct	~Correct
Both	1	1001
Box 2	1000	0

Conditional probabilities of circumstances on acts

	Correct	~Correct
Both	.9	.1
Box 2	.9	.1

Probabilities of act-circumstance practical conditionals

	Correct	~Correct
Both	$1-p$	p
Box 2	p	$1-p$

Calculations yield the same values as in the first analysis, and again make Box 2 the best news and Both the best act. Again, this is so regardless of the value of p, though this time there is no dominance in the value matrix:

$$\text{Des(Box 2)} = .9(1000) + .1(0) = 900 > \text{Des(Both)} = .9(1) + .1(1001) = 101;$$

$$\text{Util(Both)} = (1-p)(1) + p(1001) = 1 + p(1000)$$
$$> \text{Util(Box 2)} = p(1000) + (1-p)0 = p(1000).$$

Appendix V: Ratificationism[8]

"Ratificationism is the doctrine that the choiceworthy acts – realtive to your beliefs and desires – are the ratifiable ones" (Jeffrey 1983, p. 19). An option is *ratifiable* if and only if "on the hypothesis that *that* option will finally be chosen" its Desirability is at least as great as that of any other option (p. 19).

For precision, let probability function $\Pr_q$ come from a function $\Pr$ by conditionalization on q, $\Pr(q) > 0$, if and only if $\Pr_q(p) = \Pr(p/q)$. For p and q such that $\Pr(q) > 0$ and $\Pr_q(p) > 0$, let the *conditional desirability* of p on q, $\mathrm{Des}(p/q)$, be the weighted average of the values of p's worlds in which the weight for the value of a world w is $\Pr_q(w/p) = \Pr_q(w \& p)/\Pr_q(p)$. Let a^* be a *decision* to do a. And, for an agent with probability function $\Pr$, let an action a (such that $\Pr(a^*) > 0$, $\Pr_{a^*}(a) > 0$, and, for every alternative a' to a, $\Pr_{a^*}(a') > 0$) be *ratifiable* if and only if, for every alternative a', $\mathrm{Des}(a/a^*) \geq \mathrm{Des}(a'/a^*)$.

Ratificationism is a revision of Jeffrey's logic of decision, not a defense. It replaces the rule "maximize Desirability" with the rule "make ratifiable decisions" or, "romantically: 'Choose for the person you expect to be when you have chosen'" (Jeffrey 1983, p. 16). Ratificationism is a relatively conservative revision of the logic of decision. It avoids the primitives that Jeffrey most wanted to avoid – causal primitives, including kinds of causal conditionals – though it does have one new primitive: namely, *decisions* as things distinct from and prior to actions.

The question is whether Ratificationism is not only an attractive revision to Humean-minded theorists, but a successful one that solvs all Newcomblike problems. It is not. Jeffrey has come himself to this view. He thought otherwise in 1981, when he wrote "The Logic of Decision Defended," but had changed his mind by 1983, when he was putting finishing touches to the second edition of *The Logic of Decision*. I will comment on the reason that turned Jeffrey against ratificationism before detailing my own objection to it.

The rule "make ratifiable decisions" agrees with the rule that it would replace, namely, "maximize Desirability,"

> in ordinary cases, but gives what I take to be the right solutions in many (alas! [Jeffrey writes] not all) of the bothersome cases where agents see their decisions as merely symptomatic of states of affairs that they would promote or prevent if they could. (p. xii)

For the remainder of this appendix, all page references are to Jeffrey (1983). The aside "(alas! not all)" was, I believe, something like a "Stop the press!" insertion.

We consider first a bothersome case – more specifically, a bothersome Prisoners' Dilemma – in which the new rule does improve on the old rule, giving the right answer where the old rule gave the wrong one. For definiteness we assume the following Desirability matrix for "my" (a prisoner's) problem.

	C'	$\sim C'$
C	2	4
$\sim C$	1	3

Confessing is dominant, and since the other's actions are certainly causally independent (even though not evidentially independent) of mine, confessing is what Jeffrey is sure I should do.

But the Bayesian principle [the old rule] advises each prisoner *not* to confess, if each sensibly sees his own choice as a strong clue to the other's and therefore assigns high subjective probabilities . . . to the other prisoner's doing whatever it is . . . that he himself chooses . . . . (p. 16)

Assuming, for simplicity, extreme conditional probabilities and improbabilities, we have the following probability matrix.

	C'	$\sim C'$
C	[1]	[0]
$\sim C$	[0]	[1]

Des($\sim C$) = [3] > Des(C) = [2], and the old rule, "maximize Desirability," prescribes not confessing – bad advice, Jeffrey thinks (where, e.g., "[1]" denotes "nearly 1"). However, if we add to our story that "on each hypothesis about his *decision,* each prisoner . . . sees his own *performance* as predictively irrelevant to the other's" (p. 17) then the new rule, "make ratifiable decisions," prescribes confessing, which is what Jeffrey wants prescribed.

For this version of our bothersome Prisoners' Dilemma: $\Pr_{C}\cdot(C'/C)$ and $\Pr_{(\sim C)}\cdot(\sim C'/\sim C)$ should still be nearly 1; the stipulation that "on each hypothesis about his *decision* . . . each prisoner . . . sees his own *performance* as predictably irrelevant to the other's" *aligns* [1]s and [0]s; and we have the following matrices for conditional-on-my-decisions conditional probabilities for the other's actions on my own actions.

	$\Pr_{C^*}$			$\Pr_{(\sim C)^*}$	
	C'	$\sim C'$		C'	$\sim C'$
C	[1]	[0]	C	[0]	[1]
$\sim C$	[1]	[0]	$\sim C$	[0]	[1]

"As probabilities are independent of acts in each matrix, my dominant performance (confess) will have the higher estimated desirability on each hypothesis about my final decision" (p. 17). That is, given that C dominates $\sim C$ in the Desirability matrix and that the other's actions are probabilistically independent of mine in both of these matrices: although $\mathrm{Des}(C/C^*) > \mathrm{Des}(\sim C/C^*)$ (in numbers, [2] > [1]) and C is ratifiable, still $\mathrm{Des}[C/(\sim C)^*] > \mathrm{Des}[\sim C/(\sim C)^*]$ ([4] > [3]) and $\sim C$ is not ratifiable. In this version of our bothersome Prisoners' Dilemma, the new rule gets the right answer and so as promised improves on the old rule.[9]

But, alas, it is not always so. The new rule, Jeffrey came to be convinced, does not improve on the old one in every version of our bothersome Prisoners' Dilemma. It fails in what Jeffrey describes as a "more plausible variant . . . (suggested by Bas van Fraassen) [in which] the sorts of extraneous influences that would prevent me from confessing when I had decided to are likely to work on him, too, and to the same effect" (p. 20), and presumably (though Jeffrey does not spell this out) in which the sorts of extraneous influences that would move me to confess notwithstanding a final decision *not* to do so are likely to work similarly on him. In Jeffrey's version of his bothersome Prisoners' Dilemma, the other's actions, though probabilistically dependent on mine, are, *conditional on my two possible decisions,* probabilistically independent of mine. In the variant due to van Fraassen, the other's actions are probabilistically dependent on mine both simply and conditional on my two possible decisions. Making this version definite, we have the following conditional-on-decisions conditional probability matrices.

	$\Pr_{C^*}$			$\Pr_{(\sim C)^*}$	
	C'	$\sim C'$		C'	$\sim C'$
C	[1]	[0]	C	[1]	[0]
$\sim C$	[0]	[1]	$\sim C$	[0]	[1]

Confessing is not ratifiable in this case:

$$\mathrm{Des}(\sim C/C^*) = [3] > \mathrm{Des}(C/C^*) = [2].$$

Though confessing is, Jeffrey is sure, still the choiceworthy act, not confessing is the ratifiable one in this case:

$$\text{Des}[\sim C/(\sim C)^*] = [3] > \text{Des}[C/(\sim C)^*] = [2].$$

The new rule does not improve upon the old one in this "more plausible" bothersome Prisoners' Dilemma.

Whether or not van Fraassen's variant is more plausible than Jeffrey's original one is, I think, not important. What matters (though Jeffrey might disagree about this) is only that this is a logically possible variant, and that ratificationism gives the wrong answer in it. For the record, and even if this is not important, it is however very doubtful that van Fraassen's variant is more plausible. It is part of this case that "I am nearly certain that my performance will accurately indicate my final choice," that I am nearly certain that I will do what I finally decide to do (p. 16). So we ask: What sort of extraneous factors could I then believe might get in the way and prevent me from confessing when I had made a final decision to do so? What might news at that point that I was not going to confess after all tell me? There are, of course, many things that I could think might prevent me from carrying out a final decision to confess, including several mentioned by Jeffrey in another connection:

> Death or a nonfatal cerebral hemorrhage might prevent me from carrying out a decision to confess, and a surreptitiously administered dose of sodium pentothal might set me to confessing in spite of my decision not to. But changing my decision does not count as a way of not managing to perform my chosen action. [Ratifiability involves suppositions of *final* decisions.] (p. 18)

It is, however, clearly not plausible that I should think that things such as these would happen to me if and only if they happened to the other prisoner. And it is not easy to see how the stipulation could be *made* plausible that "the sorts of extraneous influences that would prevent me from confessing . . . are likely to work on him . . . to the same effect" (p. 20), given that these must be influences that I think might explain my failing to act on a final decision to confess and thus cannot be things that I think might work by moving me to change my mind.

For another objection to ratificationism, one based on what may be found to be a more plausible case, we consider popcorn problems. Suppose, as Jeffrey would have us do, that I am nearly sure that I will do whatever I decide finally to do. Then, were I to become sure that I will decide to go for popcorn, I would be nearly sure that there was popcorn there, and the improbable but still possible news that I would not be going after all might very well provide me with no evidence of whether there was popcorn there. It is natural to suppose that I would be sure that whatever would account for my failing to act on what I would be sure was a final decision to go for popcorn would then, in my mind, not have anything to do with whether or not there actually was popcorn. What might I then think could cause a slip between my decision to go for popcorn and the action pursuant to it of actually going? Remember that what is at issue is the evidential bearing of my actions, given a certainty on my part that I have made a final decision to go for popcorn and that – whatever happens – I am not going to change my mind. Well, as Jeffrey might depressingly suggest, there is always the possibility of being hit by a falling chandelier. It would not be easy to tell stories in which I found such an occurrence to be evidentially linked to the presence or absence of popcorn. It would not be easy to tell plausible stories, though mad-manager scenarios do come to mind.

So much for the effect on my credences, and on evidential bearings of my actions, of news that my decision will be to go for popcorn. This effect could well be to quite neutralize these bearings. What, however, about news of a final decision *not* to go? It is natural to suppose that, given news that my decision will be not to go – given a subjective certainty of this, and not just my present *near* certainty – subsequent news that I was *even so* going for popcorn would remain, as it presently is, good evidence that there was popcorn to be had after all. It is natural to suppose that such further news would in this case be for me evidence that, under the influence of signals on the screen, my final decision was destined to come undone – that I was, notwithstanding a final decision not to go for popcorn, going for popcorn after all *under the influence of these signals*. It is natural to suppose that news that (notwithstanding my final decision not to go) I was going after all would provide me with evidence that in one or another of many possible ways the signal was going to make me do it, and thus with evidence that there was popcorn there in the lobby.

Matrices for my "*conditionalized* on decisions G^* and $(\sim G)^*$" conditional probabilities for circumstances given actions could in a popcorn problem very well be as follows where G abbreviates "go for popcorn" and P abbreviates "there is popcorn."

$\Pr_{G^*}$

	P	$\sim P$
G	[1]	[0]
$\sim G$	[1]	[0]

$\Pr_{(\sim G)^*}$

	P	$\sim P$
G	[1]	[0]
$\sim G$	[0]	[1]

These probabilities can obtain given stipulations all of which are natural and plausible. Given these probabilities, given the following Desirabilities in a popcorn problem,

	P	$\sim P$
G	4	1
$\sim G$	2	3

and assuming that conditional-on-G^* and conditional-on-$(\sim G)^*$ Desirabilities are the same: although going for popcorn is ratifiable, not going for popcorn is not – $\text{Des}(G/G^*) = [4] > \text{Des}(\sim G/G^*) = [2]$, but $\text{Des}[G/(\sim G)^*] = [4] > \text{Des}[\sim G/(\sim G)^*] = [3]$. And yet, since I am nearly sure that there is no popcorn to be had, it remains plain that not going for popcorn is my only sensible course. Ratificationism can give the wrong answer in a popcorn problem, just as it does in certain bothersome Prisoners' Dilemmas. In partial contrast, however, ratificationism gives wrong answers in popcorn problems even when (indeed, precisely when) they are elaborated by natural and plausible stipulations.

Ratificationism gives wrong answers in some popcorn problems. I note that in other popcorn problems, and in some Prisoners' Dilemmas, both actions are ratifiable, and that ratificationism, while not prescribing wrong actions, fails to permit only right ones. For a Prisoners' Dilemma of this type – arguably a more plausible one than either Jeffrey's or van Fraassen's – tell a story for the following matrices (which come from the popcorn problem just discussed).

$\Pr_{C^*}$

	C'	$\sim C'$
C	[1]	[0]
$\sim C$	[1]	[0]

$\Pr_{(\sim C)^*}$

	C'	$\sim C'$
C	[1]	[0]
$\sim C$	[0]	[1]

For a popcorn problem in which both actions are ratifiable (I make no claim for its possible plausibility), tell a story for the following matrices (which come from van Fraassen's Prisoners' Dilemma).

$\Pr_{G^*}$

	P	$\sim P$
G	[1]	[0]
$\sim G$	[1]	[0]

$\Pr_{(\sim G)^*}$

	P	$\sim P$
G	[0]	[1]
$\sim G$	[0]	[1]

Evidential independence of the states P and $\sim P$ from my possible actions, conditional on my possible decisions, is of course not sufficient for ratificationism to solve a popcorn problem and select only going for popcorn, because this action – the right action in a popcorn problem – is not under these states a dominant action.

NOTES

1. [This note is from Sobel 1991a, n. 14, with permission of *Synthese*.] For the exact idea let the value of a person's *prospects* (his or her personal or subjective prospects) be the weighted average value of all possible worlds in which weights are this person's probabilities for worlds:

   prospects = $\sum_w [\Pr(w) \cdot (\text{the value of } w)]$.

   Let the value of this person's *prospects conditionalized on p* be this person's prospects were he or she, without changing values for worlds, to revise all probabilities by conditionalizing on p:

   prospects conditionalized on $p = \sum_w [\Pr_p(w) \cdot (\text{the value of } w)]$.

   Then the evidential Desirability of p (roughly and often, its desirability as a possible item of news) equals prospects *conditionalized* on p, which may be contrasted with

   prospects *conditional* on $p = \sum_w [\Pr(p > w) \cdot (\text{the value of } w)]$,

   which is, on one simple causal theory (with ">" a Stalnaker causal connective), equal to the desirability of p as a possible fact.

2. Regarding the apparently limited character of his defenses, Eells writes: "Notice that this defense relies essentially on the agent's believing that there *is* a common cause.... In some of the decision problems that are taken to be prima facie counterexamples to the traditional theory, no common cause is explicitly mentioned.... It is arguable, however, that if the agent believes that there is a correlation between an act and some outcome, where he believes that the act doesn't cause the outcome and the outcome doesn't cause the act, then he should postulate a common cause of the two things – even if he cannot say what that causative factor is" (Eells 1984b, p. 75). I think that while arguable it is not true of, for example, Prisoners' Dilemma Newcomblike problems that common causes must be postulated. For an agent can reasonably think that his and another's acts are correlated without thinking that either are caused. But I will not press this point. My concern is to show that Eells's "metatickle" defenses are not good even against the common-cause Newcomblike problems to which they are explicitly addressed.

3. Jeffrey (1988b; 1992) works toward these results, but similarly falls short of a defense of the fully general theory of the desirabilities of all propositions (and not just of all action propositions) as not only possible items of news but also as possible facts – the theory to which he presumably aspires.

4. [This note is with permission of *Synthese* drawn from Sobel 1991a, pp. 159–60, wherein both Eells 1989a and 1989b are taken into account.] Eells has said that it "is peculiar and unreasonable that going to the doctor provides *further* evidence for being ill, evidence *beyond how one [Uncle Albert] feels*" (Eells 1989b). I hold with Skyrms (1982, p. 700) that "this need not be ... irrational" and that there is nothing peculiar or odd about the case. I do not think it is

even unusual for actions we may perform to constitute, if performed, evidence concerning our biological states. Here is a brief personal memoir of relevance to these points: "I attended a Cub Scout camp one Monday through Friday of the summer of 1940. Back home Saturday morning I went up from Woodlawn Avenue to the Del Prado hotel to collect autographs from the visiting team on their way to batting practice at Comiskey Park. It was hot. I was tired and feeling strange (I had no name for my feelings, having at that time almost no experience with illness), and when I reached my courtyard, rather than continue upstairs I lay down in the sun on the sidewalk. I can't remember what I was thinking when I did that, but I believe it was something like, 'I think I'll just lie down for a minute and rest.' I do, however, remember what I thought as I lay there taking in the scene, not as the agent in it but as an observer: I thought, 'Gee, I must be sick.'" The news of my action taught me something about the state of my health. I really did learn (no scare quotes) something from it, for I was in fact sick with typhoid fever that shortly thereafter registered over 103° and then went up. And I'm sure that my action of lying down already harbored this evidence in potential when it (the action) was still only a possibility, though I did not then take conscious notice of its evidential potential for me.

There was, I think, nothing peculiar or unreasonable about that episode. My action was not unreasonable. I had reasons for thinking it would do me some good to lie down, though I did not consciously think them through. I'm sure it really did do me some good. But was my lying down a bona fide action? Well, yes. I didn't fall down. I knew what I was doing, and lay down of my own free will, intentionally, I should say. And the beliefs and potential evidential bearings of things for me that have been mentioned were not unreasonable. My preferences were standard. (I never went back to Cub Scout camp. We believed it was the water.)

5. With regard to the relevance of this case to ideas of Price and Talbott, note that reflections of mine on my efforts to make up my mind would not threaten to unsettle the credences on which initial tentative sentiments would be based in ways that Eells would lead one to fear. Neither would they need to threaten conditional credences in ways that Price might have us fear. For I could realize that my own credences had nothing to do with what the agent was to do, as distinct from what I was hoping he would do. Nor would reflections on my efforts to make up my mind and to settle my desires need to threaten my confidence in my competence in ways that Talbott might have us fear. For in third-person problems (causal or acausal), the first person need have no unsettling views concerning the causes of *his* credences, preferences, or ways of thinking and making up his mind. And so, regarding the complex issue of what I should hope to learn is true concerning what someone other than myself will do in a Newcomblike problem, after giving the issue every thought that could make a difference to my attitudes, I can still be of two minds *without* being at any time confused, inductively irresponsible, or in doubt concerning my competence or the causal credentials of my credences, preferences, or ways of thinking.
6. This conclusion has been reached for Eells's efforts to stay evidential, and can be extended to Jeffrey's (see note 3 and Appendix V).
7. This appendix comes with revisions by permission of Kluwer Academic Publishers from *Synthese* 86 (1991), pp. 146–7 (sec. 3 of Sobel 1991a).
8. This appendix comes with revisions by permission of Kluwer Academic Publishers from *Synthese* 75 (1988), pp. 115–23 (the second section of Sobel 1988a).

9. Jeffrey tacitly, and possibly without explicitly noticing that he is doing so, assumes in his argument that conditional Desirability matrices – conditional on C^* and conditional on $(\sim C)^*$ – are, in the case he is dealing with, the same as the unconditional Desirability matrix. Decisions in this case are assumed not to make "independent" contributions to Desirabilities. The need for this assumption is stressed in Rabinowicz (1985, pp. 172–3, p. 180, and p. 198, n. 11).

3

Not every prisoners' dilemma is a Newcomb problem

Abstract. Some prisoners' dilemmas are Newcomb problems, and some are not; whether any important real-life ones are is an open question of little theoretical significance. Whether agents in a dilemma are in side-by-side Newcomb problems depends on whether the opinions of each regarding himself are epistemically dependent on his opinions regarding the other, and if so to what extent. It is not enough that they are sure they are like-minded.

0. INTRODUCTION

Newcomb's Problem and the Prisoners' Dilemma are alike in several ways. In each problem there is a dominant action whose choice-relevant value excels in all possible circumstances. And in each problem, possible circumstances are certainly causally independent of possible actions. Furthermore, in connection with each problem there is a sense (though not the same sense) in which agents who rejected the dominant action could expect to do better than agents who employed it. And finally, it can seem that maximizers *would* reject the dominant action not only in Newcomb's Problem, which was designed for this result, but also in the Prisoners' Dilemma, or at least in one where "each prisoner knows that the other thinks in much the same way he does" (Gibbard and Harper 1978, p. 160). It can seem that these problems are not only very similar, but also that such prisoners' dilemmas *are* Newcomb problems, or rather two Newcomb problems side by side, inessential trappings aside (Lewis 1979a, p. 239). But this is not quite right, for although some such prisoners' dilemmas are Newcomb problems, others are not.[1] Whether or not one is a Newcomb problem will depend in large part on why each prisoner is convinced that the other thinks in much the same way he does, on why

---

Reprinted with revisions by permission from *Rationality, Cooperation, and Paradox: The Prisoner's Dilemma and Newcomb's Problem* (Richmond Campbell and Lanning Sowden, eds.), Vancouver: University of British Columbia Press. Copyright © 1985 University of British Columbia Press; all rights reserved.

   Some ideas in this chapter took form in a conversation with David Gauthier. I am indebted to Leslie Burkholder and Richmond Campbell for comments on early versions.

he is convinced that – as far as ways of thinking are concerned – the other and he are near perfect replicas of one another. Whether or not a prisoners' dilemma is a Newcomb problem will depend on what opinions lie behind this conviction.

It is mainly, though not only, prisoners' dilemmas in which each prisoner is nearly sure that the other will act as he does that have been thought to be Newcomb problems; they have been thought to be "near-certainty" Newcomb problems. I first set conditions for such Newcomb problems and prisoners' dilemmas, and then discuss how these are related.

1. NEAR-CERTAINTY NEWCOMB PROBLEMS

Let a Newcomb problem be defined by the following conditions: (1) An *expected utility matrix*[2] of a Newcomb problem has the structure

	Circumstances	
	c_1	c_2
a_1	x	w
a_2	z	y

(Actions)

wherein $x = U(a1\,\&\,c1)$, $w = U(a1\,\&\,c2)$, ..., and where $w > y > x > z$. A *news value matrix* has the same structure, so that $x = N(a1\,\&\,c1)$, $w = N(a1\,\&\,c2)$, and so on. In Newcomb's Problem itself, if values are linear with monetary payoffs then expected utility and news value matrices can both be represented as follows.

	There is nothing in Box 2	There is $M in Box 2
Take Both boxes	1	1001
Take only Box 2	0	1000

(2) Circumstances $c1$ and $c2$ in a Newcomb problem are certainly causally independent of actions $a1$ and $a2$. (3-N) In a Newcomb problem, circumstances are not to be epistemically independent of actions. Indeed, in a *near-certainty* Newcomb problem circumstances will be nearly maximally dependent on actions epistemically:

$$\Pr(c1/a1) \simeq 1 \simeq \Pr(c2/a2)$$

(where $\simeq$ denotes near-certain equality), and since $\Pr(c1\,\&\,c2) = 0$ and $\Pr(c1 \vee c2) = 1$,

$$\Pr(c2/a1) \simeq 0 \simeq \Pr(c1/a2).$$

In a near-certainty Newcomb problem, $a1$ and $a2$ will be near-certain potential signs of $c1$ and $c2$, respectively. And finally, mainly because of the epistemic dependence of circumstances, (4) in a Newcomb problem, the *news value* of $a2$ exceeds that of $a1$:

$$\Pr(c1/a2)N(a2\,\&\,c1) + \Pr(c2/a2)N(a2\,\&\,c2)$$
$$> \Pr(c1/a1)N(a1\,\&\,c1) + \Pr(c2/a1)N(a1\,\&\,c2).$$

In Newcomb's problem itself,

$$[\Pr(\text{nothing/only Box 2}) \times 0] + [\Pr(\$M/\text{only Box 2}) \times 1000] \simeq 1000,$$

while

$$[\Pr(\text{nothing/Both}) \times 1] + [\Pr(\$M/\text{Both}) \times 1001] \simeq 1.$$

In a Newcomb problem, $a2$ would maximize news value, or what Richard Jeffrey has termed *desirability*. In contrast, by conditions (1) and (2) – the dominance of $a1$ and the certain causal independence of circumstances from actions – $a1$ would maximize expected utility. According to a simple causal theory of expected utility,[3] using "$\square\!\!\rightarrow$" as the causal conditional connective,

$$U(a1) = \Pr(a1 \,\square\!\!\rightarrow c1)U(a1\,\&\,c1) + \Pr(a1 \,\square\!\!\rightarrow c2)U(a1\,\&\,c2)$$

and

$$U(a2) = \Pr(a2 \,\square\!\!\rightarrow c1)U(a2\,\&\,c1) + \Pr(a2 \,\square\!\!\rightarrow c2)U(a2\,\&\,c2);$$

condition (2) entails that

$$\Pr(a1 \,\square\!\!\rightarrow c1) = \Pr(c1) = \Pr(a2 \,\square\!\!\rightarrow c1)$$

and

$$\Pr(a1 \,\square\!\!\rightarrow c2) = \Pr(c2) = \Pr(a2 \,\square\!\!\rightarrow c2).$$

By (1),

$$U(a1\,\&\,c1) > U(a2\,\&\,c1) \quad \text{and} \quad U(a1\,\&\,c2) > U(a2\,\&\,c2),$$

from which it follows that

$$U(a1) > U(a2).$$

In Newcomb's Problem itself, taking both boxes would maximize expected utility, whereas taking only Box 2 would maximize news value. And so a temptation to wishful willing is apt to obscure the authority, in this problem, of the fact that the agent can be sure (and presumably is sure) that he would do better by taking both boxes than by taking just the second – exactly $1000 better.

2. NEAR-CERTAINTY PRISONERS' DILEMMAS

Let a near-certainty prisoners' dilemma be defined by three conditions. (1) An expected utility matrix of a prisoners' dilemma has the structure displayed in Section 1, as does a news value matrix. For the Prisoners' Dilemma itself, if values are linear with negative inverses of prison terms then expected utility and news value matrices can be as follows (see Jeffrey 1965b, pp. 11–12; 1981b, pp. 484–5; 1983, p. 15).

	The other prisoner will confess	The other prisoner will not confess
Confess	−5	0
Do not confess	−10	−1

(2) Circumstances $c1$ and $c2$ in a prisoners' dilemma are causally independent of actions $a1$ and $a2$. (3-P), in a near-certainty prisoners' dilemma, it is nearly certain that the other will do as I do, or, more abstractly, that

$$\Pr[(a1 \& c1) \lor (a2 \& c2)] \simeq 1;$$

and, since $\Pr(c1 \& c2) = 0$ and $\Pr(a1 \& a2) = 0$,

$$\Pr[(a2 \& c1) \lor (a1 \& c2)] \simeq 0.$$

Condition (3-P) is specific to near-certainty prisoners' dilemmas. Neither it nor anything like it is essential to the classical Prisoners' Dilemma, which was designed to dramatize the possibility of dominance/optimality confrontations. Condition (3-P) is not needed for that purpose; in fact, no other-regarding epistemic condition is needed for that purpose. It *is* a feature of standard versions of the Prisoners' Dilemma that the prisoners are in separate cells, but this feature is designed to ensure they cannot enter into agreements that would make a difference, not to ensure that neither will know what the other is up to, or know that the other knows what he is up to.

3. NEAR-CERTAINTY NEWCOMB PROBLEMS AND PRISONERS' DILEMMAS COMPARED

Near-certainty Newcomb problems and prisoners' dilemmas, though alike in their first two conditions, differ in their third; furthermore, there is no fourth condition for prisoners' dilemmas. Even so, every near-certainty Newcomb problem is a near-certainty prisoners' dilemma. Since $(a1, a2)$ and $(c1, c2)$ determine a news value matrix for a decision problem,

$\Pr(a1 \vee a2) = 1$, $\Pr(a1 \& a2) = 0$, $\Pr(c1 \vee c2) = 1$, and $\Pr(c1 \& c2) = 0$. So from (3-N),

$$\Pr(c1/a1) \simeq 1 \simeq \Pr(c2/a2), \quad \text{and} \quad \Pr(c2/a1) \simeq 0 \simeq \Pr(c1/a2),$$

it follows that (3-P),

$$\Pr[(a1 \& c1) \vee (a2 \& c2)] \simeq 1.$$

Every near-certainty Newcomb problem is a near-certainty prisoners' dilemma. But not every such prisoners' dilemma is a Newcomb problem. The important point here is that (3-P) does not entail (3-N). Even if I am nearly certain that the other prisoner will do what I do, my conditional probabilities for actions like mine on his part need not *both* be high. Whether or not they are both high will depend on just why I am nearly certain that we will act alike.

3.1. Prisoners' dilemmas that are not Newcomb problems

Let me be nearly certain that I will confess,

$$\Pr(a1) \simeq 1,$$

and on quite *independent* grounds nearly certain that you will confess,

$$\Pr(c1) \simeq 1.$$

Perhaps, for example, I have decided to confess, and have heard through channels I consider reliable that you have come to the same decision. I could then be nearly certain that you too were going to confess, even if I had no idea what sort of person you were or why you were going to confess. And if I arrived at my own decision without regard to what you were going to do, my certainty concerning what you would do could possibly have nothing to do with my certainty concerning what I was going to do.

In the case under construction I am to be nearly certain that I will confess and, on quite independent grounds of *some* sort (not necessarily the sort described so far), nearly certain that you will confess. Suppose further that, as a consequence of these near certainties, though I am not quite so confident of the conjunction I am still nearly certain that we will both confess:

$$\Pr(a1 \& c1) \simeq 1.$$

I am to come to this near certainty entirely by way of the first two. It follows from this last near certainty that I am nearly certain that we will act alike:

$$\Pr[(a1\,\&\,c1) \vee (a2\,\&\,c2)] \simeq 1.$$

Continuing with stipulations for this case: though I am to think it very unlikely that I will not confess, I am to consider it much more unlikely that both of us will not confess than that I alone will not: let "$\gg$" mean "is much greater than," so that for positive numbers j and k, $j \gg k$ if and only if $j/(j+k) \simeq 1$; it is to be part of the case that

$$\Pr(a2\,\&\,c1) \gg \Pr(a2\,\&\,c2).$$

Why? – because I have independent grounds for my opinions about what you and I will do, and so consider it much more likely that I am mistaken just about myself than that I am mistaken about both of us. It follows, given $\Pr(c1 \vee c2) = 1$ and $\Pr(c1\,\&\,c2) = 0$, that my probability for your not confessing conditional on my not confessing is very low:

$$\Pr(c2/a2) = \frac{\Pr(a2\,\&\,c2)}{\Pr(a2\,\&\,c1) + \Pr(a2\,\&\,c2)},$$

so that

$$\Pr(c2/a2) \simeq 0,$$

contrary to (3-N).

Are such cases logically possible? Are the constraints imposed on probabilities jointly consistent? Yes. One class of cases of the kind described here is determined by the following plainly consistent probabilities:

$$\Pr(a1\,\&\,c2) = 0,$$
$$\Pr(c1\,\&\,c2) = 0,$$
$$\Pr(a1\,\&\,c1) = .9,$$
$$\Pr(a1\,\&\,c2) = (\sqrt{.9} - .9),$$
$$\Pr(a2\,\&\,c1) = (\sqrt{.9} - .9),$$
$$\Pr(a2\,\&\,c2) = 1 - [.9 + 2(\sqrt{.9} - .9)].$$

Given these probabilities, circumstances are probabilistically independent of actions:

$$\Pr(c1) = .9 + (\sqrt{.9} - .9) = \sqrt{.9},$$

$$\Pr(c1/a1) = \frac{.9}{.9 + (\sqrt{.9} - .9)} = \sqrt{.9},$$

$$\Pr(c1/a2) = \frac{\sqrt{.9} - .9}{(\sqrt{.9} - .9) + (1 - [.9 + (\sqrt{.9} - .9)])}$$
$$= \frac{\sqrt{.9} - .9}{1 - \sqrt{.9}}$$
$$= \sqrt{.9};$$

$$\Pr(c2) = (\sqrt{.9} - .9) + (1 - [.9 + 2(\sqrt{.9} - .9)])$$
$$= 1 - \sqrt{.9},$$
$$\Pr(c2/a1) = \frac{\sqrt{.9} - .9}{.9 + (\sqrt{.9} - .9)}$$
$$= 1 - \sqrt{.9},$$
$$\Pr(c2/a2) = \frac{1 - [.9 + 2(\sqrt{.9} - .9)]}{(\sqrt{.9} - .9) + (1 - [.9 + 2(\sqrt{.9} - .9)])}$$
$$= \frac{1 + .9 - 2\sqrt{.9}}{1 - \sqrt{.9}}$$
$$= 1 - \sqrt{.9}.$$

Given these probabilities, the other conditions of our case hold if we assume that $x \simeq 1$ if and only if $.9 < x \leq 1$, and that $y \simeq 0$ if and only if $0 \leq y < .1$:

$$\Pr(a1) = \Pr(c1) = \sqrt{.9} \simeq 1,$$
$$\Pr(a1 \& c1) = .9 \simeq 1,$$
$$\Pr[(a1 \& c1) \vee (a2 \& c2)] = [1 - 2(\sqrt{.9} - .9)] \simeq 1,$$
$$\Pr(a2 \& c1) = (\sqrt{.9} - .9) \gg (1 - [.9 + 2(\sqrt{.9} - .9)]) = \Pr(a2 \& c2),$$
$$\Pr(c2/a2) = (1 - \sqrt{.9}) \simeq 0.$$

We may consider two theoretically interesting types of dilemma of the present sort. First, we have a dilemma in which I am nearly certain that I am rational, and on quite independent grounds nearly certain that you are rational. In this case I am confident that it would be rational to maximize *expected utility,* and confident that confessing would be utility maximizing for both you and me, as it would be. For a second type of dilemma of the present sort, we add to its general specifications (which include that I am nearly certain that I will confess and, on quite independent grounds, nearly certain that you too will confess) the further specifications that I am nearly certain I am rational, on quite independent grounds nearly certain that you are rational, and entirely confident that it would be rational to maximize *news value.* In this second type of dilemma I will be nearly certain (indeed, I am to be entirely confident) that confessing would maximize news value for you and for me, as it would in this type of case; I am nearly certain that we will both confess and that I will get five years. And, given the independence of my opinions about you and me, if in this case I were to learn for sure that I was not going to confess, then I would nonetheless be nearly certain that you *were* going to confess, and so I would be nearly certain that I was going to get *ten* years, which is worse than five. Bad news! News that I was not going to confess would be bad news.[4]

In dilemmas of the two types just described, I am nearly certain that we will do the same thing, nearly certain that we are both rational, and nearly certain that we will both do the rational thing. But even so, since your actions are by hypothesis probabilistically independent of mine, confessing maximizes (for me) both utility and news value. These prisoners' dilemmas are thus not Newcomb problems.[5]

3.2. Prisoners' dilemmas that are Newcomb problems

Let me be nearly certain that we are psychological twins and that we will act alike, and let this near certainty of mine be independent of my opinions concerning what I will do and what you will do. Thus I am nearly certain that we will do the same thing,

$$\Pr[(a1 \& c1) \vee (a2 \& c2)] \simeq 1;$$

furthermore, as (3-N) would have it, my possible acts would both be for me near certain signs that you were acting similarly:

$$\Pr(c1/a12) \simeq 1 \quad \text{and} \quad \Pr(c2/a2) \simeq 1.$$

Perhaps, for example, I am nearly certain that we are identical biological twins, that throughout our lives we have been constant companions, and that we have always thought and acted alike in situations as similar as our present ones seem to be. Perhaps, as a consequence of these convictions, I am nearly certain that we will act alike in the present case.

In this kind of case, any news of my action would for me be excellent evidence that you were acting similarly. My opinions concerning our actions, whatever these opinions are, do not in this case rest on quite independent grounds; on the contrary, there is near maximal dependence of these opinions one on the other. In this kind of dilemma I am nearly certain that you will do what I will do; further, regardless of what I did, I would remain nearly certain that you would do the same thing. Dilemmas of this sort satisfy condition (3-N), and we now stipulate that they also satisfy (4). Such dilemmas are logically possible. For a numerical illustration, let probabilities and news values include the following:

$$\Pr(a1 \& a2) = 0, \quad \Pr(a2 \& c1) = 0,$$
$$\Pr(c1 \& c2) = 0, \quad \Pr(a2 \& c2) = .1,$$
$$\Pr(a1 \& c1) = .9, \quad N(a1 \& c1) = -5,$$
$$\Pr(a1 \& c2) = 0, \quad N(a2 \& c2) = -1.$$

Given these probabilities and values,

$$\Pr(c1/a1) = 1 = \Pr(c2, a2), \quad \Pr(c2/a1) = 0 = \Pr(c1, a2),$$

and

$$\Pr(c1/a1)N(a1 \& c1) + \Pr(c2, a1)N(a1 \& c2) = -5$$
$$< -1 = \Pr(c1/a2)N(a2 \& c1) + \Pr(c2/a2)N(a2 \& c2).$$

A prisoners' dilemma can be of the present sort[6] even if I am nearly certain that we are both rational, quite certain that it would be rational to maximize expected utility, and quite certain that confessing would do that. For it is consistent with these conditions that news that I was *not* confessing should be excellent evidence that I was not being rational, and that you were similarly deficient and would thus also not be confessing. And a dilemma can be of the present sort even if I am nearly certain that we are both rational, quite certain that it would be rational to maximize news value, and nearly certain that we are not going to confess. For again, surprising news concerning my action – in this type of dilemma, that I am or will be confessing – could be excellent evidence that you too are or will be confessing. It could be excellent evidence not only that I am not doing the rational thing as I see it, but that you too are not doing the rational thing.

Finally, a dilemma can be of the present sort if, though I do not yet have any strong views concerning what we will do, I am still confident that we will behave rationally because (uncritically and without putting this idea into so many words) I take for granted that – in simple situations such as the present one seems to be – most people do behave rationally. However, though a dilemma can be Newcomblike in this way, it is very unlikely that it should be so "just before" the perceived time of action: one supposes that by then I *would* have strong views concerning what I will do. Furthermore, it is very likely that any dilemma that was Newcomblike in this way at some early stage of my deliberations would, by the perceived time of action, devolve into a non-Newcomblike dilemma. For it is likely that by then I would have views regarding what it would be rational to do, views that were probabilistically independent (or largely so) of propositions concerning my own possible actions. It is likely by then I would be confident that some particular action was the rational thing for me to do and that I was going to do it. But it would be strange indeed if prominent among my reasons for so thinking was the (for me) near-certain fact that I was going to do it. Hence it is likely that by the perceived time of action I would have views regarding what you were going to do, since I would presumably still be taking for granted that you were going to do the rational thing – views that were themselves independent of my views concerning what I was going to do. Incidentally, it is thus likely that news value maximizers (at least sufficiently sophisticated, self-conscious, and reflective ones) in a prisoners' dilemma that was Newcomblike in the present way early on would in the end - or in the event - *not* cooperate.[7]

In dilemmas of the present sort (of which I have commented on three types, the third being unlikely as a moment-of-choice dilemma), your actions are, in terms of my probabilities, probabilistically dependent on mine. In these dilemmas, confessing would maximize expected utility, whereas not confessing would maximize news value. Such dilemmas are logically possible, and news maximizers who were, at their moments of choice, in such dilemmas would cooperate. Dilemmas of this sort *are* Newcomb problems.[8]

4. Conclusions

Some logically possible prisoners' dilemmas are Newcomb problems, and some are not. The character of a particular prisoners' dilemma will depend on the *grounds* participants have for their opinions concerning one another's actions. Participants in problems broadly like the Prisoners' Dilemma may be nearly certain that they will act alike, and they may be nearly certain that they are rational and are utility maximizers, or that they are rational and are news value maximizers. But either way, they still may or may not be in side-by-side Newcomb problems. Whether or not they *are* will in large part depend on what makes them nearly certain that they will act alike, on how they come by these opinions, and, in short, on whether or not each prisoner's opinions regarding his own action and the action of the other are epistemically dependent or independent - and, if dependent, to what extent.[9]

Some logically possible prisoners' dilemmas are Newcomb problems, but I am not sure whether many real-life, moment-of-choice prisoners' dilemmas are Newcomb problems, or whether any important ones are. Whether or not *any* important real-life interactions feature (like the Prisoners' Dilemma) dominance/optimality confrontations and also (like Newcomb's problem) confrontations involving moment-of-choice utility maximization and news value maximization is a hard, partly empirical, question on which I do not have a confident opinion. Given my uncertainty here, I take some comfort in the belief that nothing important to the theory of rational agency turns on the answer to this question.[10]

NOTES

1. The title of his paper notwithstanding, it is plain that Lewis does not think that *every* prisoners' dilemma is a Newcomb problem. But he does observe that

"Prisoners' Dilemmas are deplorably common in real life," and clearly implies that many real-life prisoners' dilemmas are Newcomb problems (Lewis 1979a, p. 240). Whether or not many real-life prisoners' dilemmas are in fact Newcomb problems is, at least in Richard Jeffrey's view, an issue of importance to theory. For reasons that I hope will be clear, I think that it is not obvious (though it may be true) that all real-life prisoners' dilemmas are, in the end, Newcomb problems.

2. I set out a Jeffrey-style theory of matrices for a causal decision theory in Sobel (1985a).
3. This simple theory is for the agent who does not believe in objective chances: the agent for whom, for any propositions p and q, $[(p\,\square\!\!\rightarrow q)\vee(p\,\square\!\!\rightarrow \sim q)]$. When this agent is certain that actions a and $\sim a$ are open (and so causally possible), he is certain that a circumstance is causally independent of a if and only if he is certain that $[(a\,\square\!\!\rightarrow c)\leftrightarrow c]\,\&\,[(\sim a\,\square\!\!\rightarrow c)\leftrightarrow c]$ and so only if, for him, $\Pr(a\,\square\!\!\rightarrow c) = \Pr(c)$.
4. In a dilemma of this second type, I am nearly certain (1) that we will confess and (2) that we are both rational; I am entirely confident (3) that it would be rational to maximize news value and (4) that my confessing would maximize news value. But how could I come to such opinions? Using i for me and y for you, I could begin by being nearly certain that I was rational ($2i$), and entirely confident that it would be rational to maximize news value (3). I might then learn through what I considered to be highly reliable channels that you were going to confess ($1y$), and on reflection I then might realize that my confessing would maximize news value (4), decide to confess, and become thereby nearly certain that I was going to confess ($1i$). I might *then* learn through the same channels that you were rational ($2y$). Indeed, I might learn through these channels that you have in similar ways come to the same opinions, and that in all ways relevant to what I take to be our present situation we are doxically indistinguishable. This describes one way in which I could come to be in a dilemma of this second type, in which we could be in side-by-side dilemmas of this type. Probably there are many other ways. Since the near certainties and total confidences (1)–(4) are consistent with the axioms of probability, one supposes there must be ways whereby they can come to co-exist, though perhaps not (imprecise notion) many ways.
5. For other dilemmas that are not Newcomb problems, start with the stipulation that the agent is nearly certain that he will not confess and, on quite independent grounds, nearly certain that the other agent also will not confess.
6. Nozick's identical-twin prisoners' dilemma is of this sort (Nozick 1969, pp. 130-1). Nozick's case, in contrast with the cases I sketch in the text, does not include assumptions concerning prisoners' views regarding their rationality.
7. "One might attempt to develop a theory of the *Sophisticated Rational Decision Maker* wherein sophistication is characterized stringently enough in terms of self-knowledge and causal reasoning so that for [sophisticated] decision makers the two paradigms [for deliberation, causal and evidential] are in approximate agreement" (Skyrms 1982, p. 698; see Sobel 1985a, n. 7).
8. Gauthier (1975) has held that utility maximizers who expected to find themselves in prisoners' dilemmas would have reason to "revise their conception of rationality" so that when interacting with like-minded agents they would of their own rational accords not confess. The revision he thought would recommend itself would be to a form of practical thinking he calls "constrained"

utility maximization. We now see that another revision could recommend itself, at least for anticipated Newcomblike prisoners' dilemmas. Utility maximizers who expected to find themselves in such situations would have a reason to revise their ways of practical thinking (their ways of "translating" beliefs and preferences into actions, with and without explicit thought and calculation), so that when interacting with like-minded agents in such situations they would think and, most importantly, act as news value maximizers. Hard questions include: Are agents capable of such mind-bending self-administered psychic surgery? How would they do it, when and if they were capable of it? Under what conditions (if any) would utility maximizers have sufficient reasons for revising their ways of practical thinking? What would be the exact forms of their revised conceptions of rationality? And what would all this imply for the true nature of rationality, if such there be, as distinct from various conceptions thereof? See Gauthier (1975) and Sobel (1975) for remarks on these matters.
9. Roughly stated: Problems that I count as prisoners' dilemmas, Levi counts as versions of Newcomb's Problem – see Levi (1975, esp. pp. 162, 164) – and my reasons for holding that not every prisoners' dilemma is a Newcomb problem are like his for holding that not every version of Newcomb's Problem (in his sense) is a Newcomb problem (in my sense). Conceding terminology to Levi, every occurrence in the present paper of Newcomb's *Problem* and its cognates could be replaced by an occurrence of (a cognate of) Newcomb's *Paradox,* with of course occurrences in the present note excepted.
10. Jeffrey thinks that Newcomb's Problem is "a prisoner's dilemma for space cadets," a problem too artificial to mandate radical revisions in an otherwise satisfactory theory of rational decisions. But he is convinced that radical revisions are in fact necessitated by prisoners' dilemmas, "the prisoners in which are paradox enow" (Jeffrey 1983, p. 25). I think that Newcomb's Problem, despite its artificiality, demonstrates the need for radical revisions in Jeffrey's logic of decision, and that there may not be enough real-life Newcomblike problems of *any* sort to convince Jeffrey that radical revisions in his theory are justified. Our differences include a difference on a methodological issue the full character of which is not clear to me.

4

Some versions of Newcomb's Problem are prisoners' dilemmas

Abstract. Having maintained that some but not all prisoners' dilemmas are side-by-side Newcomb problems, I now argue similarly that some but not all versions of Newcomb's Problem are prisoners' dilemmas in which 'taking two' and 'predicting two' make an equilibrium that is dispreferred by both the box chooser and the predictor to the outcome in which only one box is taken and this is predicted. Questions are raised concerning what varieties of prisoners' dilemma versions of Newcomb's Problem can be – whether they can be dilemmas that their participants recognize to be dilemmas, and whether they can be dilemmas that are side-by-side Newcomb problems. I also comment on the opportunities that these results may open for kinds of cooperative reasoning in versions of Newcomb's Problem.[1]

0. INTRODUCTION

The Prisoners' Dilemma is a game "attributed to A. W. Tucker" (Luce and Raiffa 1957, p. 94) that dramatizes the possibility of nonoptimal equilibria; more particularly, it dramatizes the possibility of several independent actions that are dominant and so "individually" rational but, one is tempted to say, not "collectively" rational. The Prisoners' Dilemma would be an occasion for explicit cooperation were it possible for the parties to communicate and make a binding agreement, but such is not the case for parties to that problem.

Newcomb's Problem is a decision problem "constructed by a physicist, Dr. William Newcomb" (Nozick 1969, in Campbell and Sowden 1985, p. 129) "in 1960 while meditating on . . . the Prisoner's Dilemma" (Gardner 1973, p. 104). This problem may have recommended itself as a dramatization of possible conflicts between dominance and maximizing principles (Nozick 1969, pp. 110–11). It has, however, come to be viewed as staging most fundamentally not a conflict between dominance and maximization, but rather one between alternative ways of maximizing – evidential ones

© 1991 Kluwer Academic Publishers. Reprinted with revisions by permission of Kluwer Academic Publishers from *Synthese* 86 (1991), pp. 197–208.

that call for calculations of expected values, and causal ones that can be applied without calculations given the dominance of an option under certain causally independent conditions. (See Sobel 1988a for the distinction between evidential and causal maximizing, and discussion of this assessment of the theoretical significance of Newcomb's Problem.)

Though different in structures and purposes, these problems are historically intertwined and similar in certain ways. David Lewis has written that the "Prisoners' Dilemma *is* a Newcomb Problem – or rather, two Newcomb Problems side by side, one per prisoner" (Lewis 1979a, in Campbell and Sowden 1985, p. 251). I have maintained (Sobel 1985a) that in at least some versions of the Prisoners' Dilemma, "each [can see] his own choice as a strong clue to the other's" (Jeffrey 1983, p. 16) in a way that makes his problem a Newcomb problem. This is of interest because it is plain that such Newcomb problems can be acausal ones, in which agents are not required (bothersomely) to view their actions as things that will be caused in certain ways.[2] This first side of the story contributes to the seriousness of the Newcomb challenge. The present chapter is about the other half of the story, which (as it turns out) is similar. Though not all versions of Newcomb's Problem are prisoners' dilemmas, at least some are. This is of interest because it opens the possibility that at least some versions of Newcomb's Problem can be solved to the mutual satisfaction of their parties if they are given to nonmaximizing cooperative reasoning. This second side of the story may serve to ameliorate some Newcomb predicaments.

1. Preliminaries

Let a *Newcomb problem* be any situation that features at least one agent, Row, and that meets the following three conditions: (i) from Row's point of view the situation has the following ordinal structure,

		Column	
		C1	C2
Row	R1	3	1
	R2	4	2

wherein the larger an outcome's number, the more it is preferred; (ii) Row's choices R1, R2 and column conditions C1, C2 are, Row is sure, causally independent of one another,[3] and finally (iii) Row's choices are, for him, sufficiently good evidence for corresponding column conditions to make R1 of greater evidential expected value (or, in Jeffrey's sense, of greater "Desirability") than R2.

For example, here is a money-in-the-bank variant of Newcomb's Problem (the one-box-or-two problem of Nozick 1969) that when understood as intended is a Newcomb problem:

there is $1000 on the table. Your problem is whether or not to take it. [A] being [in the reliability of whose predictions you have great confidence] has put $M into your bank account or not depending on whether he has predicted that you will leave the money on the table or take it. What do you do? Leave the money on the table? [Or do you take it and run (perhaps to the bank)?] (Sobel 1985a, pp. 198-9, n. 6)

This problem has the following consequence matrix in which "row headings... represent possible acts,... column headings... conditions which might affect... outcomes [and] entries... are... notes... to help determine [cardinal, or at least ordinal] numerical [evidential] desirabilities [as well as causal utilities]" (Jeffrey 1983, p. 2).

	Leave predicted	Take predicted
Leave the money	$1,000,000	$0
Take the money	$1,001,000	$1,000

Let a *prisoners' dilemma* be any situation that features two agents and that meets the following conditions: (i) it has the two-dimensional ordinal structure

	C1	C2
R1	3, 3	1, 4
R2	4, 1	2, 2

wherein the first number of each pair represents Row's preferences and the second number Column's; and (ii) choices of Row and Column are, each is sure, causally independent. The following is a prisoners' dilemma if each detainee is interested only in minimizing his jail time.

Two men are arrested for bank robbery. Convinced that both are guilty, but lacking enough evidence to convict either, the police put the following proposition to the men and then separate them. If one confesses but the other does not, the first will go free (amply protected from reprisal) while the other receives the maximum sentence of ten years; if both confess, both will receive lighter sentences of four years; and if neither confesses, both will be imprisoned for one year on trumped-up charges of jaywalking, vagrancy, and resisting arrest. (Jeffrey 1983, p. 15)

This problem has the following consequence matrix, and thus the ordinal structure required of a prisoners' dilemma.

	Do not confess	Confess
Do not confess	1 year for each	10 years for Row, 0 years for Column
Confess	0 years for Row, 10 years for Column	4 years for each

2. SOME, BUT NOT ALL, VERSIONS OF NEWCOMB'S PROBLEM ARE PRISONERS' DILEMMAS

Some Newcomb problems are prisoners' dilemmas, given that some prisoners' dilemmas are side-by-side Newcomb problems. And it is clear that some Newcomb problems are not prisoners' dilemmas, for some Newcomb problems involve only a single agent – for example, smoking-gene problems such as "Sir Ronald lights up" (Jeffrey 1981b, p. 476), and Solomon's problem (Gibbard and Harper 1978, in Campbell and Sowden 1985, p. 141). Such Newcomb problems cannot be prisoners' dilemmas. But all versions of Newcomb's Problem itself involve two agents – a predictor and a predictee – and so these problems can be prisoners' dilemmas. Whether or not a problem qualifies depends on the interests of the predictor. Many such interests and motives are imaginable, but for simplicity I consider just three: (i) a desire to make true predictions; (ii) an interest in money, and in paying out as little as possible; and (iii) an antipathy toward takers, combined with a desire that the predictee not be one. Cases to follow culminate in one that is a prisoners' dilemma Newcomb Problem.

Case 1 – Newcomb Problem$_1$. Suppose that the predictor wants only to be right.[4] If any assumption regarding his motives can be said to be "standard," it is this minimal one. A consequence matrix of help in determining expected values under this assumption is:

	Predict leave	Predict take
Leave the money	Right	Wrong
Take the money	Wrong	Right

(Here column headings represent possible acts, and row headings represent conditions that can affect outcomes.) It is obvious that the problem of this case has the 2-dimensional ordinal structure.

	Predict leave	Predict take
Leave	3, 1	1, 0
Take	4, 0	**2, 1**

and is therefore not a prisoners' dilemma. It is noteworthy, however, that Newcomb's Problem$_1$ does nonetheless share an important property with prisoners' dilemmas. Like them, it features a unique equilibrium, (Take, Predict take), that is not Pareto optimal: the outcome (Leave, Predict leave) would be better for the predictee and not worse for the predictor.[5]

Case 2 - Newcomb Problem$_2$. Suppose that the predictor not only wants to be right but cares also about the money, though he puts being right "lexically first." Then a helpful consequence matrix is:

	Predict leave	Predict take
Leave the money	Right, and out $1,000,000	Wrong, and out $0
Take the money	Wrong, and out $1,001,000	Right, and out $1,000

and Problem$_2$ has the non-prisoners' dilemma structure

	Predict leave	Predict take
Leave	3, 3	1, 2
Take	4, 1	**2, 4**

Given the lexical priority of correctness, money enters the predictor's consideration only to break ties of correctness and of incorrectness present in Newcomb Problem$_1$. (Note that, in contrast with Newcomb Problem$_1$, the unique equilibrium in Newcomb Problem$_2$ is not suboptimal.)

Case 3 - Newcomb Problem$_3$. Suppose still that the predictor wants to be right, but that he cares mainly about the money; suppose in particular that being right is not worth $1,000 to him. The consequence matrix for Case 2 serves here, and it can be seen that the Problem of the present case has the following non-prisoners' dilemma ordinal structure:

	Predict leave	Predict take
Leave	3, 2	1, 4
Take	4, 1	2, 3

This structure also suits the simpler assumption that would have the predictor concerned exclusively with money and not at all with being right or making true predictions.

Case 4 – Newcomb Problem₄. Suppose (just as in Case 3) that the predictor wants to be right but cares mainly about money, and that being right is not worth $1,000 to him. But suppose further that the predictor dislikes takers, and would prefer that the predictee not be one. (This antipathy would explain – by making veridical – the appearance that predictors are concerned that takers should get far less money than leavers.) The effect of this additional motive is to make outcomes on the Take row less attractive to the predictor. If his antipathy toward takers is intense, so that he would rather be out $999,000 than have the predictee be a taker, then his preference in Newcomb Problem₃ for (Take, Predict take) over (Leave, Predict leave) will be reversed:

	Predict leave	Predict take
Leave	3, 3	1, 4
Take	4, 1	2, 2

This money-in-the-bank Newcomb Problem is a prisoners' dilemma in which the sole equilibrium (Take, Predict take) is nonoptimal with a vengeance, being bettered for both parties by the outcome (Leave, Predict leave).

3. Further questions

So Newcomb's Problem can be a prisoners' dilemma. Questions remain, however, concerning the kinds of prisoners' dilemmas that versions of Newcomb's Problem can be. I will comment briefly on two such questions.

Can a version of Newcomb's Problem be a "mutually recognized" prisoners' dilemma? Can Newcomb Problem₄, for example, be recognized by its players for the prisoners' dilemma that it is? There is a difficulty here on the predictee's side. For suppose he realized that the ordinal structure of the situation was symmetrical in the manner of a prisoners' dilemma. Then he could know that the predictor was not concerned solely with being right (see Case 1), or even first with being right (Case 2), and he could know that the predictor was not - in the manner of Selten and Leopold's case (see note 5 - obsessed derivatively with being right as a necessary means to the only things sought for themselves in the case.[6] But the predictee, since he confronts a Newcomb problem, must view potential news of each of his actions as good possible evidence that the predictor predicted that action. It is difficult to imagine explanations of such views on the predictee's part if he thinks that the predictor is not particularly concerned that he should make true predictions. To explain such views under the condition that the predictee realizes that the situation is symmetrical in the manner of a prisoners' dilemma, one might suppose some confusion or lack of perspicuity on the predictee's part: One might suppose that he believes things about the predictor's motives that he, the predictee, is in a position to know are not true. I will not try to deal with this difficulty, or to explore other ways in which the predictee could view potential news of his possible choices as good clues to the predictor's predictions. I do think that a version of Newcomb's Problem can be a mutually recognized prisoners' dilemma, and indeed that a Newcomb Problem$_4$ can be one. But I suspect that this possibility (suppositions of confusions on the predictee's part aside) is realized only in farfetched versions of Newcomb's Problem, the details of which would prove of limited theoretical interest. I note that even though unconfused versions of Newcomb's Problem that are mutually recognized prisoners' dielmmas seem barely imaginable, it is easy to frame mutually recognized non-prisoners' dilemma versions that feature nonoptimal equilibria (e.g. Newcomb Problem$_1$) and even 'doubly' nonoptimal equilibria (see note 5 for two examples).

Can a prisoners' dilemma version of Newcomb's Problem be a prisoners' dilemma that is a side-by-side Newcomb problem? The question, related to Newcomb Problem$_4$ and put somewhat differently, is whether it can be a Newcomb problem not only for the predictee, but also for the predictor. For that condition to be satisfied, the predictor would need to view potential news of each of his possible predictions as good evidence for the action it predicted, and so view each possible prediction as good evidence for its own correctness. He would need to view news (were he to receive it) that he will make a prediction other than the one he thinks he is about to make as evidence good enough to counterbalance his presently sufficient evidence against that prediction. It would, I think, be very difficult to tell stories of theoretical interest that made sense of such views on the predictor's part. This difficulty relates to all versions of

Newcomb's Problem, and not especially to versions that would be prisoners' dilemmas.[7]

4. CONCLUDING REMARKS

Not every Newcomb problem is a prisoners' dilemma. For one thing, some Newcomb problems feature only one agent and so cannot be prisoners' dilemmas. For another, even Newcomb problems that feature two agents, as all versions of Newcomb's Problem itself do, are prisoners' dilemmas *only* given special patterns of interests on the predictor's part. If, for example, he is interested only in being right, or only in this and the money, then his preferences will not order outcomes as required for a prisoners' dilemma. However, if he has an interest also in takers – a negative interest, of course, so that he has an aversion to them – then, if this interest is sufficiently intense, his preferences can order outcomes as required for a prisoners' dilemma. Newcomb's Problem makes this point.

Thus it can seem that agents who are given to one or another kind of cooperative reasoning might, in some versions of Newcomb's Problem, see profitable scope for their way of thinking. It seems that cooperative reasoners might sometimes manage to avoid the equilibrium (Take, Predict take) to which causal maximizers would be doomed,[8] and to meet instead in the optimum (Leave, Predict leave) that they would both prefer to that equilibrium. An exploration of these possibilities would raise questions concerning: the details of various modes of cooperative reasoning;[9] possible limitations of cooperative reasoning to mutually recognized problems (and so only to very farfetched versions of Newcomb's Problem that are prisoners' dilemmas); potential gains and losses of dispositions to modes of cooperative reasoning; and the disputed reasonableness of various modes of cooperative reasoning. Such questions, however, all lie beyond the scope of this chapter, which has been to show – and to elaborate only briefly upon – this fact: although most versions of Newcomb's Problem are not prisoners' dilemmas, some are.[10]

NOTES

1. Susan Hurley discusses the relevance of a kind of cooperative reasoning to Newcomb problems in a manuscript I read in July 1989, entitled "Newcomb's Problem, Prisoner's Dilemma, and Collective Action." Her discussion features a predictor-parent/predictee-child problem in which not only the predictee's but also the predictor's preference rankings are as in a prisoners' dilemma. It was while thinking about Hurley's manuscript that the idea for this chapter took form. The final version of her paper appears in *Synthese* 86, (1991), pp. 173–96.

2. Kinds of Newcomb problems are surveyed, and challenges to the coherence especially of causal ones are discussed, in Sobel (1990b [Chapter 2 in this volume]).
3. Row is sure not only that his actions cannot affect those conditions, but that those conditions cannot affect his actions. For example, in versions of Newcomb's Problem in which the column conditions are certain possible predictions, the agent is sure not only that he cannot affect those predictions, but also that those predictions cannot affect his actions as they might do if the predictor, after making a prediction, was in a position to make it true. (See Sobel 1989a, p. 9 [Chapter 5, Section 1] for a science-fiction hypothesis concerning self-fulfilling predictions.) I am defining what Talbott (1987) might term a "*standard* Newcomb problem."
4. Compare: "Newcomb's Problem can be viewed as a game in which Row is interested only in the money and Column is interested only in making true predictions" (Sobel 1988e, p. 251). I do not in that paper discuss versions of Newcomb's Problem in which Column is not interested only in making true predictions.
5. Selten and Leopold (1982, p. 198) observe that it is possible "to describe [Newcomb's Problem] as a two-person game, even if this is not done in the literature." They reach utilities of 1 and 0 for the predictor, arranged as in Newcomb Problem$_1$. "Presumably he prefers to observe *A* [that the transparent box that holds $1,000 is left, and only the opaque box is chosen] if he has put M into the opaque box [which he does if and only if he predicts that the transparent box will be left] and to observe *B* [that both boxes have been chosen] if he has put nothing into the opaque box [which he does if and only if he predicts that both boxes will be chosen]. Accordingly, we model his payoff as 1 in these two cases and as zero for the other two possible outcomes" (p. 199). Their grounds for this structure are, however, somewhat different from mine. For though they "assume that money does not matter" (p. 199) to the predictor, they say positively only that he is concerned to reward and punish *A* and *B* respectively (p. 199). Their predictor does not, as does mine, want only to be right; he wants primarily to reward *A* and punish *B*. It seems that he wants to be right not merely as an end in itself, but only or also as a necessary means to rewards and punishments wanted for themselves.

Their predictor is indifferent between rewarding *A* (leaving the transparent box that holds $1000) and punishing *B* (taking that box), and also indifferent between not punishing *A* and rewarding *B*. He evidently does not favor leavers over takers. If their predictor's primary motives consisted of that general favor added to a desire to reward and punish accordingly, his utilities would be not as in my Newcomb Problem$_1$ but rather, depending on the relative intensities of that general favor and his desires to reward and punish, as in one of the following structures:

	Predict leave	Predict take
Leave	3, 3	1, 1
Take	4, 0	2, 2

	Predict leave	Predict take
Leave	3, 3	1, 2
Take	4, 0	**2, 1**

Note that these structures, though not prisoners' dilemmas, are nearer to them than is the structure of Newcomb Problem$_1$. In contrast with it, these structures feature unique equilibria that are 'doubly' nonoptimal: in each, (Take, Predict take) is preferred – not only by the predictee, but also by the predictor – to (Leave, Predict leave).

6. If a predictor makes being right lexically prior to every competing consideration, then both a and d must exceed both b and c, whereas in a prisoners' dilemma b exceeds both a and d:

	Predict leave	Predict take
Leave	(Right) w, a	(Wrong) x, b
Take	(Wrong) y, c	(Right) z, d

7. For a particularly challenging exercise, one might try to construct a prisoners' dilemma version of Newcomb's Problem (such as Newcomb Problem$_4$) that is both a mutually recognized problem and a side-by-side Newcomb problem.
8. The interaction (Take, Predict take) is an equilibrium in, and for causal maximizers is the solution of, every version of Newcomb's Problem considered here and every version that comes to mind. Consider that in every money-in-the-bank version of Newcomb's Problem, "take" is strongly dominant and predictions are certainly causally independent of Row's actions. So, as the Predictor should know, if Row is a causal maximizer then he will 'take'. Barring bizarre motivations, the predictor will prefer (Take, Predict take) to (Take, Predict leave), and so will predict 'take'.
9. Evidential maximizers "cooperate" in prisoners' dilemmas that are side-by-side Newcomb problems: they attain the cooperative outcome that each prefers to the unique equilibrium in which causal maximizers must meet. But I do not count evidential maximizing as a mode of cooperative reasoning. I think of cooperative reasoners as agents who, in certain prisoners' dilemmas (e.g., in mutually recognized ones in which each realizes that the other is like-minded), are intent not on maximizing, either causally or evidentially, but on "optimizing," and who in this way attain the cooperative outcome in these prisoners' dilemmas even without aid of preplay communication and explicit binding agreements. Evidential maximizing and various modes of cooperative reasoning should amount to the same thing in some prisoners' dilemmas, but by different deliberative means. It can be expected that these ways of reasoning would amount to different things in some other situations, and not be always (even "extensionally") equivalent.

10. For some comments on theories of cooperative reasoning – in particular, Donald Regan's and David Gauthier's theories – see Sobel (1985c,d, 1987b, 1988b [Chapter 5]). Also of relevance, though I have not had an opportunity to examine it, is Hurley (1989).

5

Infallible predictors

Abstract. One-boxers sometimes argue: "If you were certain that the predictor is infallible, you would see yourself as having a choice between one box and a $M for sure and two boxes and a $T for sure, and, if rational, you would go for the certainty of a $M. Similarly, were your confidence in the predictor nearly complete, nearly nearly complete, and so on to merely middling confidence." In response, I stress the difference between absolutely infallible and merely never-erring predictors, caution against confusing interpretations of the emphasized conditional — some of which are true and some of which are relevant, but none of which is both – and comment on cases in which the first box has in it not a $T but a $M, just as the second box may.

0. INTRODUCTION

Two-boxers in Newcomb problems are persuaded by dominance arguments. They are convinced that it would be irrational to leave on the table the $T that is there for the taking. They are convinced that choiceworthiness and evidential expected utility – "Desirability," as Richard Jeffrey terms it – diverge in Newcomb problems.

One-boxers disagree. Many think that the error of two-boxers is particularly plain in limit cases, where the agent's confidence in the predictor is not merely great but complete. Such one-boxers addressing two-boxers might argue as follows.

> If your confidence in the predictor were complete and you considered him to be infallible in cases like yours, then you would see yourself as having a choice between a $M for sure and a $T for sure. You would be ordering money from a menu.[1] Taking only the second box in that case would be not only most Desirable, but also most choiceworthy.
>
> What holds for the case in which your probability for the $M (given that you take only the second box) is 1 must also,

Reprinted with revisions by permission of the publisher from *Philosophical Review* 97 (1988), pp. 3–24. Copyright 1988 Cornell University.

however, hold for the case in which it is very nearly 1; what
holds for the case in which it is very nearly 1 must hold for
the case in which it is very nearly, very nearly 1; and so on
down a slope of all values for this conditional probability.
Desirability coincides with choiceworthiness in every case. At
the top down to about its midpoint they join in recommending
that you take only the second box. At about the midpoint
they cross over to recommend that you take both boxes. But
at the top, and all the way down, they agree in their
recommendations.[2]

This argument, in particular the initial assessment of the limit case, is I think ill-served by a certain confusion, and compromised by the ease with which it is possible to indulge in certain equivocations. We consider these in turn.

1. PREDICTORS WHO ARE CERTAINLY INCAPABLE OF ERROR

It is possible to confuse cases in which an agent has complete confidence in a predictor and is sure that he is infallible, in the sense that he is unerring and in fact never mistaken, with cases in which the agent is sure that the predictor is infallible in the sense that the predictor is quite incapable of being mistaken.[3] But cases of this latter sort, wherein the predictor's infallibility is taken to the limit, are not cases in which the agent would be confronted with a decision problem of what to do. For it is a necessary condition of a decision problem that, consistent with everything else he thinks, the agent be sure of what he can do. And it is a necessary condition of a nontrivial decision problem that the agent be sure he has a *choice* – that he be sure that he can do each of at least two actions, that both actions are open to him. These are prerequisites of decision problems. Before one can wonder and deliberate *what* to do, it is necessary that one be settled in one's mind regarding what one *can* do, what one's options are. But in extreme Newcomb problems in which the agent is sure the predictor is absolutely incapable of error, the agent cannot be sure of what he can do unless he is already sure what he is going to do, and in no extreme case can he even think that he has a choice and that each of the two actions in the problem is something he can do. Let me explain.

It is part of any Newcomb problem that the agent is sure that the predictor has made a prediction. More fully, the agent in any Newcomb problem is sure that it has either been predicted that he will take both boxes, or that he will take just one box. And the agent in any Newcomb problem is sure that the prediction made is now, at the time of his choice, quite beyond his or anyone else's control – there is no sense in which it can be unmade – that whatever he or anyone were now to do, the prediction would remain unchanged. But this means that, *if* the agent in a Newcomb problem were sure that the predictor was incapable of error, that a false prediction was quite out of the question and something which could not be so much as coherently "entertained" (cf. Lewis 1973, pp. 16, 22–3) for purposes of practical counterfactual speculation; *then,* if the agent is consistent, he cannot be sure that both actions are open to him, and thus cannot be sure that they are things that can be entertained for purposes of practical counterfactual speculation. Suppose, for purposes of indirect argument, that the agent *were* sure that both actions are open to him. Things of which he was sure, including (by the current supposition) that both actions were open to him, would imply a contradiction. For from these things it would of course follow that the *un*predicted action was open to him, and so was possible in at least the very weak sense of being entertainable by anyone who knew (as he may not) all the facts. And it would follow that, if he were to perform this action, the prediction made that he would perform the other action would still stand; for he (the agent in any Newcomb problem) is sure that the prediction, whatever it was, is causally independent of his current choice and action. And so it would be a consequence of things of which he is certain that a false prediction on the predictor's part was itself entertainable, and at least in this sense a possibility. But in the present quite extreme case the agent is to be sure that a false prediction on the predictor's part is *not* a possibility, not even in the weak sense that would allow hypotheses that would have it false to be (by an all-knowing person) entertained for purposes of practical counterfactual speculation. Our supposition for argument would have the agent certain that both actions are open to him. We now see that this, taken together with other things of which he is to be certain, leads to a contradiction.[4] It is now clear that the agent in an extreme Newcomb problem cannot consistently so much as even think that both actions are open to him.[5] He must, if consistent, think that only one is (though, if he is consistent, he cannot be sure which action that is, unless and until he is sure what he is going to do). The present case of an agent who is sure that the predictor is quite incapable of error – sure that his erring is quite out of the question and not, for practical purposes, an entertainable hypothesis – is not a case in which the agent could consistently take himself to have a choice. This extreme case is thus not "on the slope" of decision problems.

Suppose for the moment that this extreme case *were* a decision problem. It would then start us down the slope. For the agent in this extreme case is sure that the predictor is absolutely incapable of error, and this means that cells in a diagonal in the first consequence matrix (Jeffrey's term) for the case, and in a column in the second one, can be left empty and can be ignored. (I stress, *pace* Jacobi 1993, p. 5, that the problem is not which of these partitions to use; each is applicable without prejudice, as explained in Sobel 1988d; see Chapter 2, Appendix IV, in this volume.)[6]

Extreme Newcomb problems
in which the agent is sure the predictor
is absolutely incapable of error

	(Prediction: only B2 will be taken) There is a $M in B2	(Prediction: both will be taken) There is not a $M in B2
Take both		$T
Take only B2	$M	

	Correct prediction	Incorrect prediction
Take both	$T	
Take only B2	$M	

Indeed, in this extreme case these cells *must* be left empty and ignored, for the conjunctions associated with them are, in the agent's view, not merely completely improbable but impossible, in the very strong sense of being (by an all-knowing person) not even entertainable, or even so much as coherently supposable for purposes of counterfactual causal speculation. This means that the agent cannot think that they have consequences in the sense of things that would happen if (that is, *supposing*) these acts were to be done; salient features of consequences could be noted in these cells.[7]

If the agent in an extreme Newcomb problem had a decision problem – if he could, consistent with his confidence in the predictor, think that he had a choice between taking both and taking only B2 – then his problem would reduce to ordering from a menu as follows:

| Take both | Get a $T |
| Take only B2 | Get a $M |

"The other outcomes," one might say, "are null"; Seidenfeld (1985, p. 203) does say this. However, given the agent's beliefs about the predictor – given that the agent is to believe, fantastically, that the predictor is absolutely infallible, so that error on his part is not even entertainable – the agent in an extreme Newcomb problem could not consistently think that he had a problem regarding whether to take both boxes or take only one. In this extreme case the agent, if consistent, must think that he cannot do these things, not *both* of them; the question of which to do cannot arise for him. This extreme case thus has little if any relevance to standard cases in which the agent does have a problem concerning which to choose.

But how could anyone think that a predictor was quite incapable of making a mistaken prediction? There are at least science-fiction ways that may satisfy some of us. One of these would have the agent think that a certain predictor, in making a prediction in the case, unavoidably (and indeed without even suspecting that he was doing so) initiates a monitoring device that would detect threatened failures of his prediction and, if need be, initiate manipulating devices. For example, the agent could believe that if the monitor detected a weakening of resolve on his part then a cosmic force would be made to beam into his brain and get him to do what it had been predicted he would do. An agent who believed these things could, on account of them, believe that a false prediction on the predictor's part was impossible in the strong sense that would make errors on this predictor's part not even entertainable.

I do not mean by the phrase "on account of" that this last opinion is entailed by the earlier ones; it is not. It goes beyond them, and amounts to viewing the monitoring and correcting devices and mechanisms mentioned in these background opinions as themselves absolutely infallible (which they are not said to be, in these earlier opinions on which the final opinion is based). To make sense of what is problematic about the opinion that some predictor is absolutely infallible in the present strong sense – to make sense of this in terms of other opinions that are in themselves more understandable – is I think to make sense of it in terms of background opinions that are free of, and do not *entail,* modalities of the kind that make this target opinion problematic and something a person might want explained. If such explanations are not good enough for a person, then – since I think such explanations are the best we can do here – I doubt that any sense that *would* satisfy this person can be made of strongly infallible predictors, or of extreme problems in which agents would believe in such predictors.

"But can't individuals be made infallible by definition? And shouldn't this satisfy anyone who wants to understand how strong infallibility is possible?" My answers are "yes" to the first question, but "no" to the second. It is possible, of course, to make it part of the definition of a kind of being that an individual of that kind is unerring (that is, weakly infallible), and so to make necessary by definition that any individual of that kind is unerring. There is no difficulty in that. But this constructive possibility does not even begin to throw light on how it could be *de re* true of an individual that he is necessarily unerring (that is, somehow strongly infallible). Making "strong infallibility" (whatever these words are to mean) a part of the definition of a kind of being clearly cannot in itself contribute to an understanding of these words.

2. Predictors who are certainly unerring

2.1

We now set aside strange predictors who would be quite incapable of error, and attend to equivocations possible in non-extreme Newcomb problems in which agents are supposed to be sure that the predictor is only in fact never mistaken. Let these be called "limit" Newcomb problems. Relevant to these problems are several closely related possible equivocations, of which we will concentrate on one in particular.

It can seem that the following sentence would say to the agent in a limit case something that was decisive for taking one box:

> If your confidence in the predictor is complete, then your choice is between a certainty of a $M and a certainty of a $T.

But this sentence can be understood in two ways. In the first it becomes:

> If your confidence in the predictor is complete, then your choice is between doing something that *would make you certain you had gotten* a $M and doing something that would make you certain you had gotten a $T.

Understood this way, and supposing that the agent will know what he has done and will conditionalize on this knowledge, the claim is true. After he has taken a box (or boxes), and before opening it (or them), if he has complete confidence in the predictor then he will be in no doubt about what he will find when he opens his box(es). But this fact about how he can control his mind and manipulate his possibly ephemeral prospects – how he can, if he chooses, manufacture for himself a pleasant subjective state of expectation – is, given that he is interested mainly in the money, not very relevant to what he should do.

Understood another way, however, our sentence comes to the claim that:

> If your confidence in the predictor is complete, then your choice is between doing something that *you can now be sure would net you* a $M (something that you can now be sure

would have a $M for you among its consequences – no more, no less) and doing something that you can now be sure would net you a $T.

If the agent were sure of this conditional then he would be sure of something that was very relevant to his problem, and he would have an easy problem. It would be clear that he should do the thing that he was sure would make him a millionaire. By this conditional, given in our limit case that his confidence in the predictor *is* complete, the agent could be sure that one choice (taking only B2) would net him a $M and that the other (taking both boxes) would net him a $T. And so, were he sure of this conditional, it would be clear that he should do: take only B2 and be a millionaire. However, while this conditional would be very relevant, the agent could be sure it is false. For he is sure that there is either a $M in B2 or nothing, and that in any case there is a $T in B1. And so he should be sure that his choice is either between something that would net him a $M plus a $T and something that would net him a $M only (see the first column of the following matrix), or between something that would net him a $T and something that would net him $0 (the second column).

Standard Newcomb problems
in which the agent is not sure the predictor
is absolutely incapable of error

	(Prediction: only B2 will be taken) There is a $M in B2	(Prediction: both will be taken) There is not a $M in B2
Take both	$M + $T	$T
Take only B2	$M	$0

To summarize briefly, there is a sense in which the agent in a limit problem can, by taking only B2, be certain of a $M, and also a sense in which possibly he cannot be. It is true that he can, by taking only B2 and not opening it, render himself certain that he has a $M in the box he has taken. But, unless he is sure there is a $M in B2, it is false that he can be certain that taking only B2 would net him a $M; likewise for taking both boxes and certainty of a $T.[8]

The agent in any Newcomb problem should be sure that of the following conditionals, interpreted in ways that make them relevant to his choice, one is true and the other false.

If I were to take only B2, I would end up with a $M.
If I were to take both boxes, I would end up with only a $T.

We suppose that he does not know which is true and which is false. But, unless he is very confused, he must know that whichever is true, the other is false.[9]

2.2

"But surely these last two conditionals are open to natural interpretations that make both of them true." Yes, I think they are.[10] With the predictor's prowess in mind, I might naturally reflect that, if I were to take only B2, then (backtracking) it would have to be (or more simply, it would be) that the predictor had predicted this action, that he had put the $M in B2, and that this was what I was going to get. With his prowess and consequent prediction in mind, I might think essentially the same thought in the more succinct words, "If I were to take only B2, it would be that I got a $M," or think it in the very words of interest to us, "If I were to take only B2, I would end up with a $M." Similar reasoning holds for taking both boxes with the words "If I were to take both boxes, it would be that I got a $T" or the words "If I were to take both boxes, I would end up with a $T." And these conditionals may be true,[11] but they would not, I think, be relevant. These conditionals would be saying how – while holding on to certain propositions I accept, including certain propositions about the predictor (that he is an extraordinarily good one) and about the past (that he has made a prediction regarding what I will do and acted on it in a certain way) – I can most easily make room for certain possibly counterfactual propositions. But such belief-revision conditionals that do not pretend to trace out potential consequences, while of possible theoretical relevance in some contexts (for example, in contexts in which I am wondering what – given one or another piece of news concerning an action – I should believe, and if rational would believe and, if I am a betting man, be prepared to bet on), are I think not of any practical relevance to what I should do: take both boxes or take only the second box.[12]

2.3

Let me digress from issues specific to limit cases in order to deepen and broaden the claims made just now about the practical irrelevance of backtracking. We now consider Newcomb problems in general, including especially nonlimit ones in which the agent has not certainties but only various non-extreme probabilities, and in which the issue is not what conditionals to rely on but rather what probabilities to use as weights in calculations of measures of choiceworthiness. (The issue in limit cases can be understood in this way, too.)

If weights in calculations in a Newcomb problem were probabilities of backtracking conditionals, these weights would reflect not probable real bearings of actions on things but instead evidential bearings of actions. Terence Horgan is, I think, committed to this view. I assume, for this interpretation, that a conditional probability $\Pr(q/p)$ measures the possible evidential bearings of p on q.[13] Horgan holds that in exceptional decision problems such as Newcomb's Problem, backtracking act-state conditionals are appropriate. He also holds that, in all decision problems, *probabilities of appropriate act-state conditionals are equal to corresponding conditional probabilities* (Horgan 1985a, p. 172).

Horgan maintains both the identity of probabilities of certain backtracking conditionals and corresponding conditional probabilities in all cases, as well as the practical relevance of these probabilities in even such exceptional cases as Newcomb's Problem. Horgan maintains these views even though they commit him to the decisiveness for choice of evidential bearings in all cases, including cases where such bearings differ from probable real bearings measured by probabilities of causal, nonbacktracking conditionals.[14] He is committed to the view that choiceworthiness of acts is determined by values of news of them,[15] even in cases where "news of an act [would] furnish evidence of a state . . . which the act itself is known not to produce" and where, "though the agent [would] indeed [as he always does] make the news of his act, he [would] not [as in the ordinary run of cases he does] make all the news his act bespeaks" (Gibbard and Harper 1978, in Campbell and Sowden 1985, p. 145). It is, however, implausible that choiceworthiness should be determined by news values in these cases.

Horgan's position has two parts. In the first he stipulates a certain identity; in the second he maintains the practical relevance of the things identified. Of these, the first part is plausible: it is plausible that probabilities of backtracking conditionals should equal corresponding conditional probabilities. Backtracking conditionals seem to be belief-revision propositions that state what else, given a belief in the antecedent, one would have to believe. It is thus plausible to stipulate, as Horgan does, that a mark of backtracking (or of a particular mode of backtracking) is that probabilities of conditionals that are backtracking (in this manner) equal corresponding conditional probabilities.[16] What is implausible is that measures of evidential bearings should be of practical relevance and decisive for choice precisely in cases in which these bearings plainly differ from probable causal bearings. Stressing the implausibility of its implications is, I think, the best way to oppose Horgan's position on the character of practically relevant conditionals. Not everything can be proved from antecedently shared premises. Not everything that can be obvious can be deduced from premises that are antecedently obvious to all parties.

This ends the digression of this section; we return now to the assertion that led to it. According to that assertion, backtracking conditionals, while of possible relevance to theoretical belief-revision issues, are not of practical relevance, or of relevance in contexts of deliberation in which the question is what I, an agent, should do.

2.4

In the very simple (even though fantastic) situation of Newcomb's Problem, the issues that are relevant to what I should do - take both boxes, or just the second one - are plainly as follows. (1) What *would not be affected;* that is, what would be left as it is, however it is, were I to perform the first (or second) action. (2) What *would be changed* were I to perform the first action, supposing that this action is not the thing that I am in fact going to do; and similarly, what would be changed were I to perform the second action. Settle these issues and it will be clear what to do. And they are easy to settle. What would be left as it is, whichever action I did, includes the money. The causal independence from these actions of the money in the boxes is given. This independence is as much a part of limit, probability-of-1 Newcomb problems as it is of any Newcomb problem. Among the things that depend on which action I perform are whether I get only what is in B2 or what is in both boxes. All this is to say what has been plain all along - namely, that my choice is between taking what *is* in B2 and taking what *is* in both boxes.

But then, given that this is my choice, what is my *problem?!* Before pressing this rhetorical question and substantive point regarding Newcomb problems, let us observe that at least one thing is clear. In a Newcomb problem (even in a limit Newcomb problem), since the $M is in B2 or not regardless of what I do, I can be sure that I do not have it in my power to take both boxes and get exactly a $T and also to take only B2 and get exactly a $M. I should be sure that I can do one or the other of these things, even if I am not sure (as I may well not be) *which* I can do (because I am not sure what is in B2). Yet it should be quite clear to me that not both of these things is in my power. My choice (I should be prepared to lament) is certainly *not* between two actions, one of which would net me exactly a $T and the other exactly a $M!

2.5

No one can deny these claims and say that I do have it in my power both to obtain a $T and to obtain a $M, or say that my choice is between a $T and a $M - except "for the sake of maintaining an argument." But then, although many things are clear, almost nothing is incontrovertible; almost anything can be denied or said for the sake of maintaining an argument. Much has been said of limit Newcomb problems. Indeed this very thing has been, if not said, at least seriously entertained by Horgan, who has claimed that it is at any rate very natural to say that the agent does in the limit case have it in his power both to obtain a $T and to obtain a $M, and that these are indeed precisely the things he can do (Horgan 1985b, p. 230). Horgan has written that

> it must be admitted that there is something very natural about saying that the agent simply cannot falsify the being's prediction. After all, he knows that he is bound to fail, no matter how hard he tries. Furthermore, pre-assured failure will not be a matter of simple bad luck; luck does not enter into it at all, since the probability that the being will be wrong is dead zero.
>
> The naturalness of using "can" in this way, I suggest, reflects the fact that our ordinary notion of power is actually more flexible than it initially appears to be. One way of employing this notion – in numerous contexts the only appropriate way – is to build in causal efficacy and to deny such a thing as power over the past. But another possible usage, one which has a strong air of appropriateness in the present context at least, is to employ the term in such way that the conditions of the limit case of Newcomb's Problem do indeed imply that the agent cannot falsify the being's prediction and do indeed imply that he has the power to obtain for himself either the $1000 outcome or the $1 million outcome. (Horgan 1985b, p. 231)

There are, according to Horgan, two ordinary senses of "power," and the issue is which sense is relevant for practical purposes: "The two-boxer claims that the agent ought to perform the act which yields the better of the outcomes that are within his *standard* [causal] power; while the one-boxer says he ought to choose the act with the better of the outcomes that are within his *backtracking* power" (Horgan 1985b, p. 232).

I think the presently (i.e., practically) relevant sense of "power" is a causal one, and that there is no ordinary backtracking sense. I concede, of course, that by hypothesis the agent is sure that he will not falsify the being's prediction. And we can assume that, at least in the agent's mind, "luck does not enter into it": The agent can be assumed to be sure, for example, that it is not the case that, though both he and the predictor are flipping coins, to his bad luck they will land alike. Furthermore, the agent should be sure that "no matter how hard he tries" to falsify the being's prediction, and no matter what he the agent is going to do – that is, no matter what he actually *does* – he is going to fail to falsify that prediction. That is, he will in fact fail to falsify it. For the agent is sure that this prediction is true, and he like everyone else should be sure that – no matter how hard he has tried, tries, or will try – he never has and never will make false anything that is true.[17] The agent should be sure that he will fail to falsify the prediction, whatever it is.

But it is not given that he should be sure that he *cannot* falsify the prediction, or that he knows that (despite his efforts) he is *bound* to fail, unless "bound" here means merely "certain," an epistemic notion, and not "constrained," a notion involving some kind of objective necessity. And even if these things were given, it would not follow that the agent was sure that he had the power (at the later date of his action) to change the prediction, and for its truth the contents of the boxes. Even if he were sure that he was bound to fail - that he was objectively bound to do whatever it is that the being predicted - it would not follow that he was sure that he had both the power to gain exactly a $T and the power to gain exactly a $M. It would not follow that he had this or any other choice in the case. Far from it!

I can understand the temptation to say of a limit case that the agent cannot falsify the being's prediction, and to say that he can obtain for himself either a $T or a $M - that he has both of these things really in his power. I can understand how natural it could be for an agent in a limit case to think these things himself. I can appreciate "the strong air of appropriateness" in a limit case of the words "I have it in my power to obtain exactly $1000, for there is something I can do that would make me certain of that" and "I have it in my power to obtain exactly $1 million, for there is another thing I can do that would make me certain of *that*." But I think that it is plain - really, quite plain - that there is not any natural and ordinary sense of these words in which they would actually be appropriate. It is not for us to "build . . . causal efficacy" into the notion of power and, consonant with that stipulation, to deny that we have power over things we cannot affect. Notwithstanding temptations to careless (and confused) ordinary speaking to the contrary, such efficacy is already part of our ordinary notion of power, and it is a tautology that we do not have power over things we cannot affect.

Will it now be suggested that "affect" has an ordinary backtracking sense? The word "affect" can have a *contrived, acausal,* backtracking sense defined in terms of backtracking conditionals. So can "power" and, making a small leap, "choice"; so can "cause"! They can have contrived senses of all sorts, as can all words. There is no law against it and no particular difficulty in it, but in the present case, I am sure, no profit in it either.

3. Conclusions

In a limit Newcomb problem, as in every standard Newcomb problem, the agent should be sure that he has it in his power either to obtain just what is in the second box, or to obtain what is in it *along with* the $T that he knows is in the first box. The agent's choice in every non-extreme or standard Newcomb problem is just that – to take the second box alone, or to take both boxes. His choice is to act so that he gets only what is in the second box, or so that he gets that and in addition what is in the first box. And thus, if it has not been all along (given that he likes money), the agent's rational choice is now settled and clear. Perhaps we can now agree that in all standard Newcomb problems, including limit ones, *the agent should take both boxes,* thereby making certain that, whatever money there is on the table, he gets *all* of it.

However, on the chance that we are not yet agreed on this, I cannot forbear offering for reflection one further case before we all become no-boxers. Let us consider a "fully inflated" Newcomb problem. This is to be a limit, probability-of-1 Newcomb problem with this difference: The first box in a fully inflated problem has in it not a thousand dollars but a cool million, just as the second box may.

Fully inflated Newcomb problems
in which the first box contains a million,
just as the second box may

	There is a $M in B2	There is not a $M in B2
Take both	$M + $M	$M
Take only B2	$M	$0

	Correct prediction	Incorrect prediction
Take both	$M	$M + $M
Take only B2	$M	$0

Here one "naturally argues" (to use Nozick's words) as follows. "I can have, if I choose, a chance for two million. And I have nothing to lose by taking both boxes and giving myself that chance. So of course I should go for it!" If, as I hope, we find ourselves inclined by this or any other line of reasoning to favor taking both boxes rather than only one in this problem,[18] let us consider next a limit Newcomb problem that is *almost* fully inflated. The bank draft in the first box is in this problem drawn not for a million dollars but rather for $999,999.99. Does one then say: "Now I do have something – a penny – to lose; so, as in all Newcomb problems except the fully inflated one, it is just one box for me, thank you." Would anyone really wish to argue that a monetary difference, however minute, could make a difference here in what was rational to do? If not, and one still wishes to say that a difference of $999,000 (the difference that leaves only the standard $1000 in the first box) *would* make such a difference, then I wonder where – between one cent and $999,000 – and for what reasons anyone would draw the line and say, "Enough; now I have too much to lose."[19]

NOTES

1. I owe the menu metaphor to Teddy Seidenfeld. He writes, "Consider an infallible demon.... Then... the problem is one of choice under certainty.... It is certain that the demon predicts correctly; hence, the decision matrix is

	Demon predicts correctly
Option₁ [one box]	$1,000,000
Option₂ [two boxes]	$1000

   where the other outcomes are null" (Seidenfeld 1985, p. 203). Seidenfeld has said in discussion that in this case the problem reduces to choosing from a menu.

2. For the record, and before proceeding to discussion, I note that Robert Nozick entertains the first part of this objection but seems not to endorse it. He writes that if the probability is 1 then "one naturally argues: I know that if I take both, I will get $1000. I know that if I take only what is in the second, I get $M. So, of course, I will take only what is in the second" (Nozick 1969, p. 141). After saying how one "naturally" argues in this case, Nozick asks, rhetorically it seems: "And does a proponent of taking what is in both boxes in Newcomb's example (e.g., me) really wish to argue that it is the probability, however minute, of the predictor's being mistaken which makes the difference?" Nozick then asks, argumentatively it seems: "And how exactly does the fact that the predictor is certain to have been correct dissolve the force of the dominance argument?" (p. 141). Rather than take a stand on probability-of-1 cases, Nozick intends his comments on them as reasons for thinking (as he has said earlier in his paper) that "an adequate solution to this [Newcomb's] problem will go much deeper than I have gone or shall go in this paper" (p. 135). (Postscript: "That's exactly right, all of it, and you can quote me," or words to that effect, said Nozick to me on December 30, 1986, at meetings in Boston of the Eastern Division of the American Philosophical Association.)

3. It can seem that cases of the former sort are not bona fide cases of agents who are sure that predictors are infallible. Doesn't "infallible" mean precisely and only "incapable of error"? No.

   "Infallible," we are told, sometimes means "not fallible," that is, "not liable or apt to err." "Fallible" means "liable to err"; dictionaries, at least the most authoritative ones, do not testify to this word's ever meaning merely "capable of erring." Possibly when they say that "infallible" sometimes means "not liable to err," they intend that it sometimes means "liable, or apt, or disposed not to err" or "unerring by nature or disposition." "Infallible," we are told, also sometimes means "unerring," which would seem to be stronger than "not apt to err," stronger than "apt or disposed not to err," and somewhat weaker than "unerring by nature or disposition." We are also told that sometimes, especially in theological contexts, this word means "incapable of erring" (thus it may be said that the Pope is incapable of erring when speaking *ex cathedra* on a matter of faith or morals).

Evidently "infallible" is ambiguous, and has the two senses I stress in addition, perhaps, to others. So say *The Oxford English Dictionary* - "Not fallible. 1. Of persons . . . : Not liable to be deceived or mistaken; incapable of erring" - and *Webster's Third New International Dictionary* - "1: not fallible: incapable of error: unerring."

4. The argument concerning implications of what the agent would be sure of begins with the supposition that he would be sure both actions are open and thus entertainable. Using "$\Diamond$" to denote "entertainability" (for purposes of practical, possibly counterfactual, speculation), steps in the argument are as follows. (Statements that the agent in any Newcomb problem would be sure of are marked with asterisks, and truth-functional inferences are indicated by "TF".)

1. $\Diamond(\text{Both}) \,\&\, \Diamond(\text{B2})$ assumption for indirect argument
2. $\sim\text{Pred}(\text{Both}) \lor \sim\text{Pred}(\text{B2})$ *
3. $[\sim\text{Pred}(\text{Both}) \,\&\, \Diamond(\text{Both})] \lor [\sim\text{Pred}(\text{B2}) \,\&\, \Diamond(\text{B2})]$ 1, 2, TF
4. $\{\sim\text{Pred}(\text{Both}) \to (\text{Both} \,\square\!\!\to [\text{Both} \,\&\, \sim\text{Pred}(\text{Both})])\}$
   $\& \{\sim\text{Pred}(\text{B2}) \to (\text{B2} \,\square\!\!\to [\text{B2} \,\&\, \sim\text{Pred}(\text{B2})])\}$
            * (causal independence of predictions from actions)
5. $\{\Diamond(\text{Both}) \,\&\, (\text{Both} \,\square\!\!\to [\text{Both} \,\&\, \sim\text{Pred}(\text{Both})])\}$
   $\lor \{\Diamond(\text{B2}) \,\&\, (\text{B2} \,\square\!\!\to [\text{B2} \,\&\, \sim\text{Pred}(\text{B2})])\}$ 3, 4, TF
6. $\Diamond[\text{Both} \,\&\, \sim\text{Pred}(\text{Both})] \lor \Diamond[\text{B2} \,\&\, \sim\text{Pred}(\text{B2})]$
            5, $[\Diamond p \,\&\, (p \,\square\!\!\to q)] \to \Diamond q$
7. $\sim\{\Diamond[\text{Both} \,\&\, \sim\text{Pred}(\text{Both})] \lor \Diamond[\text{B2} \,\&\, \sim\text{Pred}(\text{B2})]\}$

Line 7, which of course contradicts line 6, is by the stipulation of our extreme case: the agent is sure that the predictor is infallible, in the strong sense that predictor errors are not even entertainable for purposes of practical counterfactual speculation.

5. Suppose he were to think this. Then $\Pr[\Diamond(\text{Both})] > 0$ and $\Pr[\Diamond(\text{B2})] > 0$. So, given certainties stipulated for any problem, it would follow (as in note 4) that $\Pr(\Diamond[\text{Both} \,\&\, \sim\text{Pred}(\text{Both})] \lor \Diamond[\text{B2} \,\&\, \sim\text{Pred}(\text{B2})]) > 0$. However, by the stipulation for an extreme problem we have that $\Pr(\Diamond[\text{Both} \,\&\, \sim\text{Pred}(\text{Both})] \lor \Diamond[\text{B2} \,\&\, \sim\text{Pred}(\text{B2})]) \not> 0$. *Contradiction.*
6. Nozick, in his first statement of Newcomb's Problem, suggests that it matters which partitions one uses: "setting up the situation in the first way leads me to [take both]; setting it up in the second way leads me to [take only B2]" (Nozick 1963, p. 225). His considered view (Nozick 1969, p. 124, n. 10) is that it only *seems* to matter which partition one uses.
7. If the conjunctions associated with these cells were for the agent merely completely *improbable,* he could (and presumably would) have views concerning what would happen were they to be realized. Nor would it be necessary or reasonable to ignore these consequences. For completeness, I note that from the perspective of Jeffrey's (1965b) logic of decision - which is, of course, a theory for one-boxers - entries in these cells can be ignored, and might as well not be

made, even when associated conjunctions are merely completely improbable: When these conjunctions are completely improbable, corresponding cells in associated conditional probability matrices contain zeros. In contrast, corresponding cells in associated probability-of-causal-conditionals matrices (suited to two-boxer theories of expected utility) do not all contain zeros.

Consequence matrices reduce to menus for both evidential and causal decision theorists in extreme Newcomb problems, where the agent is sure that certain conjunctions are not even entertainable. They reduce to menus for evidential (but not also for causal) theorists even in limit Newcomb problems, where these conjunctions are for the agent completely improbable though *not* certainly non-entertainable.

8. Seidenfeld would have us consider "an infallible demon ... a special-case version of Newcomb" (Seidenfeld 1985, p. 203). Notwithstanding Seidenfeld's move to delete what he terms "null outcomes," I take him to intend a case in which the agent is sure that the predictor is in fact not mistaken, but not also convinced that he is absolutely incapable of being mistaken. I take him to intend a case where the agent's opinions generate, for him, conditional probabilities of 1 for states of Box 2 on his acts. Regarding his special case, Seidenfeld (1985, p. 203) asks: "Do the causal decision theorists advocate option$_2$ [two boxes] when it is certain to yield $1,000 whereas option$_1$ [one box] is certain to yield $1,000,000?" Plainly, however, this is a complex question. It presupposes that option$_2$ is certain to yield $1,000 and that option$_1$ is certain to yield $1,000,000 in this probability-of-1 case. The question presupposes that these certainties are both true even though "the causal relationship between the choice and the 'fixed and determined' state of the opaque box are unchanged by this special-case version of Newcomb" (Seidenfeld 1985, p. 203). But ... ?

9. Some distinctions need to be stressed in connection with Nozick's argument. He writes that if I know that the predictor is correct then "I know that if I take both, I will get $1000. I know that if I take only what is in the second, I get $M" (Nozick 1969, p. 141). However, I could in fact not know both of these things and still know that the $M is either there or not quite independently of what I do. For example, it seems that, these sentences – "I know that if I take both, I will get $1000"; "I am sure that if I take both, I will get $1000"; "If I take both, I get $1000 for sure"; "If I take both, I will be sure of getting $1000"; and "If I take both, I will be sure that I am getting $1000" – are apt to be run together, even though at least the first and second express very different ideas from that expressed by the last. Similarly, "If I take both boxes, I will do so in the full knowledge that I will be executed tomorrow" and "If I take both boxes, I will as I do so be completely certain that I will be executed tomorrow" express very different ideas. The first (but not the second) sentence includes the idea that if I take both boxes I *will* be executed tomorrow. Consider Horgan (1985b, p. 232).

I digress to comment on the problem just broached in which Box 1 contains a delicious meal and Box 2 contains, depending on the being's prediction, either nothing or a note that will – whatever the agent's choice – go to the authorities and cause them to cancel his execution which is scheduled for tomorrow. If the agent can, he might in this problem opt for the compound action of taking only the second box and having its contents (if any) transmitted *without is being informed of their ch. acter* to the authorities. In contrast with Newcomb's Problem, where the agent is interested mainly in the money, one

115

assumes that in Horgan's problem the agent will be interested not only in his life and in food but also in his state of mind through the night, and thus in the possibility of choosing in a way that would give him hope for and indeed confidence in his future.

But then the agent might (I think *I* would) value knowledge, whether of good news or bad, more than what he could view from his prechoice epistemic perspective as possibly false hope. An agent might, while still trying to make up his mind what to do, be averse to creating for himself what he thinks could be only a night (and his last night at that) in a fool's paradise. If so, the compound option described would not recommend itself to him. He would want to open the second box and know what was in it, whether he had taken it alone or along with the first box. If the news he would get is good, he would want to enjoy both it and the meal. And, we are assuming, he would also rather eat in despair than be both assured of his execution and hungry. So he, this lover of knowledge and food, would have a causal dominance argument for taking both boxes and examining the contents of the second one without delay.

10. There are in fact several ways. A way different from the one discussed in the text would have our conditional sentences express not bona fide propositions (truth bearers) but rather "conditional beliefs," that is, "states of high conditional credence" (cf. Adams 1975; Falk 1985; Gibbard 1981).

Another approach would have our conditional sentences, in contexts of perfect-predictor problems, stand for certain sentences which, it has been maintained, make a plausible argument for taking only one box. For example: "'If I were to take both boxes when the predictor is not mistaken than I would receive but $1000. If I were to take only the one box when it was predicted that I would, I would receive a million dollars. Since I prefer a million dollars, I should take only the one box.' This argument has considerable appeal" (Hubin and Ross 1985, p. 444). The argument corresponds, we are told, to a dominance argument for taking just one box (p. 445). We are furthermore assured that "it makes no difference whether the puzzle stipulates that the predictor's infallibility is coincidental or a matter of physical or logical necessity" (p. 444). Although the agent in a perfect-predictor case will view these two conditionals as true, it is clear (contrary to Hubin and Ross) that he should not view them as expressed by our conditional sentences, even in the context of perfect-predictor problems; it is clear also that the agent cannot correctly view them as decisive for choice. At the risk of belaboring the obvious, I note that these conditionals correspond to two of the four cells in a consequence matrix for the problem. For a set that is decisive to choice, these two need to be joined by conditionals for the other two cells (all four cells are relevant as long as we are dealing with perfect-predictor cases that are bona fide decision problems). That set of conditionals generates the matrix of Section 2 and so, given certain causal independence, supports a dominance argument. However, it supports the familiar dominance argument for taking both boxes, not a dominance argument for one box.

11. Why "may"? – because I wish to leave open the possibility that the best theory of belief-revision conditionals is nonpropositional. (See Adams 1975; Gibbard 1981.)

12. I have thought for some time that, when making room for a contrary-to-fact antecedent, "what counts as a minimal accommodating change depends

(perhaps) upon the purpose for which the hypothesis is being entertained" (Sobel 1970, p. 431). One thing I had in mind was a nonprobabilistic version of Newcomb's Problem retold over lunch one day by Edmund Gettier, and how it seemed to trade on the possibility of taking conditionals both in ways appropriate to theoretical purposes and in other ways appropriate to practical purposes.

13. I assume that the signed difference $\Pr(q/p) - \Pr(q)$ measures the potential evidential bearing of p on q. The conditional probability $\Pr(q/p)$ can then be said to measure the possible evidential bearing - the greatest possible evidential bearing of p on q. It measures what the potential evidential bearing of p on q would be were $\Pr(q)$ vanishingly small but the evidential relevance of p to q unchanged; $\Pr(q/p)$ is, of course, the limit of $\Pr(q/p) - \Pr(q)$ as $\Pr(q)$ approaches zero and $\Pr(q/p)$ remains constant. Rather than say that $\Pr(q/p)$ measures the possible evidential bearing of p on q, one could say that it measures the evidential relevance of p to q, where this is understood in a way that makes it independent of the probability of q.

14. Strictly speaking, I should say that probabilities of causal conditionals measure probable *possible* real bearings, or probable *real relevance,* since I say that conditional probabilities measure possible evidential bearings or evidential relevance. The present contrast would be with the signed difference $\Pr(p\,\square_c\!\!\to q) - \Pr(p)$, which I should say measures the probable potential real bearing of p on q.

15. Horgan reports, with apparent approval and concurrence, that V (expected utility calculated in Jeffrey's way, using conditional probabilities as weights) measures "the *value of the act as news* - that is, the welcomeness to the agent ... of learning that he is about to perform [it]" (Horgan 1985a, p. 170). He thus, by easy implication, holds that this is what is measured by U_c (calculated expected utility, using as weights probabilities of conditionals - probabilities of what Horgan considers to be "appropriate" conditionals), for he holds that "U_c and V never diverge" (p. 172).

16. Horgan describes how backtracking would work in Newcomb's Problem, illustrating without stating (one might say) principles for this way of resolving the vagueness of a conditional (Horgan 1985a, p. 164). But the identity of probabilities of such backtracked conditionals with corresponding conditionals, when (p. 172) he comes to assert it, is maintained without any concerted effort to relate it to earlier, descriptive passages on backtracking. The main thing Horgan does at this point is to declare that the mode of resolution of vagueness of act-state conditionals appropriate in any decision problem is *conditionalized* resolution, which he then defines in part by the condition that probabilities of conditionals thus resolved should equal corresponding conditional probabilities (p. 172). The identity of the probabilities of appropriate backtracking conditionals with corresponding conditional probabilities follows of course directly from this stipulated general identity.

17. "No power in heaven or earth can render false a statement that is true. It has never been done, and never will be" (Taylor 1974, p. 66). Substituting "has rendered or ever will render" for "can render," I agree. I discuss logical fatalisms, including Richard Taylor's, in Sobel (in press).

18. I note that Jeffrey's (1965b) logic of decision does *not* endorse this favor. According to that theory, in a limit problem the agent is by hypothesis *sure* that his action has been correctly predicted, so Desirabilities of the two actions -

take both, or take B2 only – are equal in a fully inflated problem. According to that theory, two cells in each consequence matrix can be ignored, since corresponding cells in the associated conditional probability matrices contain zeros.

19. Horgan's considered opinion is that, "though two-boxers seem to predominate among those who are currently working on the foundations of decision theory" (Horgan 1985b, p. 233), the debate between one-boxers and two-boxers is a "hopeless stalemate" (pp. 227, 229, 232, and 234, n. 3). He says (p. 234, n. 3) that David Lewis expresses this view in "Causal Decision Theory" (1981a); presumably Horgan here adverts to, "I will not enter into debate with [one-boxers], since that debate is hopelessly deadlocked and I have nothing new to add to it" (Lewis 1981a, p. 5). However, Lewis here implies not that the debate is hopelessly deadlocked "and that's the end of it," as Horgan seems to think he does, but only that the debate was, when he spoke, hopelessly deadlocked. Lewis says, "I have nothing new to add," implying that there may be useful things to add that he has not thought of. It is perhaps relevant to note that when stalemate is achieved in chess, the game is over; but, when a door is secured with a deadbolt or a debate is deadlocked, it still can be (and indeed generally is) possible with a key, or by argument, to unlock it. Stalemates are final; deadlocks need not be. "Hopeless stalemate" (despite journalistic usage) is redundant, whereas "hopelessly deadlocked" is not.

Arthur E. Falk reports that, according to Lewis in "Why Ain'cha Rich?" (1981b), 'the debate with us one-boxers is at a standoff for lack of sufficient common ground to resolve issues" (Falk 1985, p. 465, n. 20). In fact, Lewis, in the paper Falk cites, expresses explicitly only the more limited view that, as regards the "why ain'cha rich?" taunt, "it's a standoff" (Lewis 1981b, p. 378).

I think that it is too early to say whether or not considerations will in time prove sufficient to incline intellects and secure even wider agreement than Horgan thinks now exists on points of practice and theory that are at issue in Newcomb's Problem. Perhaps Horgan in fact agrees, and intends to claim only that "there is no way [and will never be any way] to establish definitively" (Horgan 1985b, p. 229), and in a manner sufficient to coerce intellects, precisely what is rational in Newcomb's Problem, or just what the true merits of competing decision theories are. If so then I agree, for I think that these issues are no different than most issues in philosophy.

6

Kent Bach on good arguments

Abstract. An argument to show that some action is rational in a situation may be good (i) in that it establishes rationality, (ii) in that doing the action would probably lead to a maximum payoff, or (iii) in that a disposition to such arguments in such situations would probably lead to maximum payoffs. Causal decision theorists claim their arguments are always good in the first two ways, and concede that they may sometimes not be good in the third.

I take two passages in Bach (1987) as occasions for several observations about practical arguments and senses in which they may "work" and be "good."

First passage

One can only be amused by those advocates of BOTH who . . . realize that takers of BOTH almost always get but $1K whereas takers of ONE almost always get $1M, and proceed to bemoan the fact that rational people do so much worse than irrational ones. Despite their logical scruples, they seem to have a curiously low standard of what constitutes a good argument. Evidently they would rather be right than rich. One would think that a solution requires not merely a seemingly irrefutable argument but an argument that *works,* one whose use is likely to pay off to the tune of at least $1M. (p. 412)

In response, I say that what is true of advocates of BOTH, expressed in a nonprejudicial way, is not that they have a low standard for good arguments, and far less that they have a low opinion of their own arguments. Rather, they have a low opinion of the material profitableness in some circumstances of being the kind of person who has a settled and predictable disposition to employ arguments that, as advocates of BOTH, they think are good arguments in Newcomb problems. It is possible that this low opinion might move an advocate of BOTH – that is, one who

---

Reprinted with revisions by permission of the University of Calgary Press from *Canadian Journal of Philosophy* 20 (1989), pp. 447-53.
    I thank Paul Abela for questions that led to this chapter, and Richmond Campbell for comments on its penultimate version.

believes that BOTH is rational – not to advocate the kind of thinking that leads to BOTH, and not to say that arguments leading to it are good, lest he influence others to what he could fear would be their considerable loss. It is possible that this low opinion might even move him, out of fear of personal loss, quite literally somehow to change his own mind about these arguments, and to cease thinking of them as good. It is possible that this low opinion might move an advocate of BOTH to change, if he can, how own way of thinking and deliberating, and to make himself into a person who has a settled and predictable disposition to think in a way that would have him take ONE. He might, for material profit, be moved (for example) to make himself over into a person who would maximize "desirability" or "evidential expected value" in Richard Jeffrey's sense.

Whether or not an advocate of BOTH would be moved in these ways would depend on circumstances (and, in particular, on what kinds of decision problems he anticipates being involved in), but in any case, as an advocate of BOTH he must quite despair of solutions to Newcomb's Problem that both work and are irrefutable arguments. He must believe that in this situation any argument that works (i.e., any argument "whose use is likely to pay off to the tune of at least $1M") is a bad argument, just as anyone must believe that in some situations any opinion that works is a false one. Suppose that a mad millionaire will give you $1M if and only if you believe that he will not give you $1M.

Second passage

A good argument for BOTH cannot rely on luck. A good one, giving you excellent chances of getting $1M + $1K, must be such that despite your using it, PR [the predictor] will not have anticipated your using it . . . . Indeed, it must include a lemma to that effect. However, there seems to be no way to establish such a lemma. . . . So you must conclude that there is no good argument for BOTH. (Bach 1987, p. 414)

Several distinctions are relevant to this passage, and others in which Bach writes of "good arguments." We distinguish three senses in which practical arguments to the effect that certain actions would be rational can be good arguments and arguments that work.

(1) A practical argument may be a sound argument that really proves its conclusion, and works in that it does show that the action addressed in it is rational as claimed.

(2) A practical argument may be such that using it and doing what it says is rational would probably work, in that it would lead to material payoffs at least as great as would acting on the conclusion of any other argument in the situation.

(3) A practical argument addressed to a situation may be of a kind that works, in that a settled disposition to use practical arguments of this kind in certain kinds of situations (including this) and to do what they deem rational, would work and would probably lead to payoffs that on balance are at least as great as would follow from a settled disposition to act on practical arguments of any other kind in these situations.

A causal decision theorist claims that his arguments are good and work in the first two ways, but – at least with regard to material payoffs (we have set aside the value that may be attached to being rational) – possibly not good in the third way nor related as there indicated to a disposition that *would* work. Why not necessarily good in the third way? Because, for just one thing, an agent may think it very likely that he will get into Newcomb problems. Why good in the second way? Because if one maximizes expected utility – that is, if one maximizes expected value of probable causal consequences – in a case, then that is precisely what one does![1]

But surely I have propounded a paradox, and something is wrong. For how can it be (i) that a policy and disposition to causal expected value maximizing in certain situations might not have on balance (probably) best payoffs, and yet *also* be (ii) that in every situation causal expected value maximizing has (probably) best payoffs? Isn't this simply *impossible?*

No. The key to the possibility here challenged is that an agent with a causal maximizing disposition that extends to absolutely all kinds of situations might, on account of his disposition and given the character of decision problems he anticipates, expect to be faced with poorer choices than would an agent who was disposed to think differently – for example, to maximize evidential expected value at least in certain kinds of situations. Compare this with the relatively familiar point that a "rational egoist" or "straight maximizer of personal satisfactions" might in a conventional society, since he could not be trusted, be faced with poorer options than a rule-bound person. And consider that, given this possibility, although by hypothesis a rational egoist would in any choice situation do as well as anyone could in that situation, and would do better in some situations than a rule-bound person would do, it could still be that a rule-bound person would do better on balance through a period punctuated with choice situations than would a quite unconstrained rational egoist: the rule-bound person could have better options amongst which to choose in his choice situations.

The point I am making about causal utility maximizing, and that I have just compared with a possible feature of rational egoism, is in form the same as the main point of the passage to follow regarding "act-utilitarians" – "straight maximizers" who on each occasion of choice are out to do what will have the best consequences and produce the most satisfaction, not just for themselves but for everyone. In the passage to be quoted, D. H. Hodgson – who was, I think, the first to make the point of present interest concerning a kind of maximizing – sees act-utilitarians as prone to problems like those that I have said are possible for rational egoists. Hodgson has argued that act-utilitarians might well not be trusted by those who knew them nor counted on by them to keep promises, for act-utilitarians keep promises not as a matter of principle but only if, when the time for keeping them comes, it is in fact best in terms of consequences to keep them. And this, Hodgson maintains, means that act-utilitarians can find themselves handicapped in some societies and not able to do as much good as nonutilitarians can do, even though act-utilitarians are at all times out to do as much good as they can, and even though a successful act-utilitarian always does as much good as he can. At the very least, act-utilitarians could find themselves cut off from intimate cooperation and coordination conducted with complete motivational candor and openness:

Even though an act-utilitarian might consistently be successful in choosing the act which would have best consequences, this only means that the consequences are the best possible *in the circumstances of his being an act-utilitarian;* that is, in the circumstances of his being . . . unable to be both sincere and effective in making a promise . . . .

"It is therefore not absurd to assert that the adoption and (even) correct application of act-utilitarianism by individuals in a non–act-utilitarian society might not have best consequences. There are alternatives which are open to those who accept personal rules approximating to the conventional moral rules of their society, but which are not open to those whose only personal rule is the act-utilitarian principle. Even though the former persons might not always choose, from the acts open to them, those with the best consequences, nevertheless the consequences of the acts which they do choose might be better than the best consequences which the latter persons could bring about through the alternatives open to them. (Hodgson 1967, pp. 58–9)

Hodgson is concerned in this passage with handicaps possible for act-utilitarians in predominantly non–act-utilitarian societies. He is prepared, I assume, to make a complementary claim about possible advantages that could be enjoyed in predominantly utilitarian societies by certain non–act-utilitarian, "qualified" promoters of the good of all – by certain "constrained" act-utilitarians who pursue the good of all in their actions subject to constraints of morality (e.g., to constraints of genuine trustworthiness and honesty).

The point I am stressing is analogous to "the essential point" of David Gauthier's straight maximizing argument for a policy of constrained maximizing: "It might seem that a maximizing disposition to choose would express itself in maximizing choices. But we have shown that this is not so. The essential point in our argument is that one's disposition to choose affects the situations in which one may expect to find oneself" (Gauthier 1985a, p. 89; 1986, p. 183). Relating this essential point to evidential and causal forms of unconstrained maximizing, we see that it is possible – in some conceivable circumstances it is true – that an agent committed to evidential expected value maximizing (always, or only in certain cases including all Newcomb problems) might, depending on circumstances, expect to have better choices to make and better problems to solve than the choices and problems an agent committed (always, or only in those certain cases) to causal expected value maximizing would expect to confront. For perhaps there are Newcomb problems in the offing, problems that he anticipates. The former agent – the (sometimes) evidential maximizer – would figure, in any Newcomb problem he got into, to have a choice between $M and $M + $1000, whereas the latter would figure to have a choice only between $0 and $1000 in any Newcomb problem that *he* faced. And so, even though the former agent would not make the best choices – he would leave $1000 on the table in his Newcomb problems – he would figure to do better in these problems than the latter (causal maximizing) agent would figure to do in *his*. In fact, the first agent figures to do [($M less $1000) × (the number of Newcomb problems he anticipates)] better in these problems.

The first agent – the evidential maximizer – could say to a causal maximizer: "Sure, you would do better in some choice situations I figure to get into than I would do in them, but you will almost certainly not get into them and so can expect to be poorer than me." To this a causal expected value maximizer might reply: "Sadly, I do expect to be poorer than you. Budweiser."[2]

A causal maximizer could say he was wiser, and say this sincerely. After all, as a causal maximizer he would believe in the rationality of causal maximizing and in the irrationality of evidential maximizing. And what the evidential maximizer says (as the causal maximizer might concede) could be true. A causal maximizer could – in some circumstances, given certain anticipations concerning pending decision problems – figure to be on balance poorer, on account of his way of thinking and choosing, than if he were an evidential maximizer.

What is to be said here about competing claims to wisdom and the goodness$_1$ of causal and evidential maximizing arguments? (See note 1 for an explanation of these numerical subscripts.) I say that the causal maximizer may be wiser and his arguments better$_1$, even if he would probably be poorer and his arguments worse$_3$. His arguments, which are certainly always good$_2$, may also be (as he thinks they are) always good$_1$. They may be good$_1$ even if and when they are not good$_3$. Whether or not, in certain circumstances, these arguments are good$_3$ is, I think, quite irrelevant to whether or not they are always good$_1$. Of course, not everyone agrees; one gathers from what he has written that David Gauthier, for one, does not agree. Additional "evidence" (as David Hume might say) – arguments of sorts – for the irrelevance always of the possible goodness$_3$ of causal maximizing arguments (and of whether or not unrestrained dispositions would work$_3$) to the goodness$_1$ of these arguments can be gathered from Sobel (1988e [see Chapter 7 in this volume]), wherein I oppose Gauthier's "pragmatic turn."[3]

NOTES

1. Though they are somewhat beside the object of this paper, I interject here comments on Bach's purported "$1,000,000 solution," which is said to be specifically for Newcomb problems where the predictor "makes his predictions by plugging [a] *psychological profile* [for the agent] . . . into a powerful psychological theory" (Bach 1987, p. 413).

    Bach's "meta-argument for ONE" (p. 412) is for the conclusion that in any such problem there is not for BOTH, though there is for ONE, a "solution" in the sense of a "not merely irrefutable argument but an argument that *works,* one whose use is likely to pay off to the tune of at least $1M" (p. 412). By a "solution" we can gather that Bach means an argument that is "good" in a sense additional to the three I have distinguished, one that is furthermore not a simple compound of these three. Being "seemingly irrefutable" is not the same as being "good$_1$". And an argument "whose use is likely to pay off to the tune of at least $1M" *may* be an argument whose use on the occasion would yield *evidence for* such a payoff, which is different from being either good$_2$ or good$_3$. (Subscripts denote the three senses recently enumerated.)

    In any case, and regardless of the exact details of Bach's solution concept, there is a problem with his meta-argument. For though he presents reasons for the conclusion "that there is no good argument for BOTH" (p. 414), he does not present reasons for thinking that there *is* (in his sense) a good argument for ONE. Even if all arguments for ONE work in his sense – as he implies, "on the assumption that PR anticipated you would use it, you can use *any* argument for ONE" (p. 424) – Bach gives no reason for the conclusion that there is even one argument (first-order, meta-, meta-meta-, or what have you) for ONE that is "seemingly irrefutable." He has in fact not addressed this issue, or done anything to counter even the extreme opposite view (held by most two-boxers, even if rarely expressed) that every argument for ONE, once clarified and reduced to its fundamentals, will be found to contain some *"obvious* error" or other (Swain 1988, p. 391, emphasis added). Bach has not taken seriously the possibility that there may be no solution in his sense to Newcomb's Problem. Subject to uncertainty concerning his sense of "solution," that there is no solution to Newcomb's Problem is what all two-boxers believe.

124

2. The pun on "Budweiser" is lifted from *Letters of E. B. White,* collected and edited by Dorothy Lobrano Guth (New York: Harper & Row, 1976), p. 537.
3. If there is implicit in Gauthier's writings a meta-argument for ONE in certain Newcomb problems, it is a simpler argument than Bach's (see note 1). If there is such an argument implicit in Gauthier's writings then it is one that claims, quite without regard for conditions like that of "seeming irrefutability," that there is a one-box rational solution but not a two-box one – an argument that makes this claim solely on the ground that there is a one-box (first-order) argument that works in a certain sense in these problems, whereas there is no two-box argument that does. Rationality, at the level of particular choices, is (or can at least seem to be) in Gauthier's writings "whatever works."

7

Maximizing and prospering

Abstract. I take up and oppose David Gauthier's equation of practical rationality with that way of thinking a firm disposition to which would probably pay on balance. Implications are developed for Newcomb problems, particularly nonstandard ones that feature prior options to see what is in the second box, and Gauthier's pragmatism concerning practical rationality is opposed.

0. INTRODUCTION

A central feature of David Gauthier's philosophy of practice is his argument for a "reinterpretation of the maximizing conception of rationality" (Gauthier 1986, p. 15). "Our argument," he writes, "identifies practical rationality with utility maximization at the level of dispositions to choose and carries through the implications of that identification in assessing the rationality of particular choices" (Gauthier 1985b, p. 93; 1986, p. 187). In fact, only some relatively pleasant implications concerning prisoners' dilemmas are carried through. It is reassuring to be told that, at least in certain not unlikely circumstances, individual rationality (notwithstanding initial appearances to the contrary) entails cooperation and makes possible mutually beneficial interactions without hidden, heavy, Hobbesian hands. But, as I try to bring out, Gauthier's pragmatism regarding practical rationality has other implications that are not so pleasant. Included here are some for Newcomb problems, both standard ones as well as nonstandard ones that feature prior options to look and see what is in the second box.

According to Gauthier's pragmatism, practical rationality for an agent, at least at the level of particular choices, is simply that way of thinking a firm disposition to which would in this agent's view probably pay on balance. In opposition to this interesting and important perspective on the subject, I offer the idea that practical rationality is in fact not so variable

Reprinted with revisions by permission from *Dialogue* 27 (1988), pp. 233-4 and 248-60. This is Part Two of Sobel (1988e).

but has at all levels a fixed form (namely, straight maximization) that accommodates variable contents (credences and preferences). Practical rationality at the level of choice is, I think, not perfectly plastic – it is not for an agent simply whatever ratiocinative disposition would in his view probably pay. Rather, it consists of a certain definite way of making up one's mind, given one's beliefs and preferences, which may or may not probably pay on balance in what one takes to be the conditions of one's life. I maintain that it is not a contradiction in terms for a person to say, "I would be better off if I were not always so damned reasonable."[1]

1. GAUTHIER'S PRAGMATISM

For text I take the following:

A person who chooses in such a way that he maximizes the satisfaction of his desires does not always choose to maximize their satisfaction. This is the deep lesson of the Prisoner's Dilemma.
    Let us spell it out. Consider a person who chooses to maximize satisfaction. Whenever he is in a Prisoner's Dilemma . . . he cannot expect the outcome to be optimal, unless he is interacting with a sucker. . . . Now suckers are unlikely to do well in the struggle for survival. . . .
    Consider another person . . . who chooses on the basis of a principle satisfying the optimality condition when she expects others to do so as well . . . but is no sucker. . . . Now if she is usually able to form correct expectations about others and to enable them to form correct expectations about herself, she may expect optimal outcomes in most PD [Prisoner's Dilemma] situations in which others are like-minded. . . .
    The first person chooses to maximize the satisfaction of his desires. The second person chooses in such a way that she maximizes the satisfaction of her desires. . . . [T]he first suffers except in a world of suckers; the second benefits except in a world where her fellows are not like-minded. We take the second person to be fully rational. . . .
    We must now withdraw our previous proposal that the principle of maximizing happiness . . . is a practical law. Kant is right; such a law would be destructive, in leading to suboptimal, mutually disadvantageous outcomes in PD situations. . . . The correct law must ensure that the person who conforms to it may expect to maximize her happiness. (Gauthier 1985b, pp. 85–6)

Gauthier proposes to reinterpret

the utility-maximizing conception of practical rationality. The received interpretation . . . identifies rationality with utility maximization at the level of particular choices. . . . We identify rationality with utility maximization at the level of dispositions to choose. A disposition is rational if, and only if, an actor holding it can expect his choices to yield no less utility than the choices he would make were he to hold any alternative disposition. We shall consider whether particular choices are rational if, and only if, they express a rational disposition to choose. (Gauthier 1985a, p. 89; 1986, pp. 182–3)[2]

This reinterpretation of utility-maximizing rationality makes no difference, Gauthier thinks, for situations of, "*parametric* choice, in which the actor takes his behaviour to be the sole variable in a fixed environment [and] regards himself as the sole centre of action" (Gauthier 1986, p. 21). "In parametric contexts, the disposition to make straightforwardly maximizing choices is utility-maximizing" (Gauthier 1985a, p. 90; 1986, p. 183). The reinterpretation can, however, make a difference for situations of "*strategic* choice, in which the actor takes his behaviour to be but one variable among others, so that his choice must be responsive to his expectation of others' choices, while their choices are similarly responsive to their expectations" (Gauthier 1986, p. 21).

We are told that agents, given their beliefs and preferences, would on reflection think that they could get on better with one another if they were disposed, when interacting with one another, not to maximize. From this it follows, by Gauthier's reinterpretation of utility-maximizing rationality, that maximizing is not rational when one is interacting with others. It is said that, in contrast, agents would on reflection think that it is best for them to be (by their dispositions to choose) straight utility maximizers when acting alone. Of course, agents may be mistaken. Their expected utilities for various policies of deliberation and dispositions to choose may derive from ill-founded credences. But if, as credences and preferences stand, straight maximizing is subjectively best for private projects but not for situations in which one is interacting with others, then by Gauthier's pragmatism it is a rational disposition for situations when one acts alone but not for those in which one interacts with others.

I think that there is purchase for Gauthier's pragmatism not only sometimes (though certainly not always; Sobel 1993b) with respect to interaction situations, but also sometimes with respect to situations in which one acts alone. Primarily to embarrass (if I can) this pragmatism, and only incidentally to relate it even to agents acting alone, I propose to change from the usual subject in its connection of prisoners' dilemmas to Newcomb problems.[3]

2. Implications for Newcomb problems

Newcomb's Problem can be viewed as a game in which Row is interested only in the money and Column is interested only in making true predictions.

	Predict 1	Predict 2
Take 1	1000, 1	0, 0
Take 2	1001, 0	1, 1

(T1, P2) is the only equilibrium. Only (T1, P1) and (T2, P1) are optima. I now adapt Gauthier's remarks concerning prisoners' dilemmas to this game.

The deep significance of Newcomb's Problem, we might say (mocking Gauthier), is that there is a sense in which maximizers can expect not to maximize in such problems, and in which "a person who chooses in such a way that he maximizes the satisfaction of his desires" will not, in Newcomb problems, "choose to maximize their satisfaction" (Gauthier 1985b, p. 85).

Consider a maximizer whose decisions are informed by what he takes to be the probable consequences of his possible acts. He cannot, when in Newcomb problem [NP] situations, expect an optimal outcome unless he is interacting with a bad predictor. For he will take 2, and so optimality would require a mistake – namely, predict 1. Bad predictors "are unlikely to do well in the struggle for survival, so that our maximizer may not expect to find many around. He must then expect suboptimality" (Gauthier 1985b, p. 85). This first agent of ours must expect not to do well in such problems.

Consider another agent who, at least when she does not think she is dealing with a bad predictor, chooses in Newcomb problems as an evidential desirability maximizer would – she can be counted on to take 1. When she does not think she is acting in a situation set by a bad predictor, she chooses as she wishes the predictor predicted that she would. She can expect to do quite well in NP situations. "She may . . . expect to do better, in terms of overall satisfaction, than those who choose . . . to maximize" causal utility (Gauthier 1985b, p. 85).

The first person chooses in such a way as to maximize *in his choices*. The second person chooses in such a way that she maximizes *in being such a chooser*. It is, in her case, as if someone or some mechanism had compared decisionmaking strategies and kinds of practical personalities, and had chosen the one that it is best for her to have. Lucky her, if someone else (or some selective mechanism) has chosen the best for her.

"We take the second person to be fully rational," Gauthier might say (Gauthier 1985b, p. 85). Why? Because "the correct law [of practical reason] must ensure that the person who conforms to it may expect to maximize her happiness" (p. 86). As against the first person's way of thinking we may ask: If he is so smart, why ain't he rich? For people can have it both ways; they can be rational *and* rich. Indeed, being rational cannot call for sacrifice and so entails at least expecting to be rich. To repeat, "The correct law must ensure that the person who conforms to it may expect to maximize her happiness" (p. 86).

3. POSSIBLY DISTURBING ASPECTS OF THESE IMPLICATIONS

According to Gauthier, it follows from the nature of practical rationality – specifically, that it is utility maximizing at the level of dispositions, though not necessarily at the level of choices – that, other things being equal, the prospects for rational beings always at least match (and sometimes exceed) those for beings who are not rational. One consequence of this theory, however, is that a rational person would actually leave the $1000 on the table in a standard Newcomb problem, even though she knew that it was there for the taking and that it would cost her nothing in the long run or the short run to take it. If this consequence of Gauthier's reinterpretation of maximizing rationality is not disturbing enough (it is for me), then perhaps implications for the conduct of fully rational persons in some nonstandard Newcomb problems will be found to be.

Suppose that a fully (or sufficiently for the case we now consider) evolved rational being is in a Newcomb problem with this difference: She can, before opting for take 1 box or take 2, inspect the second box and see what it holds in store for her. (See Skyrms 1982, p. 699.) She is going to take this box and get what is in it; the only issue is whether she is also going to take the first box, and get as well the $1000 that she already knows for sure is in *it*. Suppose further that she is not only fully evolved in her principles, but sufficiently sophisticated to appreciate how they would work were she to choose boxes in a standard Newcomb problem without prior proof of the contents of the second, and how they would work when it came to the boxes in a nonstandard NP supposing she had opted to inspect and thus *had* such proof. Will she avail herself of the opportunity to find out what is in the second box before opting for one box or two? Will she opt to inspect?

By hypothesis she opts for take 1 in standard Newcomb problems – she maximizes evidential Desirability. And, in her sophistication, she knows that this is what she will do when it comes to the boxes in our nonstandard problem *if* she decides not to inspect, for that decision gives rise to a standard problem. Whether or not she wants prior proof of what is in the second box – whether or not she opts to inspect in our nonstandard NP – depends therefore on what she realizes she would do when it came to the boxes, given such prior proof and if she were to have inspected. She will not inspect if, having inspected, she would take two boxes. For a fully evolved being could be counted on – and by a good predictor would be predicted – to take one box even in nonstandard Newcomb problems. A fully evolved rational being chooses in all cases in such a way that she maximizes by being such a chooser, that is, by being a person who is disposed to choose in that way.

130

What then will she do? What will be the character of her practical principles? There are two possibilities here, between which we (given our critical purposes) need not decide. According to the first principle, knowing what was in the second box she would take 2. Note that this is what she would do if she were to maximize evidential expected Desirability not only in standard Newcomb problems in which she lacks prior proof of the contents of the second box, but also in nonstandard NPs in which, when it came time to choose boxes, she did have such proof. That assurance would make M (box 2 contains $M) probabilistically independent of acts, and take 2 dominates under the partition $\{M, \sim M\}$. (See Skyrms 1982, p. 699; Falk 1985, p. 453.) This first possibility for evolved prosperity-promoting principles would have her wanting and choosing *not* to know what is in the second box. According to it evolved rational animals would be like ostriches (consider Falk 1985, p. 454) – they might, if they had to, even cover their heads with sand in order not to see or hear what was in the second box. For they would want very much not to know.

According to the second possibility, even with proof of the second box's contents an evolved rational being would still take 1, just as she would in a standard Newcomb problem (in which she may be nearly sure what is in the boxes). According to this second possibility, a fully evolved rational agent who was interested only in the money could, in a nonstandard Newcomb problem, be indifferent between inspecting or not. According to the present hypothesis, these actions are alike in monetary consequences. (However, if she is interested not only in the money but also in knowing what is in the box, then she might be inclined to inspect.) According to the present possibility concerning her practical principles, even if she inspected and found that the second box was empty she would still take only it. For according to this second possibility she would, when it came to the boxes, do what she would then wish had been predicted, even if she knew for sure that it had *not* been predicted.

Evolved rational beings would, according to both possibilities, in a sense be choosing alike in standard and nonstandard Newcomb problems when it came down to the boxes. And clearly this is as it should be. For while agents in standard NPs lack prior proof of what is in the second box, by the time of choice they can (and, if sophisticated, do) have highly probable opinions on this subject. The difference between near-certain proofs and prior proofs should not make any large difference either in manner of ratiocination or in choice. These two possibilities for the character of evolved rational principles would have our evolved rational being choosing alike with regard to the boxes, but they would have her choosing alike in different senses. The first possibility would have her choosing *in accordance with the same maximizing principle* in both problems – she would maximize evidential expected Desirability in both problems when it came to the boxes. The second possibility would have her choosing *the same action* in both problems – she would choose take 1 in both problems whether or not she had opted to inspect (and so whether or not take 1 would still maximize expected Desirability). Each of these possibilities for practical principles would have her prosper in both standard and nonstandard Newcomb problems. But each entails a sense in which, when in a nonstandard NP she came to the choice of boxes, if she had opted to inspect then she would *not* choose as she would choose in a standard NP. Either (this is the first possibility again) she would take 1 in standard problems and take 2 in the nonstandard ones in which she had (mistakenly, in her view) chosen to inspect, or (this is the second possibility), though she would maximize evidential Desirability in standard Newcomb problems, she would *not* do so in nonstandard ones in which (perhaps to satisfy her curiosity) she had chosen to inspect. This is another implication of Gauthier's pragmatism – another anomalous implication. For one would have thought that rational beings (whatever their principles) would choose alike when it came to the boxes, in *both* of the senses of "choose alike." As stated, the difference – regarding the boxes' contents – between proofs (to which inspections would give rise) and very high probabilities (which can be present even in standard problems) should not make appreciable differences to the deliberations and choices of rational beings.

4. Judgmental conclusions

Which way would fully evolved rational beings take? Would they decline prior options to inspect, or would they, even having inspected, still resolutely take only one box and do so even if inspection has revealed it to be empty? Perhaps some fully evolved beings would be of one disposition. Perhaps all would be of some still other dispositions that I have not thought of. We do not know, but then I think we should not care. We should at any rate put away the ideas that currently motivate these questions, and see the prosperity of rational beings not as a condition on correct articulations of the character of practical reason, but as an *issue* or a set of questions concerning how things would probably work out for rational beings in various realistic, as well as contrived, cases.

We may wonder whether, and if so in what circumstances, rational beings could find themselves disadvantaged by their rationality. That seems an intelligible line of speculation, and an interesting one. It is possible that rational beings should find themselves disadvantaged by their rationality. It seems, for example, that expectations of prosperity in Newcomb problems might be furthered by dispositions to irrational choices. This seems the case in connection with standard Newcomb problems. What could be less rational, for one who likes money, than to leave on a table a thousand dollars that he knows is there for the taking, absolutely without penalty? And yet a known disposition to do just that could be thought to make one a good bet for millions. Turning again to nonstandard Newcomb problems, humanoid bipeds – disposed to cover their eyes and stop their ears in order not to know what was in the second box – as well as beings disposed to take only the second box even knowing it was empty, might under some believed conditions prosper. However, it seems plain that they would do so precisely by their *reputations for irrationality*. The case is similar, I think, for prisoners' dilemmas. Cooperators in dilemmas, though prosperous and perhaps nice, seem confused. What is perhaps clearer is that they *can* be confused and irrational even though prosperous, and be prosperous just because they are in certain ways confused and irrational. There is no contradiction here.

Gauthier has heard such claims. He has heard it said that "it may be rational to dispose oneself" to be irrational (Gauthier 1985a, p. 91; 1986, p. 184), just as it can be rational to bring oneself to believe things that one has every reason to think are false. Reflecting on this suggestion, which he thinks rings of paradox, Gauthier declares: "We are unmoved" (Gauthier 1985a, p. 92; 1986, p. 185).

Biting the bullet, he would be prepared to count as rational even carrying out a threat to end the world, if circumstances were such as to make rational a disposition to carry out such threats. Carrying out such a threat would be rational, we are told, if it were rational to dispose oneself to carry it out:

> Our argument shows that deterrent policies in general, and nuclear retaliation in particular, may be maximizing.... [And] I claim that if it is rational to form this conditional, deterrent intention, then, should deterrence fail and the condition is realized, it is rational to act on it. (Gauthier 1984, pp. 485-6)

> The would-be deterrer who fails to deter and who must then make good on her threat in order to carry out her conditional intention, is not maximizing at all. Her reason for sticking to her guns is not to teach others by example, not to improve her prospects for successful deterrence in the future, or anything of the sort. Her reason is simply that the expected utility or payoff of her failed policy depended on her willingness to stick to her guns. (Gauthier 1984, pp. 488-9)

"That rational lady is not for turning."

For my part, I can imagine beliefs and preferences that would make rational an agent's disposing herself to carry out even the most awful of threats. I can imagine beliefs and preferences that would make rational her disposing herself to count as a reason for carrying out an awful threat the consideration that it was reasonable (when she made it) to be disposed to carry it out should it fail in its deterring purpose. Furthermore, if such dispositions are actually established when the threat is made, if nothing happens in the interim to undo them, and if the threat fails, then of course it follows that she will be disposed to carry out the awful threat, that she will be disposed to count as a reason for doing so the fact that it was, when the threat was made, reasonable to be disposed to do so in the event the threat failed, and that if she can (if, for example, no one stops her) she will count that as a reason and will carry out that awful threat. But it does not follow that, given her present beliefs and preferences, she *should* carry out this threat; and it may or may not be a reason for her that it was reasonable, when she made the threat, to be disposed to carry it out should it fail. Whether or not she has in fact a reason for carrying out the threat, and (if so) how strong a reason it is, are issues that depend on whether or not doing what in this case was once reasonable to be disposed to do would serve any *present* purpose or preference, and (if so) how strong this preference is. This in turn depends on whether, for example, it is important to her to be consistent and a woman of her word, and, if these characteristics are important to her, *how* important they are to her. (In the event that the awful threat failed, she should want to know whether being steadfast is more important to her than life – not just her life, but all life.)

There is something paradoxical on the face of it about the idea that it is sometimes rational to be disposed to behave irrationally. But there certainly is nothing really paradoxical about this, or about the suggestion that we can have reasons for making ourselves irrational, and for being irrational. Consider that we can have practical reasons for believing things we realize at the time of belief we have good grounds for thinking false. A little imagination should convince us that a person can have reasons for making himself mad and, indeed, good practical reasons for taking on – not feigning, but really assuming – every manner of defect and irrationality. Imaginable (even if *only* imaginable) cases are ready at hand. Cases that do not involve recognitions by agents of their own already existing and uneliminable weakness and imperfections are ready at hand.[4] The problems that generate reasons for an agent's making himself irrational can be ones that have to do not so much with him (or with any other agent) as with his (or their) circumstances.

What is *really* implausible is Gauthier's idea that practical rationality for an agent consists in whatever way of thinking and choosing the adoption and employment of which would, given his beliefs and preferences, maximize expected utility. What is really implausible is the idea that there is no such thing as – no fixed-for-everyone nature to or form for – rational practical thinking. What is really implausible is the idea that "rational thinking," properly so called, is for each agent simply what*ever* would in his considered view work best for him. What is really paradoxical is the idea (implied it seems by Gauthier's pragmatism regarding practical reason) that it should be impossible ever to have good reasons for making oneself irrational in a practical sense. No theoretical reasons or objective grounds for this suggestion have, I think, ever been produced, though it is not difficult to appreciate possible motives for making and embracing it. It is a somehow agreeable fancy that, necessarily, rational agents could get on well – could negotiate Prisoners' Dilemmas, and in general coordinate and cooperate for mutual benefit – *without* the aid of systems of artificial sanctions or dependence on possibly absent civilizing sentiments, and without in any way compromising and in part alienating their rationality. Gauthier's pragmatism (his Jamesian pragmatism) would vindicate this pleasant thought. That it would do so, while also providing an understandable motive for his pragmatism and an explanation of it as a wish-fathered thought, is of course not an argument that even begins to display such pragmatism as a correct (as distinct from a pleasantly reassuring) theory of practical reason.

Expected gains from forms of irrationality can be worth the pain. I don't deny this. It is, in particular, possible that persons who are not already optimizers or evidential Desirability maximizers would be well advised to make themselves into such, so far as they can. It could even be that we should join forces in the promotion of an agreed form of constrained maximization as part of a social ideology or religion. These possibilities relate to proper, important, and very difficult questions concerning prospects of various conceptions of rationality, questions that ought to be considered. What I wish to stress here is that such questions are distinct from questions concerning the nature and correct conception of rationality. Indeed, they presuppose and depend for their answers on answers to prior nature questions if, as I think, the worth of practical rationality depends in no small part on its intrinsic importance to persons. (I sketch a case for the intrinsic importance of theoretical rationality in Sobel 1987.)

Rational thinking is not necessarily, and may not be in fact, that form of thinking which, if permanently and unreservedly embraced by us as we are and in the world as it is, can be expected to pay off. It may not excel in instrumental value over all competitors. However, even if being rational is not for us the best instrumental policy, it can still be – and I think (as Plato did) that it is – of great intrinsic importance to us in our inner beings, and it can still be best for us, all considered. The main reason for being rational in both theoretical and practical ways is not, I think, to be found in the instrumental utilities of these orientations, great though they may be, but rather in certain intrinsic and profoundly *personal* considerations that on proper reflection would, for almost anyone in almost every set of circumstances, prove powerful enough to counterbalance even considerable penalties in prisoners' dilemmas and great temptations in Newcomb problems. But then, I address myself to philosophers and should thus perhaps desist from further praise of reason, there being no point in preaching to the converted.

APPENDIX: SOCIALIZING SENTIMENTS

Gauthier underestimates and sometimes misses, I think, the problem-solving potentials of sentiments that are suited to cooperation. For example, after allowing that a person who "prefers being a woman of her word . . . may value sincerity directly" (Gauthier 1984, p. 475), he soon thinks he sees that, in order to make sense of her choosing "to make commitments to [otherwise] dispreferred courses of action," he must suppose either that "she finds masochistic satisfaction in making and carrying out such commitments" or that her concern "is with the instrumental and not the intrinsic benefits of adhering to an expressed intention" (p. 477). In fact, however, if she really does value sincerity directly – if she really does value directly being a person who does what she has said she will do – then, having said she will under certain conditions bomb away, she *will* under those conditions value doing so; she will directly value bombing away in that by bombing away she would be keeping her word. If the question is asked why she would ever give such words, an answer on a particular occasion can be, "For the deterrent effects of giving them." If the additional question is asked why she values being trustworthy and a person of her word, a good answer might well begin with the perspective-setting question, "Don't you know – don't we all know?" The point is that straight maximizing rationality is not necessarily exclusively forward-looking. It is not exclusively forward-looking for persons possessed of backward-looking values. It is not exclusively forward-looking for persons who put a premium on being persons of their words and who value directly, and not just for its usefulness, this characteristic.

Gauthier seems to underestimate the potentials of civilizing and cooperation-enabling sentiments, and not to realize the extent to which they detract from the force of his pragmatic arguments for nonmaximizing thinking at the level of particular choice. Part of this point can be seen by considering that a person who thinks that he would be better off were he not a straight maximizer but a certain kind of optimizer might also think that he would be better off were he possessed of sentiments of solidarity and of direct interests in being a person of his word.[5] He might think that each of these changes would be sufficient alone for certain goods, and that changing how he *feels* about things would in fact be far easier, and vastly less disruptive of his psyche, than would be changing how he *thinks* about them – changing *how* he thinks about them, not what he thinks about them. A person might be brought to realize, by a story of a sort different from those usually told in connection with the "compliance" problem, not only that the problem-solving sentiments would be useful if widespread, but also that they recommend themselves to him on their intrinsic merits and without regard to whether or not they are widespread. Appreciating what solving states of heart and mind in themselves are could, in time, work changes in his heart and mind, and do so without artifice or manipulation. These points are broached in the last paragraph of Sobel (1976 [Chapter 15 in this volume]). They go with the idea that (as Gauthier realizes) it is not necessary to leave "altogether open the content of human desires" (Gauthier 1986, p. 19), and that desires and preferences are amenable to modes of rational, thoughtful review and revision.

Gauthier contrasts two kinds of social artifice (1986, p. 19) for dealing with situations in which maximizers might not do well. One kind is massive (we might say, without any claim to historical accuracy, Platonic) manipulation, including perhaps genetic manipulation, designed to secure distributions of capacities and preferences of kinds that would dovetail in natural harmony. The other kind of artifice Gauthier considers would involve changes in the ways people think about practical issues and make their choices. There are (as Gauthier realizes) artifices other than these two. One would consist in the establishment, through processes of socialization (i.e., through training, indoctrination, and education), of coordination- and cooperation-enabling sentiments. Another of course is the Hobbesian political solution of coercive institutions.

Of the modes of artifice Gauthier sees fit explicitly to consider, the second – persons changing their ways of thinking and deliberating – is, he holds, probably the best we can do. I think that of the four possibilities now before us, Gauthier's favorite vies with the Platonic one for last place, both in terms of attractiveness and in terms of feasibility. The best we can do is a mix of the other two, which, as it happens (and subject to some Platonic adulteration), seems to be what we are and have always been doing.

NOTES

1. Part I of the paper (Sobel 1988e) from which this chapter is drawn discusses technical matters concerning games and universal conformities to rules. Gauthier distinguishes "maximizing given the actions others do" from "maximizing given the happiness others get," and maintains that universal conformities to the first rule invariably coincide with equilibria whereas universal conformities

to the second invariably coincide with optima. Some of his views regarding these matters are tested under a variety of interpretations and against a range of situations, including situations in which actions of agents are *inter*dependent. Limitations and needed qualifications are discovered.

2. There is a problem concerning exactly how Gauthier's reinterpretation of utility maximization is to be understood. Shall a particular choice be rational if and only if it expresses a maximizing disposition to choose that the agent actually *has*? If so, and if an agent's disposition to choose is not maximizing, then his particular choices are not rational even when they accord with dispositions that would be for him utility maximizing at the level of dispositions to choose. Or are actions to be assessed by reference to maximizing dispositions to choose whether or not he has such a disposition? My arguments in this chapter do not depend on how such details of Gauthier's pragmatism are settled.

3. It is a hard (but probably not important) question whether Newcomb problems are to be counted as situations of "parametric choice" in which agents see themselves as acting alone. Gauthier's characterization of parametric choice seems to apply, as agents in Newcomb problems take their behavior to be sole variables in fixed environments. In standard problems, the agent supposes that the predictor's work is done and that all relevant aspects of the environment – most notably, the contents of the boxes – are settled and fixed. But the point need not be pressed, for even if standard Newcomb problems are not to be counted as situations of parametric choice, certainly naturalized Newcomb problems (in which Nature is a rewarding predictor; compare with naturalized versions of Pascal's wager) must be situations of parametric choice in which agents regard themselves as sole centers of action, if there be any such situations. I note here, in connection with problems of interpretation, that there is difficulty even with the classification of prisoners' dilemmas. They are, I am sure, to be counted as situations of strategic choice. And yet maximizing agents need not when making choices in them be responsive to their expectations regarding the other's choice; dominance coupled with independence yields decisions without calculation for maximizers. Prisoners' dilemmas are not situations of strategic choice if it is required that a maximizing agent's "choice must be responsive to his expectation of others' choices, while their choices are similarly responsive to their expectations" (Gauthier 1986, p. 21). These words are also not appropriate to Newcomb problems.

4. Gauthier concedes that it can be reasonable for imperfect agents to dispose themselves to behave unreasonably, and observes that "no lesson can be drawn from this about the dispositions and choices of the perfect actor" (Gauthier 1985a, p. 91; 1986, p. 184). But then objectors such as Derek Parfit, who claim that it can be reasonable even for perfect actors to dispose themselves to kinds of irrationality, do not feel the need to make this inference. (See Parfit 1984, pp. 19-23.)

5. It would be natural to include a person's preferences in that "disposition to choose" which Gauthier would put to a maximizing test for rationality. A person's disposition to choose is made up it seems of his credences, preferences, and mode of practical ratiocination. Gauthier, for reasons that are unclear to me, confines dispositions to choose to the third of these factors. He leaves "altogether open the content of human desires" (Gauthier 1986, p. 19), and does not consider useful manipulations of these or of opinions.

III

Causal decision theory

8

*Notes on decision theory:
Old wine in new bottles*

Abstract. According to Bayesians, rational actions maximize weighted average values of outcomes. Richard Jeffrey identifies weights involved with conditional probabilities. This evidential decision theory is conceptually conservative (causality does not figure in its primitive basis), and it is simple (it works for all partitions of circumstances). However, I argue that it gets wrong answers in Newcomblike problems and is thus false.

A less conservative and more complex theory is offered that gets right answers in Newcomblike problems. It identifies weights with melds of objective conditional causal chances and subjective unconditional probabilities. It is maintained that though this theory does not work for all partitions, it works for all natural ones for which one wants it to work. Remarks toward foundations are made in conclusion.

0. Introduction: Bayesian decision theory

To judge what one must do to obtain a good or avoid an evil, it is necessary to consider not only the good and the evil in itself, but also the probability that it happens or does not happen; and to view geometrically the proportion that all these things have together. (*The Port-Royal Logic*)[1]

Actions reflect beliefs and preferences. Rational actions reflect intensities of beliefs and preferences in certain orderly ways. According to Bayesian decision theories, rational actions *maximize expected utilities,*[2] where the expected utility of an action is a kind of average of the values of its possible total upshots. More precisely, taking possible total upshots of an action a to be total ways in which it might happen – or total "worlds" that might conceivably be were it to happen:

the *expected utility* of an action a

$= \Sigma_{\text{world } w}$ [(the probability of w given a)·(the value of w)].

Applications of Bayesian decision theories usually proceed not in terms of supposed values of an action's many possible worlds but rather in terms

Reprinted with revisions by permission from *Australasian Journal of Philosophy* 64 (1986), pp. 407–37.

of its expected utilities in various circumstances or kinds of worlds, drawn from what is taken to be an adequate partition of possible circumstances or kinds of worlds in which the action might take place. A *partition* of possible circumstances is, by definition, adequate in this sense relative to an action if and only if the expected utility of this action is a kind of average of its expected utility in the various circumstances in this partition. More precisely again, a partition C of possible circumstances is *adequate* relative to an action a if and only if

the expected utility of $a = \sum_{c \in C}$ [(the probability of c given a)
$\cdot$ (the expected utility of a given c)].

The definition of expected utility makes it a weighted average of the values of worlds, whereas in the definition of an adequate partition of circumstances we have a weighted average of expected utilities of actions in circumstances. In standard matrix presentations of decision problems, members of "appropriate" partitions of options are assigned rows (for example, choose both boxes or choose only box 2 in Newcomb's Problem), and members of adequate partitions of possible circumstances are assigned columns (for example, there is a $M in box 2 or there is not).

A developed Bayesian theory explains these things. It explains the character of the probabilities in the formulas just displayed, and says whether they are conditional probabilities, probabilities of kinds of conditionals, or some other sorts of probabilities. A developed theory also explains, in other and useful general terms, what sorts of partitions of circumstances are adequate to the analyses of actions' expected utilities. A developed theory also explains the notion of an action's expected utility given certain circumstances, and says whether this notion is of the expected utility of the action in conjunction with these circumstances, of the expected utility of the action somehow conditional on these circumstances, or (again) some other notion. The first of these explanations could be said to identify the expected utility of a given c with the expected utility of an action, specifically a "version" of the action a, and thus to bring it under the general definition for the expected utilities of actions. Explanations of the second sort would presuppose an account, in terms of values of worlds, of appropriately conditional expected utilities; explanations of other sorts might presuppose other accounts. And a developed Bayesian theory would explain what partitions of options were "appropriate," and why. This last problem, however, is less connected to the others than they are to each other, and will be for the most part ignored in the present chapter. I have dealt with it at length, and proposed a solution to it, in Sobel (1983 [Chapter 10 in this volume]). For a somewhat different treatment of this last problem, see Weirich (1983).

1. Jeffrey's logic of decision

Jeffrey's "logic of decision" (1965b) is a developed Bayesian theory. It is a simple and economical Bayesian decision theory, a philosophically and formally attractive theory possessed of considerable prima facie plausibility. His theory makes probabilities in definitions of the previous section *conditional* probabilities: In his theory, for any propositions a and c and world (world proposition) w,

the probability of w given $a = \Pr(w/a)$

and

the probability of c given $a = \Pr(c/a)$,

where $\Pr(w/a)$ and $\Pr(c/a)$ are, in accordance with the usual definition of conditional probability, the ratios of unconditional probabilities $\Pr(a \& w)/\Pr(a)$ and $\Pr(a \& c)/\Pr(a)$, respectively.

It can seem from the definitions of the previous section that primitive bases of developed Bayesian theories will need to include more or less explicitly *causal* notions. "The probability of w given a" and "the probability of c given a" admit of, and in the context can seem to demand, causal interpretations – one expects them to stand for the probabilities that w and c would be upshots of a. According to Jeffrey, however, these probabilities can be explicated in entirely acausal terms; in fact, he takes "it to be the principal virtue of [his] theory, that it makes no use ... of any ... causal notion" (Jeffrey 1965b, p. 147). If rational action *can* be satisfactorily explained without primitive recourse to causal categories, then for a theory to have done this would be a strong recommendation.

The acausality of its primitive basis makes Jeffrey's theory philosophically attractive. Prominent among its specifically formal virtues is the maximal simplicity of its account of adequate partitions of circumstances. In this theory, every partition of circumstances is adequate to every action! Jeffrey's theory can be founded on the following definition in which "Des" is short for Desirability, Jeffrey's term for expected utility: For any proposition p such that $\Pr(p) > 0$,

$$\text{Des}(p) = \sum_w [\Pr(w/p) \cdot (\text{the value of } w)].$$

(See Jeffrey 1965b, pp. 67, 198.) And it is a consequence of this definition that, for any proposition p such that $\Pr(p) > 0$ and any logical partition Q of propositions,

$$\text{Des}(p) = \sum_{q: q \in Q \,\&\, \Pr(p \& q) > 0} [\Pr(q/p) \text{Des}(p \& q)].$$

This theorem states that every logical partition is adequate to the analysis of the Desirability of every positively probable proposition. I note that not only is every logical partition adequate to the analysis of a proposition's Desirability but also every probability partition is adequate, where a *probability partition* is a set of propositions such that there is a probability of 1 that exactly one member of this set is true. A deduction of a partition theorem for probability partitions can be found in the appendix to this chapter. Implicit in this theorem is, incidentally, Jeffrey's very simple explication of "the expected utility of a given c," which in this theory is the Desirability of a conjoined with c.

The principal thesis of this logic of decision, which is that *rational actions maximize Desirabilities,* is plausible. Desirabilities, based as they are on conditional probabilities, measure what can suggestively be termed "news values" of propositions, or the relative welcomeness of propositions as items of news. In Jeffrey's words: "To say that A is ranked higher than B [in the Desirability ranking] means that the agent would welcome the news that A is true more than he would the news that B is true: A would be better news" (Jeffrey 1965b, p. 72).

In fact, while "news value" and "how welcome would be the news that" are often good indications of what a proposition's Desirability measures, they are not always so. Evaluatively significant shifts in a person's expectations or prospects consequent to reception of a piece of news are in some cases due not so much to the evidential potential of the news as to how this person realizes he would act on this news, or react to it. Suppose, for example, that I do not own IBM stock, have no intention of buying any, and do not think that I will learn that it will double even if in fact it will double. It seems that for me – and for anyone who shared my credences for d (IBM stock will double), r (I will get rich on IBM stock), and $(d \& r)$ – $\Pr(r/d)$ could well equal $\Pr(r)$, so that merely conditionalizing on d would make little or no difference to my prospects. Even so, if I were to become certain that d then I would presumably be, on reflection, confident that r. The news that p would for me (even if not for all persons who shared with me the indicated credences) be very good news, very welcome news. For another relevant case, consider b (tulips are blooming in the park); suppose John is going to the park, "loves surprises," and "would be surprised to see some blooming in the park" since he is far from sure that tulips are blooming in the park. News that they are in bloom in the park would, he realizes, spoil the surprise (Weirich 1980, p. 705).

The general point is that since one would not always revise one's credences consequent to the reception of a piece of news by conditionalizing on that news, what would be one's prospects consequent to receptions of pieces of possible news are not perfectly aligned with Desirabilities of these pieces of possible news. Desirabilities do not measure precisely news values, though the two are often close enough to make news value a useful rough indication of what they do measure, and probably as good an indication as any that can be framed briefly in ordinary terms. (Closer to the general truth would be an equating of "the Desirability of p for me" with something like "the news values of p for another ego who, like a would-be guardian angel, shared my credences and preferences for p-worlds, but was not himself an actor or reactor.")

Propositions, including action propositions, do have Desirabilities, and it seems that rational actions must have (when compared with alternatives) *maximum* Desirabilities. It can seem decisive for this connection that, again in Jeffrey's words: "If the agent is deliberating about performing act A or act B, and if AB is impossible, there is no effective difference between asking whether he prefers A to B as a news item or as an act, for he makes the news" (Jeffrey 1965b, pp. 73-4, corrected in accordance with Jeffrey 1983, p. 84). In choosing to do A rather than B, the agent would make for himself the news that A, though he could instead make for himself the news that B. It can seem indisputable that, if he is rational then he will choose A over B - and thereby make for himself the news that A rather than the news that B - only if he would *prefer* news that A to news that B, and so only if the Desirability of A is greater than the Desirability of B.

Jeffrey's theory is attractive and plausible, but it is not true. Though in many cases rational actions maximize Desirability, there are cases (albeit somewhat out-of-the-way cases) in which they do not. The best known and most discussed such case is Newcomb's Problem. Because of various sorts of resistance to Newcomb's Problem, and for other reasons including not least novelty, I offer a different sort of case in which it seems plain that rationality and Desirability maximization would diverge. The case I offer is still out of the way (as I suspect all such cases must be), but is, I think, not as fantastic as Newcomb's Problem.

2. THE POPCORN PROBLEM

2.1

The story:

> I am in a cinema watching a movie. I want very much to have some popcorn, though I would not want to miss part of the movie for nothing, as I would do if I went to the lobby and found that there was no popcorn there to be bought. However, I would *really* hate to learn that I had passed up

an opportunity to get some popcorn, as I will have done if though there is popcorn to be had I do not go for it.

We pause here, though there is more to the story. The case, as so far set out, is summarized in the following consequence matrix, wherein p abbreviates "there is popcorn in the lobby to be bought" and g abbreviates "I will in just a moment go to the lobby to buy some popcorn."

Consequences

	p	$\sim p$
g	Popcorn!	Part of movie missed for nothing
$\sim g$	Regret and recrimination	None of any of that

According to Jeffrey's logic of decision, what I should do - that is, which of the actions g and $\sim g$ would be *rational* - depends upon quantities in the following Desirability and probability matrices.

Desirabilities

	p	$\sim p$
g	Des($g \& p$)	Des($g \& \sim p$)
$\sim g$	Des($\sim g \& p$)	Des($\sim g \& \sim p$)

Probabilities

	p	$\sim p$
g	Pr(p/g)	Pr($\sim p/g$)
$\sim g$	Pr($p/\sim g$)	Pr($\sim p/\sim g$)

Desirabilities of actions g and $\sim g$ are sums of rows in the matrix product of these matrices:

Des(g) = [Pr(p/g) Des($p \& g$)] + [Pr($\sim p/g$) Des($p \& \sim g$)];

Des($\sim g$) = [Pr($p/\sim g$) Des($p \& \sim g$)] + [Pr($\sim p/\sim g$) Des($\sim p \& \sim g$)].

Numbers in the desirability matrix could be (we assume that they are) as follows:

Desirabilities

	p	~p
g	4	2
~g	1	3

As will become clear, little more than the order of these Desirabilities matters to my argument.

Numbers in the probability matrix depend on the rest of the story, which we now proceed to tell.

> I am nearly sure that the popcorn vendor has sold out and closed down, and that there is no popcorn in the lobby to be bought. And so, quite sensibly it seems, I have decided not to go for popcorn, and I am nearly sure that I *will* not go for popcorn. Still, I am also nearly sure that in this particular theatre, when and only when there is popcorn in the lobby to be bought, the signal
>
> POPCORN!
>
> is flashed on the screen repeatedly, though at a speed that permits only subliminal, unconscious awareness. Furthermore – and this is where the plot begins to thicken – I consider myself to be a highly suggestible person. Indeed, given what I take to be the practices of the theatre and given that I realize that I have for good reasons decided not to go for popcorn, I am nearly sure that I will change my mind and go for popcorn if and only if I am influenced by this subliminal signal to do so, and so if and only if there actually is popcorn in the lobby to be had! But what do I imagine I would be thinking if I were to find myself going for popcorn? It is not important to the argument this case serves that we say, but to fill out the story we can say that I am not sure but consider it somewhat probable that I would be thinking: "The management made me change my mind and do this"; or "I chose to maximize Desirability, and wasn't *I* clever?"; or "What could I have been thinking when I changed my mind?"; or several of these things in conjunction or confused succession.
>
> As has been said, I am nearly certain that there is no popcorn in the lobby and that I am not going to the lobby in an effort to buy some, and I am nearly certain of the

conjunction of these things. Furthermore – and this is at least nearly implicit in what has been said – although I think it is very unlikely that I will go for popcorn and thus very unlikely that I will go for popcorn in the circumstance in which there is popcorn there ($g \& p$), I think it is much more unlikely that I will go for popcorn though there is none there to be bought ($g \& \sim p$). I can hardly imagine how *that* conjunction could come true, for I am nearly sure that its truth would involve my changing my mind and going for popcorn unprodded by that sometimes flashing signal, and despite the absence of those sometimes present subliminal suggestions. And *that,* given my conviction that there is no popcorn there to be bought, is an eventuality which I consider as near to impossible as makes no difference.

Using "$\simeq$" in the sense of "is nearly equal to," we have in this case that

$$\Pr(\sim p) \simeq \Pr(\sim g) \simeq 1.$$

And, using "$\gg$" in the sense of "is much greater than" – where (by definition), for positive j and k, $j \gg k$ if and only if $j/(j+k) \simeq 1$, we have:

$$\Pr(g \& p) \gg \Pr(g \& \sim p).$$

(Though I think it very unlikely that I will go for popcorn in the circumstance in which there is popcorn there, I think it is much more unlikely that I will go for popcorn though there is none there to be bought.) Observe that from this extreme inequality, together with the convention that, for positive j and k, $j \gg k$ if and only if $j/(j+k) \simeq 1$, it follows that

$$\Pr(p/g) \simeq 1.$$

($\Pr(p/g) = \Pr(g \& p)/\Pr(g)$ and $\Pr(g) = \Pr(g \& p) + \Pr(g \& \sim p)$.) What else holds in the case? Well, we have another extreme inequality:

$$\Pr(\sim g \& \sim p) \gg \Pr(\sim g \& p).$$

($\Pr(\sim g \& \sim p) \simeq 1$ and, since $\Pr(\sim p) \simeq 1$, $\Pr(p) \simeq 0$ and so $\Pr(\sim g \& p) \simeq 0$.) Thus we have

$$\Pr(\sim p/\sim g) \simeq 1.$$

In short, according to the story, my going for popcorn would provide me with a near-certain sign of (near-certain evidence for) there being popcorn in the lobby to be bought: $\Pr(p/g) \simeq 1$. And my *not* going would provide me with a near-certain sign that there is *not* popcorn there to be bought: $\Pr(\sim p/\sim g) \simeq 1$. Letting $[x]$ be a number that is nearly equal to x, numbers in the probability matrix for this case are thus as follows.

Probabilities

Multiplying matrices cell by cell and summing the rows in their product yields Desirabilities for g and $\sim g$:

$$\text{Des}(g) = [1]4 + [0]2 \simeq 4;$$

$$\text{Des}(\sim g) = [0]1 + [1]3 \simeq 3.$$

Hence going for popcorn excels in Desirability and so, according to Jeffrey's logic of decision, would be rational. As promised, the conclusion that $\text{Des}(g)$ exceeds $\text{Des}(\sim g)$ is supported in the main by the order of these magnitudes in the Desirability matrix. In the case at hand, we have the near equalities

$$\Pr(p/g) \simeq 1 \simeq \Pr(\sim p/\sim g) \quad \text{and} \quad \Pr(\sim p/g) \simeq 0 \simeq \Pr(p/\sim g).$$

If, on the other hand, we had the equalities

$$\Pr(p/g) = \Pr(\sim p/\sim g) \quad \text{and} \quad \Pr(\sim p/g) = \Pr(p/\sim g),$$

then the conclusion that $\text{Des}(g)$,

$$\Pr(p/g)\text{Des}(p\,\&\,g) + \Pr(\sim p/g)\text{Des}(p\,\&\sim g),$$

is greater than $\text{Des}(\sim g)$,

$$\Pr(\sim p/\sim g)\text{Des}(\sim p\,\&\sim g) + \Pr(p/\sim g)\text{Des}(p\,\&\sim g),$$

would be quite independent of the specific magnitudes in the Desirability matrix, and would be secured entirely by their order. These equalities could of course be added to the case: We could stipulate that my going and not going for popcorn would provide me *equally good* signs of popcorn and no popcorn, respectively. That would make the conclusion that $\text{Des}(g)$ is greater than $\text{Des}(\sim g)$ independent of magnitudes not only in the Desirability matrix but in the probability matrix as well.

An objection to the popcorn problem. But neither Jeffrey's logic of decision nor any other theory of rational choice is properly applicable in this case. For the agent in this case is nearly sure that, if he goes for popcorn, the signal will have made him do it. And so he cannot (if he is consistent) be sure that not going for popcorn is even *open* to him – he cannot be sure he has a choice in the case between going and not going.

A brief response to this objection. I accept the suggestion that the theories I am discussing are applicable only in cases in which the agent is sure of certain actions that each is open, but deny that the agent in the popcorn case cannot (if he is consistent) be sure that both going and not going are open to him. He can, I maintain, consistently think that if he were to change his mind and go for popcorn (as he thinks he just might – he is only nearly sure that he won't) then – though he would in his choice be under the influence of the signal, which would be making him do it – he would even so be able to resist its proddings, and not going would still be open to him and a choice that he could make.

Jeffrey, in commenting on a similar problem, has written that it challenges his logic of decision "only if it is the *choice* of smoking that the agent takes the bad gene to promote directly.... If he takes the performance itself to be directly promoted by presence of the bad gene, there is no question of preferential choice: the performance is [in the agent's view] compulsive" (Jeffrey 1983, p. 25). I note that, in my "brief response," I have the agent thinking that the signal (when present) directly influences his choices, not his actions. (I revisit this objection in Sobel 1990 [Chapter 2, Sections 3 and 4].)

2.2

Going for popcorn would maximize Desirability, but surely my going for popcorn would be irrational if anything ever is. For I am nearly sure that there *is* no popcorn, and I of course do not, in this case, believe that my going to the lobby to buy popcorn would tend to bring some forth. It has all along been a barely implicit feature of the case that – although I am nearly sure that the management can influence me and my decision – I do not think that I can, by simply going for popcorn, influence the management. I am nearly sure that when it's gone it's gone and there is nothing I can do about it. So I am nearly sure that I have something to lose and nothing to gain by going for popcorn.

Willa Fowler Freeman-Sobel has suggested a very similar case. In it a woman who – despite an extreme aversion to telephones – has, since taking a lover, answered her telephone when and only when her lover has called, even though far more often than not when her telephone rings it is not her lover who is calling. She knows all this and so, despite the fact that there is no *pattern* that she can discern to her lover's telephone calls, this woman has come to think that she knows when her lover is calling. She cannot say how she knows, and is indeed devoid of every conscious shadow or hint – of every "tickle" – of this knowledge until she finds herself actually picking up her telephone's receiver. But she is convinced that she always does somehow know when he is calling, and that she is moved to action by this secret knowledge just as she would be if, before the act, she had in its place full conscious knowledge that it was her lover who was calling.

Her telephone is ringing. ("Don't answer that, Howard, I'll get it!") As on all such occasions, (i) she is almost sure that it is not her lover who is calling, and (ii) news that she is even so going to answer her telephone would provide her with a near-certain sign that it *is* her lover who is calling. Though she thinks it would be best for her not to answer the telephone, she is in a position to change her mind on this evaluative matter, at least momentarily, by answering it. It would (she thinks) be best not to answer, but news that she was even so answering would be best news.

2.3

The trouble with Jeffrey's theory is, I think, obvious. It would in some cases have a person act with an eye to things that he thinks do not obtain and thinks he cannot affect; this theory would sometimes have one put considerable weight on such things. Jeffrey's theory would have one act on the *evidential* potentials of one's actions, even when these potentials do not coincide with what one takes to be the *causal* potentials of one's actions. The principal initial philosophic attraction of Jeffrey's theory, its acausal character, is I think precisely what is wrong with it.

Since in doing an action an agent makes for himself the news that he is doing that action, it can seem that a rational agent will make for himself the *best* news. What is wrong with this argument is that, although by acting in a certain way a person does make for himself the news that he is acting in that way, he does not necessarily make, or tend at all to bring about, everything for which this news is *evidence*. As Gibbard and Harper have written:

It should now be clear why it may sometimes be rational for an agent to choose an act B instead of an act A, even though he would welcome the news of A more than that of B. The news of an act may furnish evidence of a state of the world which the act itself is known not to produce. In that case, though the agent indeed makes the news, he does not make all the news his act bespeaks. (Gibbard and Harper 1978, p. 140)

"And so," we can add, "in that case the greater Desirability of action A may be due in part to things that are quite irrelevant to its expected utility, properly so termed. And, notwithstanding action A's greater Desirability, it may be irrational to choose it over action B."

3. A causal decision theory

3.1. Definition of expected utility

That I was going for popcorn would be for me a near-certain sign that there was popcorn to be had, but that it would *be* such a sign is not in itself a reason for my going. What would be of interest is that my going would tend to cause there to be popcorn to be had. But its being for me a *sign* of popcorn, and its being in my opinion a probable *cause* of popcorn, are different things, and the latter is not entailed by the former.

Probabilities in the definition of expected utility cannot be conditional probabilities, for conditional probabilities measure sign potentials. We must it seems interpret the weights involved in expected utilities in a way that makes them measures of probable potentials of things as causes. Toward this end, let the conditional form "$p \, \square\!\!\rightarrow q$" - if it were the case that p, then it would be the case that q - be true if and only if *either* q is the case and would remain the case even if it were the case that p, *or*, though q is not the case, were it the case that p then (as a consequence of that) it would also be the case that q. The form "$p \, \square\!\!\rightarrow q$" is to express a kind of causal conditional, but not a *purely* causal conditional. For example: if, in a standard Prisoners' Dilemma, I am sure that the other agent is going to confess, then I should be sure that [(I do not confess) $\square\!\!\rightarrow$ (he will confess)]. I should be sure of this conditional even if I think that his confession, though quite settled and in a sense determined (perhaps because he has made up his mind, is not going to change it, can confess if he chooses, etc.), is not *causally* determined. The form "$p \, \square\!\!\rightarrow q$" is to express what might be termed a *practical* conditional. (I confess that in these last sentences I pretend to greater comfort with, and a surer grasp of, the idea of causality than I in fact own. For example, it is not really clear to me that an agent's action, though already "determined" if he has made up his mind, etc., will not be a consequence - seemingly an ordinary causal notion - of his determination to do the thing if he will not *have* to do it and it is not a logical consequence of facts - about what will be at its time the past, taken together with principles of nature - that would still be true whatever, of all the things he will be able to do, he was to do.)

Jeffrey's logic of decision is an acausal decision theory. A first and simplest *causal* decision theory would make the weighting probability for the value of w in the expected utility of a not the conditional probability $\Pr(w/a)$, but rather the probability $\Pr(a \, \square\!\!\rightarrow w)$ of the corresponding practical conditional. We start with the following definition.

Definition 1. For an action a,

$$U(a) = \sum_w [\Pr(a \,\square\!\!\rightarrow w) \cdot (\text{the value of } w)].$$

Definition 1 is like one proposed by Robert Stalnaker in 1972 (see Stalnaker 1981, p. 129), and like the definition discussed favorably by Gibbard and Harper (1978, p. 128). It is, as they suggest theirs is, a useful first approximation to a correct causal account, though it is in my view limited in two ways.

Definition 1 is, I think, limited first in that it is suited only to agents who do not believe in *objective chances,* and who think that for every action there is a unique world that would definitely be realized were this action to take place, with no chance at all for the realization of any other world. (See Gibbard and Harper 1978, p. 161, n. 3.) Definition 1 is also limited in that it is suited only to actions the agent is sure are open to him. I will propose definitions that are not limited in these ways, dealing first separately with the two ways in which I think that Definition 1 is limited, and then together.

Suppose that an agent who thinks that the world is somewhat indeterministic is, for some action a and each world w, sure that it is *not* the case that were a to take place, it would (definitely) take place in w. For him, $\Pr(a \,\square\!\!\rightarrow w) = 0$ for every world w, and, by Definition 1, his expected utility for a is zero. But consistent with what has been supposed, this agent could think that were a to take place then it might take place in some world w that he considered to be a very good world – he could think that a might, with a considerable chance, take place at this very good world. The choice-relevant value – the expected utility – of a should be sensitive to such chance views. To provide for this sensitivity (while ignoring the other way in which I think that Definition 1 is limited), I would explicate weights in the definition of expected utility not in terms of probabilities of causal conditionals, but rather in terms of what I call *probable chances.* Let "$a \,\diamond_x\!\!\rightarrow c$" be a *practical chance conditional;* let it say: if it were the case that a, then it might – with a chance of x – be the case that c, where the vagueness of this conditional is resolved in the manner appropriate to contexts of decision. Using chance conditionals, I would, to eliminate just the present limitation of Definition 1, move to Definition 2.

Definition 2. For any action a,

$$U(a) = \sum_w [\text{PrCh}(w/a) \cdot (\text{the value of } w)],$$

wherein, by a preliminary definition, $\text{PrCh}(w/a)$ – in words, "the probable chance of w given a" – is a probability-weighted average of the possible conditional chance for w given a,

$$\sum_y \Pr(a \diamondsuit_y \to w) y,$$

where y takes on values from 0 to 1 and $\sum_y \Pr(a \diamondsuit_y \to w) = 1$. (To illustrate, if $\Pr(a \diamondsuit_{.1} \to w) = .5$ and $\Pr(a \diamondsuit_1 \to w) = .5$, then $\mathrm{PrCh}(w/a) = .55$.)

This definition features chance conditionals, not conditionals with chance consequents. Lewis makes use of conditionals of this latter sort. Differences between chance conditionals and conditionals with chance consequents, and between Lewis's theory and mine, are not great. Rabinowicz (1982) has a useful discussion of relations between Lewis's theory and mine. One difference between our theories is noted in Sobel (1985a, pp. 197–8, n. 2).

Lewis explains "dependency hypothesis" in terms of conditionals with chance consequents, and defines expected utility thus: For option A,

$$U(A) \stackrel{\mathrm{def}}{=} \sum_K C(K) V(AK),$$

where the Ks are dependency hypotheses and $V(AK) = \mathrm{Des}(A \& K)$. Regarding the different forms taken by our definitions, I recall that in the logic of decision the Desirability of an action is a weighted average of its Desirabilities in conjunction with members of a partition, weights being the probabilities of these members on this action. Two reactions are possible to analogs of this formula addressed to expected utility, properly so termed. One says, "Those are not the right weights." The other says, "But not just any partition." Gibbard and Harper (and I) can be interpreted as in the beginning, in our definitions of expected utility, being more insistent on the first of these reactions. Regarding the second, it is as if we go to very "fine" partitions as a matter of course, with little if any comment or even notice. I employ the ultimate, the finest, partition, – the set of possible worlds. They, in effect, employ a partition that is by hypothesis fine enough for all practical purposes, one that when applied to the action at issue makes all value-relevant distinctions. We then concentrate on getting the weights right. Lewis is in the beginning, in his definition of expected utility, more insistent on the second reaction to Jeffrey's thesis. (So is Jeffrey, when he is in a causal decision theory mood; see Jeffrey 1981b, sec. VII.)

Starting from either form of theory it is possible with some work to derive the other form. Even so, there are reasons for preferring the definitional form I employ. First, it is closer to the form of Jeffrey's logic of decision (see esp. Jeffrey 1965b, sec. 12.8), and yields quick access to parts of and adjuncts to that very useful theory – for example, to its system of decision matrices. Second, my form of definition of the expected utility of an action does not presuppose the definition of another kind of proposition value such as Desirability (though my form of definition does presuppose world values). And third, the form of definition I employ makes the theory of expected utility, from the start, a natural extension of standard modal logic, and this seems to facilitate rigorous early demonstrations of principles.

We come now to the second way in which Definition 1 is, I think, limited. This definition can, it seems, be right only for actions that are certainly *causally possible,* for if an action a is not causally possible, $\sim\Diamond a$, then *either* the proposition that were a to be then as a consequence c would be fails for want of a presupposition, and the conditional $a \:\square\!\!\rightarrow c$ is neither true nor false, *or,* though this conditional is true, it is so only by artificial stipulation and for completeness and simplicity of theory. That Definition 1 is unproblematic only for actions that are certainly causally possible is, I suppose, reasonably plain. My present contention is that, beyond this limitation to certainly causally possible actions, there is another limitation that *comprehends* it. Definition 1 is, I think, right only for actions that are not merely certainly causally possible but are furthermore certainly *open* to the agent – it is right only for options in a decision problem. For example, suppose that I am nearly sure that the financial future of some business firm is very bright, whether or not it attracts a large number of investors; nearly sure that many persons are of this view and will in fact invest in this firm; nearly sure that it is therefore not open to me to be one of a very *few* investors in this firm; and nearly sure that it is open to me to be one of a very few investors in it *only if* this firm's future is in fact not bright but dismal. I will then be nearly sure that if I were to be one of a very few investors in this firm – if that causally possible event were to come to pass – then I would benefit greatly: I am nearly sure that its future is bright, and I don't think that there being a very few investors in it including me would change that. And so, according to Definition 1, the expected utility of my being one of a very few investors in this firm is *great* (at least so far, and unless I think this action would have bad consequences not yet mentioned). But in this case I am nearly sure that being one of a very few investors in this firm is not open to me, and nearly sure that if being one of a very few investors in this firm *is* open to me – if it is something I can do, something I can bring about – then this firm's prospects are dismal; were I to invest in it – along with many others or with only a very few others – I would not benefit greatly but would instead suffer a loss. And so, contrary to Definition 1, it seems that the expected usefulness of this action for me ought (so far, at least) *not* to be great. It seems that the expected utility of this action – if this not certainly open action is even to have an expected utility, properly so termed – ought not to be great. The Desirability of this action would (so far, at least) not be great, and I think that, in the case of this action, Desirability and expected utility (if the latter is defined) should coincide. The expected utility for an agent of an action he is not certain he can do ought, it seems, to depend not on his probability that if it were to *happen* (perhaps largely thanks to choices of others) then it would have a certain effect, but rather (if this is different) on his probability that were *he* to *do* it, it would have that effect. If defined, the expected utility of an action an agent is not sure he can do should depend on his probabilities – *conditional* on its being open to him and something he can do – that, were he to do it, it would have a certain effect. To eliminate just this second limitation of Definition 1, I would favor a move to Definition 3.

Definition 3. For any a such that $\Pr(\Diamond a) > 0$,

$$U(a) = \sum_w \Pr[(a \,\square\!\!\rightarrow w)/\Diamond a] \cdot (\text{the value of } w),$$

or equivalently,

$$U(a) = \sum_w (\Pr[\Diamond a \,\&\, (a \,\square\!\!\rightarrow w)]/\Pr(\Diamond a)) \cdot (\text{the value of } w),$$

where "$\Diamond a$" abbreviates "a is open." (For an action that is open for sure, Definition 3 reduces to Definition 1.)

According to Definition 3, the expected utility of a probably open action a is its expected utility conditional on its being open – that is, its expected utility from the epistemic perspective that would result from conditionalizing on its being open for sure, in which perspective it *is* of course open for sure.

Attending further to the new idea in this definition, an action is to be *open* for an agent if and only if either it will take place or this agent can definitely do it or bring it about (that is, if and only if it will take place or, given certain conditions under the agent's full control, it would take place). Being open is a stronger condition than being causally possible: In a Prisoners' Dilemma, for example, if the other prisoner is going to confess then, though my confessing alone could happen and is causally possible, it is not something that I can do or that I (unaided) can bring about – it is *not* something that is *open* to me (it is not something that is open to *me* even if it is part of something open to *us*).

Valid principles for openness are to include, for every proposition a,

$$a \rightarrow \Diamond a,$$

and, for any propositions a and b such that a entails b,

$$\Diamond a \rightarrow \Diamond b.$$

I assume that if a is an *option* in a decision problem (an idea studied in Sobel 1983 [Chapter 10]) – for example, confessing in a Prisoners' Dilemma –

then $\Pr(\diamond a) = 1$ and, at every positively probable world, it would happen or take place if and only if he did it or it was his choice. Options in a decision problem are to be things the agent is sure are completely under his control. So I assume for any option a in a decision problem, and any proposition b, that

$$(a \mathbin{\square\!\!\rightarrow} b) \rightarrow \diamond b$$

is valid. Further, since an option (or version of one) that is open happens if and only if the agent does it, we have according to Definition 3 (see its second form) the idea that the expected utility of a probably open (version of an) option is a weighted average of the values of its possible upshots, where the weight for an upshot's value is proportional to the probability that the (version of an) option is open and would, were the agent to do it, have that upshot. I think this definition, though complicated, is in relation to options and versions of options very natural. And it is noteworthy that no expected utilities other than those of options (and versions of options) figure in applications of decision theory in decision problems; in an application, expected utilities of versions of options are set out in an expected utility matrix, and expected utilities of options themselves are computed from these using probabilities in the probability matrix.

Definition 3 provides a basis for partition theorems. Indeed, this is its reason for being. To illustrate again in terms of a Prisoners' Dilemma, the expected utility of my confessing should be a weighted average of the expected utilities of my confessing in conjunction with the other prisoner's confessing, and of my confessing alone; but, while typically each of these conjunctive actions is probably open, in no Prisoners' Dilemma will both be certainly open. Definition 3, by addressing itself to all probably open actions, and by dealing with them in the way that it does, provides a basis for relatively simple partition theorems.

Rather than replace Definition 1 by Definition 3, one could confine it to certainly open actions, add to it the following definition of conditional expected utility, and let partition theorems take a slightly more complicated form than that of straightforward analogs of Jeffrey's principle.

For any propositions p and q, if $\Pr(p) > 0$ and $\Pr_p(\diamond q) > 0$ then

$$U_p(q) = \sum_{w \in (q)} (\Pr_p[\diamond q \& (q \mathbin{\square\!\!\rightarrow} w)]/\Pr_p(\diamond q)) \cdot V(w),$$

where $\Pr_p$ comes from $\Pr$ by conditionalization on p.

(This definition was intended, but is garbled, in Sobel 1985a, p. 200, n. 10.) Let $U_p(q) = U(q/p)$. Then $U[(a\&c)/\diamond(a\&c)]$ by the preceding definition equals $U(a\&c)$ under Definition 3; for any certainly open action a, $U(a/\diamond a)$ by the preceding definition equals $U(a)$ by Definition 1; and partition theorems can take the form of conditions on partitions such that, for a certainly open action a,

$$U(a) = \Sigma_{c \in C} \Pr(a \Box \!\!\!\rightarrow c) U[(a \& c)/\Diamond(a \& c)].$$

Under this arrangement, expected utility proper – unconditional expected utility – can be confined to certainly open actions.

Having responded separately to the limitations of Definition 1, it remains only to put these responses together. This is done in Definition 4.

Definition 4. For any action a such that $\Pr(\ominus a) > 0$,

$$U(a) = \Sigma_w \{\text{PrCh}(w/a)\} \cdot (\text{the value of } w),$$

where, by a preliminary definition,

$$\{\text{PrCh}(w/a)\} = \Sigma_y \Pr[(a \Diamond_y \!\!\rightarrow w)/\ominus a] y.$$

Note that $\{\text{PrCh}(w/a)\}$ is what I call a *ramified* probable chance. If $\Pr(\ominus a) = 1$ then $\{\text{PrCh}(w/a)\} = \text{PrCh}(w/a)$, and Definition 4 reduces to Definition 2. Definition 4 employs a new notion, the operator "$\ominus$", which stands for the condition of *possible* openness. This condition is related to, but somewhat weaker than, that of openness. While an action is open to an agent if and only if either it will take place or this agent can definitely bring it about, an action is *possibly open* to an agent if and only if there is a chance that it will take place, or that this agent might possibly do it or bring it about (that is, if and only if there is a chance that it will take place, or, given certain conditions under this agent's full control, it might take place). Valid principles for possible openness are to include, for every action a, proposition b, and positive x,

$$\Diamond_x a \rightarrow \ominus a;$$

and, if a entails b,

$$\ominus a \rightarrow \ominus b;$$

and, for every option a in a decision problem, at every positively probable world,

$$(a \Diamond_x \!\!\rightarrow b) \rightarrow \ominus b.$$

For similarities and a connection between openness and possible openness, note that both

$$a \rightarrow \Diamond a \quad \text{and} \quad a \rightarrow \ominus a$$

are valid, as are

$$\Diamond a \rightarrow \Diamond b \quad \text{and} \quad \ominus a \rightarrow \ominus b$$

when a entails b, and that

$$\Diamond a \rightarrow \ominus a$$

is also valid. For one contrast between the conditions, I note that although
$$\Diamond_x a \to \ominus a$$
is valid for every positive x,
$$\Diamond_x a \to \oplus a$$
is valid only for $x = 1$. For another contrast between the conditions, note that for an option a in a decision problem, at every positively probable world,
$$(a \Diamond_x \!\!\to b) \to \ominus b$$
is valid for every positive x, but
$$(a \Diamond_x \!\!\to b) \to \oplus b$$
is valid only for $x = 1$; that is, only
$$(a \Box\!\!\to b) \to \oplus b$$
is valid for "$\oplus$". I say there is an *unconditional* chance of x for a, $\Diamond_x a$, if and only if there is a conditional chance of x on the necessary proposition T for a, $(T \Diamond_x \!\!\to a)$. Since I assume that there is a positive x such that $(T \Diamond_x \!\!\to a)$ if and only if $(T \Diamond\!\!\to a)$, it is a consequence of definitions that $\Diamond_x a$ for some positive x if and only if, in Lewis's terms, "$\Diamond a$," which says that a is an "inner possibility" (Lewis 1973, p. 18). So, the principle that $(\Diamond_x a \to \ominus a)$ for any positive x is equivalent to $(\Diamond a \to \ominus a)$, and the five modalities – $\Diamond$, $\oplus$, $\ominus$, Lewis entertainability for purposes of causal hypotheses, and logical possibility – are progressively weaker forms of possibility. I note that under the assumptions I make, a is an unconditional certainty, $\Diamond_1 a$, if and only if a is an "inner necessity," $\Box a$ (Lewis 1973, p. 18). Thus, although $(\Diamond a \to \ominus a)$ is valid for possible openness, only $(\Box a \to \oplus a)$ is valid for openness. I have said that a is open if a will take place, that $(a \to \oplus a)$ is to be valid. There may be reasons for retrenching here, and insisting only that a is definitely open if a will definitely take place – that is, $(\Box a \to \oplus a)$ is valid – while allowing that, even if a will take place, if a is not an inner necessity, $\sim\!\Box a$, then it can happen that a is not definitely open, $\sim\!\oplus a$.

In the view of an agent who does not believe in objective conditional chances as distinct from objective conditional certainties and impossibilities, the two conditions are equivalent. Such an agent is sure that an action is possibly open to him if and only if he is sure that it is definitely open to him. But the two conditions need not coincide in the views of agents who do believe in objective chances. For such agents, Definition 4 can extend beyond Definition 3, and define expected utilities for actions an agent is sure are *not* definitely open to him. According to Definition 4, the expected utility of a probably open action is its expected utility from the epistemic perspective that would result from conditionalizing on its being possibly open, a perspective in which it need not be certainly definitely open, and can indeed be certainly *not* definitely open.

3.2. A speculative interlude

While it improves upon its predecessors, Definition 4 may not be quite right. It is possible that an adjustment is called for that cancels a part of the movement to it from Definition 1. To reach another definition of expected utility that is closer to Definition 1, replace ramified probable chances in Definition 4 by what can be called "ramified probable augmented chances" – put {PrAugCh(w/a)} in place of {PrCh(w/a)}, and accept that by definition {PrAugCh(w/a)} = $\sum_y \Pr[\text{AC}(w/a) = y/a]y$. Augmented chances (AC) combine facts with chances. The *augmented chance* at world w of state s given action a is, if a is false at w, equal to the chance at w of s given a; but if a is *true* at w, the augmented chance at w of s given a is either 1 or 0, depending only on whether s is true or false at w – the simple chance at w of s given a, which may be other than 1 and 0, is in this case not relevant. To denote "the augmented chance at world w of state s given action a is x" I write $\text{AC}_w(s/a) = x$. It can be seen that: for x other than 0 and 1, $\Pr[\text{AC}(s/a) = x] = \Pr(a \diamond_x \to s)$; for $x = 1$, $\Pr[\text{AC}(s/a) = x] = \Pr(a \diamond_1 \to s) + \sum_{x<1} \Pr[a \& s \& (a \diamond_x \to s)]$; and, for $x = 0$, $\Pr[\text{AC}(s/a) = x] = \Pr(a \diamond_0 \to s) + \sum_{x>0} \Pr[a \& {\sim} s \& (a \diamond_x \to s)]$.

Reasons for using augmented instead of simple chances can be suggested (but not fully explained) by saying (1) that it seems that the expected utility of an action should turn out to be a probability-weighted average of its utilities at worlds; and (2) that it can seem there should be a certain asymmetry between the utilities at a world of actions that take place in it and the utilities at this world of actions that do not take place in it. The utility at a world of an action that takes place in it should (it can seem) be simply this world's value, and not at all a function of values of other worlds, even if there are (given this action) *chances* for other worlds. In contrast, it can seem that the utility at a world of an action that does *not* take place in it should in fact be a weighted average of the values of worlds that might obtain were it to take place, the weight for a world being its chance given this action. To establish this asymmetry one can equate the utility of an action at a world with a weighted average of the values of that action's worlds, where the weight for a world is its augmented chance given that action: $U_w(a) = \sum_{w'} \text{AC}_w(w'/a) V(a)$. Augmented chances are designed to service this asymmetry.

However, there are objections to this asymmetry and, equivalently, to using augmented chances for utilities at worlds. First, it can seem that the utility of an action should depend solely on its efficacy, and that therefore what matters – even in the case of an action that will take place – are *chances* of things on it, and not whether or not things will in fact take place. Second, it can seem that the relative utility of an action *a* that is not going to take place, when compared to the utility of the action that will take place, should depend on the difference (if any) that *a*'s taking place would make – the difference it would make not to things but rather to chances for things. For this reason, too, it can seem that the utility of an action that will take place should depend on chances of things on it, and not on their augmented chances.

I am not sure how a general account of utilities at worlds should go, in terms of chances or augmented chances, though I am inclined to favor chances and to make utilities functions of efficacies only. Since, however, the issues dividing these definitions seem irrelevant to the main issues in this chapter – and, in particular, to the issues of the next section – I think it best to set aside for now problems concerning augmented chances and proceed in terms of Definition 4, even if, contrary to my present tentative opinion, a more complicated definition (invoking augmented chances) would be a better definition of expected utility. (I have Włodek Rabinowicz to thank for many of my difficulties with chances and augmented chances.)

3.3. Adequate partitions of circumstances

In Jeffrey's logic of decision there is no problem with adequate partitions of circumstances. In that theory, every partition of circumstances is adequate to the analysis of every action; this is one thing that makes that theory attractive. Nobody likes problems. But this source of attractiveness is at the same time a source of embarrassment, for it is not really credible that every partition be adequate. Jeffrey's theory lacks a problem that one expects a decision theory to have. Our causal decision theory does have a problem here, which is thus not only bad for it but also good. And though in this causal decision theory not every partition of circumstances is in a certain sense adequate (see Sobel 1985a, sec. II for a case, reprised in Sobel 1989a [Chapter 9], and a formal model to this point), I think that in this theory every partition is adequate that one expects to be adequate and that should be adequate in this sense.[3] The problem that this theory has with adequate partitions leads (through its solution) to evidence for it – evidence which, because not planned for or a part of the theory's initial motivation, is particularly gratifying.[4]

The analog, for the theory of Definition 4, of Jeffrey's unrestricted logical partition theorem would take this form:

For any a such that $\Pr(\diamond a) > 0$ and for a logical partition C,
$$U(a) = \Sigma_{c:\, c \in C\, \&\, \Pr[\diamond(a\,\&\,c)] > 0} \{\mathrm{PrCh}(c/a)\} U(a\,\&\,c).$$
This principle is not valid in the theory of Definition 4, but certain restrictions of it are. We have, for this theory, two partition theorems: one restricts the principle just displayed to "sufficiently fine" partitions, and the other restricts it to "sufficiently exclusive" partitions. First, every logical partition that is, in relation to an option in a decision problem, *sufficiently fine* is adequate for that option, where a partition C is sufficiently fine in the intended sense relative to action a if and only if C analyzes a into parts (disjuncts) or versions whose expected utilities are the same as those of their probably possibly open parts or versions. A partition theorem for sufficiently fine partitions is proved in the appendix; this theorem extends not only to sufficiently fine logical partitions but also to what are termed sufficiently fine "probable chance partitions." (Theorems for sufficiently fine partitions are proved elsewhere in theories suited for agents who do not believe in chances: Sobel 1985a, 1989a [Chapter 9].)

Sufficiently fine partitions make all utility-relevant distinctions. It is not surprising that all such partitions are adequate, but it is also not completely reassuring since in real cases the number of utility-relevant distinctions is likely to be unmanageably large. It might be said that in practice we make only important distinctions, and that it is part of the art and wisdom of practice to know which distinctions need to be made in a case and which can be suppressed. But it would be more satisfying to learn that practice need not compromise, and that partitions that seem natural and right are quite adequate even when, as they often are, they are far from being sufficiently *fine* - even when, as is often the case in game theory, we do not even pretend that they are sufficiently fine. Practice and our intuitions are vindicated by our second partition theorem, whereby every logical partition that is in relation to an option *sufficiently exclusive* is adequate for it. A partition C is sufficiently exclusive in the intended sense, relative to an option a, if and only if C analyzes a into parts or versions such that the agent is sure of every pair of these that not *both* are possibly open - that is, if and only if, for any distinct c and c' in C,
$$\Pr(\sim[\diamond(a\,\&\,c)\,\&\,\diamond(a\,\&\,c')]) = 1.$$
A sufficiently exclusive partition need not be even nearly sufficiently fine. A partition theorem for sufficiently exclusive probable chance partitions is proved in the appendix. (Sufficiently exclusive partition theorems are proved in an appendix to Sobel 1985a and in Sobel 1989a [Chapter 9] for formally similar but conceptually different theories suited to agents who do not believe in objective chances - agents in whose views the conditions of definite and possible openness coincide.)

Brian Skyrms asks whether "all forms of causal decision theory must be stated in terms of ultimate consequences" (Skyrms, 1985, p. 609). I have given somewhat different reasons from his for saying No. I define expected utilities in terms of values of "ultimate consequences" (worlds), but partition theorems make expected utilities functions of expected utilities of proximate consequences (action–circumstance conjunctions). Skyrms (1985, p. 609) presents a Newcomb problem in which "Gibbard–Harper theory fallaciously applied to proximate consequences gets uncausal results." The fallacy, Skyrms suggests, is to apply such a theory to proximate consequences. A possibility not considered is that, although it is not a fallacy to apply such a theory to proximate consequences, it *is* a fallacy when so applying it to employ the evidential Desirabilities of these consequences rather than their (possibly different) expected utilities. One might say that the fallacy in the application of the Gibbard–Harper theory that Skyrms considers consists not in applying it to proximate consequences, but in doing so in a wrong manner.

Partitions that seem natural for decision problems are, I think, always sufficiently exclusive. In the popcorn problem I am certain that not *both* – going for popcorn when there is popcorn to be had, and going for popcorn when there is none to be had – are possibly open to me. (In this problem I am nearly sure that the second of these actions is definitely open to me, and that only it is even possibly open to me.) And in a Prisoners' Dilemma, each agent will be certain that not *both* – confessing along with the other prisoner, and confessing alone – are possibly open.

Natural partitions are, I think, always theoretically adequate, but not all theoretically adequate partitions are natural. The probable chance partition (defined in the appendix) {I take both boxes, I take only Box 2} would be an unnatural one to employ as the partition of *circumstances* in an application of a decision theory to Newcomb's problem. But this partition (which is certainly not sufficiently fine) is a sufficiently exclusive partition in the problem; for example, I (the agent) am sure that of (Both & Both) and (Both & One) only the first is open, and thus sure that not both of these versions of Both are open. So, although this partition would be quite unhelpful if employed in an application of my theory to the problem, it is "adequate."[5]

	Both	One
Both	U(Both & Both) $= U$(Both)	
One		U(One)

(Neither $U(\text{Both} \& \text{One})$ nor $U(\text{One} \& \text{Both})$ is determined by Definition 4, since neither (Both & One) nor (One & Both) is even possibly open: we have both $\Pr[\diamondsuit(\text{Both} \& \text{One})] = 0$ and $\Pr[\diamondsuit(\text{One} \& \text{Both})] = 0$.)

	Both	One
Both	{PrCh(Both/Both)} = 1	0
One	0	1

Employing the entries in these matrices in applications of Definition 4, we get the entirely unremarkable, unhelpful, and unfaultable result that $U(\text{Both}) = U(\text{Both})$ and $U(\text{One}) = U(\text{One})$. So we see that {I take both boxes, I take only Box 2} would be, as a partition of circumstances, unnatural and unhelpful, even though sufficiently exclusive and thus "adequate" in the sense in which I employ this term. Similar assessments would hold for singleton logical partitions such as {either there is or is not a million dollars in Box 2}; such degenerate partitions qualify (though only vacuously or by default) as "sufficiently exclusive." (Remarks in this and the previous paragraph are of some relevance to Levi 1982.)

4. Conclusions

Bayesians think that rational actions maximize weighted-average value or expected utility. The first task for a developed Bayesian theory is to make precise the nature of the weights in expected utilities, of the probabilities that are involved.

Jeffrey identifies these probabilities with conditional probabilities. His theory is conceptually economical, formally simple, relatively problem-free, and, alas, false. Indeed, on reflection, what seem initially to be important virtues of this theory can take on the aspect of defects. That every partition of circumstances is adequate can seem on reflection too good to be true. And, on reflection, it is not really credible – given the ubiquity and centrality of causal notions in everyday practical rhinking – that a good theory of rational action should do justice to the phenomena without aid of causal categories among its more or less explicit primitives.

In the theory of Definition 4, probabilities involved in expected utilities are what I term "ramified probable chances." Causality enters the primitive basis of this theory under the guise of chance conditionals, and in it not every partition of circumstances is in a certain natural sense adequate, though it seems that every partition of circumstances that one expects to be adequate *is* adequate in this sense in this theory.

Finally, lest its distance from Jeffrey's theory be exaggerated, I stress that the theory I endorse is fashioned very much on the model of his. While steering clear of certain of his theory's errors, it borrows extensively from its perspectives and machinery. The theory I endorse is very much in his theory's debt. I propose, in fact, to retain Jeffrey's theory, not (of course) as a theory of rational choice or of preferences for propositions as facts, but as a theory of rational preferences for propositions as items of news. I propose to combine this theory with a theory of rational preferences for facts on epistemic conditions; foundations could then combine conditions on preferences for news with conditions on preferences on epistemic conditions for facts. Given that these preferences should coincide for worlds, it may be possible - by a judicious selection of conditions on these two preference relations - to assemble purely qualitative foundations (in contrast with the foundations in Fishburn 1973 and Armendt 1986) that (in contrast with Jeffrey–Bolker foundations - Bolker 1967 and Jeffrey 1983) are sufficient to determine unique probability functions, as well as measures for these two kinds of preference that are unique to linear transformations.

My suggestion is a modification of one that Jeffrey has made. Writing of Jeffrey–Bolker foundations, he has said that the

> undetermination of u and P by the preference relation [which] may be seen as a flaw [is] removable, for example, by using *two* primitives: preference and comparative probability. With these primitives, one ought to be able to drop some of Bolker's restrictions on the algebra of prospects [and to] get significantly closer to an existence theorem in which the conditions are necessary as well as sufficient for existence of u and P, while obtaining the usual uniqueness result.... It would be a job worth doing. (Jeffrey 1974, pp. 77–8)

Rather than add comparative probability as a primitive to news preference, however, it may be possible instead and with the same effect to add as a primitive factual preference. Perhaps unique probabilities can be secured without taking probability as primitive, and the whole thing accomplished in a purely qualitative manner, without demanding restrictions on the field of propositions and while employing as axioms only necessary conditions (conditions "derivable backwards from the representation theorem" and properties of the measures being axiomatized - Armendt 1986). As Jeffrey has said, if possible it would be a job worth doing.

APPENDIX: DEMONSTRATIONS OF PARTITION THEOREMS

Definition of Desirability. For any proposition a, if $\Pr(a) > 0$ then $\mathrm{Des}(a) = \sum_w \Pr(w/a) V(w)$.

Definition of conditional probability. For any propositions a and c, if $\Pr(a) > 0$ then $\Pr(c/a) = \Pr(a \& c)/\Pr(a)$.

Definition of probability partition. A set of propositions C is a probability partition if and only if, for every positively probable world, exactly one c in C is true at this world.

Lemma 1. *For any propositions a and c and world (proposition) w, if w is a $(a \& c)$-world then $\Pr(w) = \Pr(a \& w) = \Pr[(a \& c) \& w]$.*

Theorem 1. *For any proposition a and probability partition C, if $\Pr(a) > 0$ then*
$$\mathrm{Des}(a) = \Sigma_{c:\, c \in C \,\&\, \Pr(a \& c) > 0} \Pr(c/a)\, \mathrm{Des}(a \& c).$$

Proof.

1. C is a probability partition
2. $\Pr(u) > 0$
3. $\mathrm{Des}(a) = \Sigma_w \Pr(w/a) V(w)$ 2, def Des
4. $= \Sigma_{c \in C} \Sigma_{w \in (a \& c)} \Pr(w/a) V(w)$ 1, 3, def ProbPart
5. $= \Sigma_{c \in C} \Sigma_{w \in (a \& c)} (\Pr[(a \& c) \& w]/\Pr(a)) V(w)$
        4, 2, def CondPr, Lemma 1
6. $= \Sigma_{c:\, c \in C \,\&\, \Pr(a \& c) > 0} [\Pr(a \& c)/\Pr(a)]$
   $\cdot \Sigma_w (\Pr[(a \& c) \& w]/\Pr(a \& c)) V(w)$ 5
7. $= \Sigma_{c:\, c \in C \,\&\, \Pr(a \& c) > 0} \Pr(c/a)\, \mathrm{Des}(a \& c)$
        6, 2, def CondPr, def Des

Definition of expected utility. For any proposition a, if $\Pr(\diamondsuit a) > 0$ then $U(a) = \Sigma_w \{\mathrm{PrCh}(w/a)\} V(w)$.

Definition of ramified probable chance. For any propositions a and c, if $\Pr(\diamondsuit a) > 0$ then $\{\mathrm{PrCh}(c/a)\} = \Sigma_x [\Pr(a \diamondsuit_x \to c)/\diamondsuit a] x$.

Definition of probable chance partition. For any proposition a and partition C, C is a probable chance partition relative to a if and only if the agent is certain that: if a is causally possible then, at each "nearest" a-world, exactly one member of C is true.

For example, (b, c, d) is a probable chance partition relative to a if and only if $\Pr(\Diamond a \to [a \,\square\!\!\!\to (b \vee c \vee d)]) = 1$, $\Pr(\Diamond a \to [a \,\square\!\!\!\to \sim(b \& c)]) = 1$, $\Pr(\Diamond a \to [a \,\square\!\!\!\to \sim(b \& d)]) = 1$, and $\Pr(\Diamond a \to [a \,\square\!\!\!\to \sim(c \& d)]) = 1$. Note that possible openness ($\diamondsuit$) entails causal possibility ($\Diamond$).

Definition of U-elementary. For any proposition a, if $\Pr(\diamond a) > 0$ then a is U-elementary if and only if, for every a' such that a' entails a and $\Pr(\diamond a') > 0$, $U(a) = U(a')$.

Definition of sufficiently fine. For any proposition a and partition C, C is sufficiently fine relative to a if and only if, for every c in C, if $\Pr[\diamond(a \& c)] > 0$ then $(a \& c)$ is U-elementary.

Lemma 2. *For any proposition a and probable chance (relative to a) partition C, for any distinct c and c' in C, if $w \in [a \& (c \& c')]$ then, for every positive y, $\Pr[\diamond a \& (a \diamondsuit_y \to w)] = 0$.*

If $\diamond a$ and, for a positive y, $(a \diamondsuit_y \to w)$, then w is an $[a \& (c \& c')]$-world. But the agent is certain that no nearest a-world is an $[a \& (c \& c')]$-world.

Lemma 3. *For any option a in a decision problem and any a-world w, if $\Pr(\diamond w) = 0$ then, for every positive y, $\Pr[\diamond a \& (a \diamondsuit_y \to w)] = 0$.*

This is so because, for any option a, $(a \diamondsuit_y \to w)$ entails $\diamond w$ for every positive y.

Lemma 4. *For any world w, if $\Pr(\diamond w) > 0$ then $U(w) = V(w)$.*

Lemma 5. *For any option a in a decision problem, proposition c, and $(a \& c)$-world w: If $\Pr[\diamond(a \& c)] = 0$ then, for every positive y,*

$$\Pr[\diamond a \& (a \diamondsuit_y \to w)] = 0.$$

If w is an $(a \& c)$-world then $\diamond w$ entails $\diamond(a \& c)$.

Lemma 6. *For any proposition a and sufficiently fine (relative to a) partition C: If $c \in C$, w is an $(a \& c)$-world, and $\Pr(\diamond w) > 0$, then $U(w) = U(a \& c)$.*

Lemma 7.

$$\sum_y [(\Pr[\diamond a \& (a \diamondsuit_y \to c)] / \Pr(\diamond a)) y]$$
$$= \sum_{w \in (a \& c)} \sum_y [(\Pr[\diamond a \& (a \diamondsuit_y \to w)] / \Pr(\diamond a)) y].$$

Proof.

$$\sum_{w \in (a \& c)} \sum_y [(\Pr[\diamond a \& (a \diamondsuit_y \to w)] / \Pr(\diamond a)) y]$$
$$= \sum_y \sum_{w \in (a \& c)} [(\Pr[\diamond a \& (a \diamondsuit_y \to w)] / \Pr(\diamond a)) y]$$
$$= \sum_y [(\Pr[\diamond a \& (a \diamondsuit_y \to c)] / \Pr(\diamond a)) y].$$

Theorem 2. *For any option a in a decision problem and any partition C: If C is a sufficiently fine probable chance partition relative to a, then*

$$U(a) = \Sigma_{c:\, c \in C\, \&\, \Pr[\ominus(a\,\&\,c)] > 0} \{\mathrm{PrCh}(c/a)\} U(a\,\&\,c).$$

Proof.

1. C is a probable chance partition relative to a
2. C is sufficiently fine relative to a
3. $\Pr(\ominus a) > 0$ a is an option
4. $U(a) = \Sigma_w \{\mathrm{PrCh}(w/a)\} V(w)$ 3, def ExpUt
5. $\quad = \Sigma_w \Sigma_y [(\Pr[\ominus a\,\&\,(a\,\Diamond_y\!\to w)]/\Pr(\ominus a))y]V(w)$
                                                                      4, 3, def RamPrCh, def CondPr
6. $\quad = \Sigma_{c \in C} \Sigma_{w \in (a\,\&\,c)} \Sigma_y [(\Pr[\ominus a\,\&\,(a\,\Diamond_y\!\to w)]/\Pr(\ominus a))y]V(w)$
                                                                                 5, 1, Lemma 2
7. $\quad = \Sigma_{c \in C} \Sigma_{w:\, w \in (a\,\&\,c)\,\&\,\Pr(\ominus w) > 0}$
$\quad\quad \cdot \Sigma_y [(\Pr[\Diamond a\,\&\,(a\,\Diamond_y\!\to w)]/\Pr(\ominus a))y]U(w)$
                                                                             6, Lemma 3, Lemma 4
8. $\quad = \Sigma_{c:\, c \in C\,\&\,\Pr[\ominus(a\,\&\,c)] > 0} \Sigma_{w:\, w \in (a\,\&\,c)\,\&\,\Pr(\ominus w) > 0}$
$\quad\quad \cdot \Sigma_y [(\Pr[\ominus a\,\&\,(a\,\Diamond_y\!\to w)]/\Pr(\ominus a))y]U(w)$ 7, Lemma 5
9. $\quad = \Sigma_{c:\, c \in C\,\&\,\Pr[\ominus(a\,\&\,c)] > 0} \Sigma_{w \in (a\,\&\,c)}$
$\quad\quad \cdot \Sigma_y [(\Pr[\ominus a\,\&\,(a\,\Diamond_y\!\to w)]/\Pr(\ominus a))y]U(a\,\&\,c)$
                                                                          8, 2, Lemma 6, Lemma 3
10. $\quad = \Sigma_{c:\, c \in C\,\&\,\Pr[\ominus(a\,\&\,c)] > 0}$
$\quad\quad \cdot \Sigma_y [(\Pr[\ominus a\,\&\,(a\,\Diamond_y\!\to c)]/\Pr(\ominus a))y]U(a\,\&\,c)$
                                                                               9, Lemma 7
11. $\quad = \Sigma_{c:\, c \in C\,\&\,\Pr[\ominus(a\,\&\,c)] > 0} \{\mathrm{PrCh}(c/a)\} U(a\,\&\,c)$
                                                                                 10, 3, def CondPr, def RamPrCh

Definition of sufficiently exclusive. For any proposition a and partition C, C is sufficiently exclusive relative to a if and only if, for any distinct c and c' in C,

$$\Pr[\ominus(a\,\&\,c)\,\&\,\ominus(a\,\&\,c')] = 0.$$

Lemma 8. *For any option a in a decision problem and any partition C: If C is a sufficiently exclusive probable chance partition relative to a, then for any c in C, positive y, and $(a\,\&\,c)$-world w, at every positively probable world,*

$$[\ominus a\,\&\,(a\,\Diamond_y\!\to w)] \leftrightarrow (\ominus(a\,\&\,c)\,\&\,[(a\,\&\,c)\,\Diamond_y\!\to w]).$$

Proof.

1. a is an option in a decision problem, w is an $(a\&c)$-world, $y>0$, $c \in C$
2. $\ominus a$
3. $(a \Diamond_y \to w)$
4. $\ominus(a\&c)$ 1, 3
5. $(a \Diamond_x \to c')$ for some positive x and c' other than c in C
6. $[a \Diamond_x \to (a\&c')]$ 5
7. $\ominus(a\&c')$ 6
8. $\sim\ominus(a\&c')$ 4, [C is sufficiently exclusive.]
9. $(a \Diamond_x \to c')$ for no positive x and c' other than c in C 5–8
10. $(a \Box\!\!\to c)$ 9, [There is no positive chance on a of any c' in C other than c.], 2, [C is a probable chance partition relative to a. So, given 2, if a then positive chances summing to 1 are distributed among members of C. Thus, $(a \Diamond_1 \to c)$, which is logically equivalent to $(a \Box\!\!\to c)$.]
11. $[a \Box\!\!\to (a\&c)]$ 10
12. $[a \Box\!\!\to (a\&c)] \& [(a\&c) \Box\!\!\to a]$ 11
13. $[(a\&c) \Diamond_y \to w]$ 12, [The nearest part of a is the nearest part of $(a\&c)$. Chances on a and $(a\&c)$ are thus identical.], 3
14. $\ominus(a\&c) \& [(a\&c) \Diamond_y \to w]$ 4, 13
15. $\ominus(a\&c)$
16. $[(a\&c) \Diamond_y \to w]$
17. $\ominus a$ 15
18. $(a \Box\!\!\to w)$ 15, 17, [As 5–10, given 2 and 4.]
19. $(a \Diamond_y \to w)$ 16, 18, [As 11–13, given 10 and 3.]
20. $[\ominus a \& (a \Diamond_y \to w)]$ 17, 19
21. $[\ominus a \& (a \Diamond_y \to w)] \leftrightarrow (\ominus(a\&c) \& [(a\&c) \Diamond_y \to w])$ (2, 3)–14, (15, 16)–20

Lemma 9. *For any proposition a, world w, and positive y: If w is not an a-world, then $\sim(a \Diamond_y \to w)$.*

Lemma 10. *For any option a in a decision problem and any partition C: If C is a sufficiently exclusive probable chance partition relative to a, then at every positively probable world, for every positive x less than one,*

$$\ominus a \to \sim(a \Diamond_x \to c).$$

Proof.

1. ◇a
2. $(a \diamondsuit_x \rightarrow c)$ for some positive $x < 1$
3. $(a \diamondsuit_y \rightarrow c')$ for some positive y and c' in C other than c
                                        2, [C is a probable chance partition.]
4. ◇(a & c)   2
5. ◇(a & c')   3
6. ~◇(a & c')   4, [C is sufficiently exclusive.]
7. $(a \diamondsuit_x \rightarrow c)$ for no positive $x < 1$ 2-6
8. ~$(a \diamondsuit_x \rightarrow c)$ for every positive $x < 1$ 7

Lemma 11. *For any option a in a decision problem and any partition C: If C is a sufficiently exclusive probable chance partition relative to a, then at every positively probable world,*

$$[\diamond a \& (a \sqcup\!\!\!\rightarrow c)] \leftrightarrow \diamond(a \& c).$$

Proof.

1. ◇a
2. $(a \square\!\!\!\rightarrow c)$
3. ◇(a & c)   2
4. ◇(a & c)
5. ◇a   4
6. $(a \square\!\!\!\rightarrow c)$
           4, 5, [As 5-10 given 2 and 4 in the proof of Lemma 8.]
7. $[\diamond a \& (a \square\!\!\!\rightarrow c)]$ 5, 6

Lemma 12. *For any option a in a decision problem and any partition C: If C is a sufficiently exclusive probable chance partition relative to a, then*

$$\{\text{PrCh}(c/a)\} = \Pr[\diamond(a \& c)/\Pr(\diamond a)].$$

Proof.

1. $\{\text{PrCh}(c/a)\} = \sum_x [\Pr(a \diamondsuit_x \rightarrow c)/\diamond a] x$ def RamPrCh

2. $\quad = \dfrac{\sum_x \Pr[\diamond a \& (a \diamondsuit_x \rightarrow c)] x}{\Pr(\diamond a)}$ Def CondPr

3. $\quad = \dfrac{\sum_{x \in (0,1)} \Pr[\diamond a \& (a \diamondsuit_x \rightarrow c)] x}{\Pr(\diamond a)}$ 2, Lemma 10

4. $\quad = \Pr[\ominus a \& (a \square\!\!\!\rightarrow c)]/\Pr(\ominus a)$

   3, $(a \diamondsuit_1 \!\rightarrow c)$ is logically equivalent to $(a \square\!\!\!\rightarrow c)$.]
5. $\quad = \Pr[\ominus(a \& c)]/\Pr(\ominus a)$ 　　　　　　　　4, Lemma 11

Theorem 3. *For any option a in a decision problem and any partition C: If $\Pr(\ominus a) > 0$ and C is a sufficiently exclusive probable chance partition relative to a, then*

$$U(a) = \Sigma_{c:\, c\in C \,\&\, \Pr[\ominus(a\&c)]>0} \{\mathrm{PrCh}(c/a)\} U(a\&c).$$

Proof.

1. C is a sufficiently exclusive partition relative to a
2. C is a probable chance partition relative to a
3. $\Pr(\ominus a) > 0$
4. $U(a) = \Sigma_w \{\mathrm{PrCh}(w/a)\} V(w)$ 　　　　　　3, def ExpUt
5. $\quad = \Sigma_w \Sigma_y ([\Pr(a \diamondsuit_y \!\rightarrow w)/\ominus a] y) V(w)$ 　　4, 3, def RamPrCh
6. $\quad = \Sigma_w \Sigma_y \left(\dfrac{\Pr[\ominus a \& (a \diamondsuit_y \!\rightarrow w)]}{\Pr(\ominus a)} y \right) V(w)$ 　　5, 3, def CondPr
7. $\quad = \Sigma_{c\in C} \Sigma_{w\in(a\&c)} \Sigma_y \left(\dfrac{\Pr[\ominus a \& (a \diamondsuit_y \!\rightarrow w)]}{\Pr(\ominus a)} y \right) V(w)$

   　　　　　　　　　　　　　　　　　　　　6, 2, Lemma 2
8. $\quad = \Sigma_{c:\, c\in C \,\&\, \Pr[\ominus(a\&c)]>0} \Sigma_{w\in(a\&c)}$

   $\quad \cdot \Sigma_y \left(\dfrac{\Pr[\ominus a \& (a \diamondsuit_y \!\rightarrow w)]}{\Pr(\ominus a)} y \right) V(w)$ 　　7, Lemma 5
9. $\quad = \Sigma_{c:\, c\in C \,\&\, \Pr[\ominus(a\&c)]>0} \Sigma_{w\in(a\&c)}$

   $\quad \cdot \Sigma_y \left(\dfrac{\Pr(\ominus(a\&c) \& [(a\&c) \diamondsuit_y \!\rightarrow w])}{\Pr(\ominus a)} y \right) V(w)$

   　　　　　　　　　　　　　　　　　　　　8, 1, 2, Lemma 8
10. $\quad = \Sigma_{c:\, c\in C \,\&\, \Pr[\ominus(a\&c)]>0} (\Pr[\ominus(a\&c)]/\Pr(\ominus a)) \Sigma_w$

    $\quad \cdot \Sigma_y \left(\dfrac{\Pr(\ominus(a\&c) \& [(a\&c) \diamondsuit_y \!\rightarrow w])}{\Pr[\ominus(a\&c)]} y \right) V(w)$

    　　　　　　　　　　　　　　　　　　　　9, Lemma 9
11. $\quad = \Sigma_{c:\, c\in C \,\&\, \Pr[\ominus(a\&c)]>0} (\Pr[\ominus(a\&c)]/\Pr(\ominus a)) U(a\&c)$

    　　　　　　　　　　　　3, 10, def CondPr, def Ram PrCh, def ExpUt
12. $\quad = \Sigma_{c:\, c\in C \,\&\, \Pr[\ominus(a\&c)]>0} \{\mathrm{PrCh}(c/a)\} U(a\&c)$

    　　　　　　　　　　　　　　　　　　　　3, 11, Lemma 12

NOTES

1. Jeffrey uses these words from *La logique, ou l'art de penser,* Paris 1662 (the work of Antoine Arnauld, Pierre Nicole, and others associated with Port Royal des Champs, published anonymously) as a motto for Chapter 1, "Deliberation – A Bayesian Framework," of Jeffrey (1983). See *La logique,* part. IV, chap. XV, p. 384.
2. This is Bayesian decision theory in its simplest form, where maximization of expected utility is for every action both necessary and sufficient to rationality. There are other forms that make the rationality of at least some actions depend not on their own expected utilities but rather on the expected utilities of other, more "basic," actions. There are forms of Bayesian decision theory that make an action's rationality depend not on the maximization of actual expected utilities "just before" its time, but on the maximization of hypothetical or conditional expected utilities. And there are forms that make rationality depend not only on actual but also on hypothetical or conditional expected utilities.

    Ratificationism, a theory described but not unreservedly endorsed in Jeffrey (1983), departs from simple Bayesianism in the second of these ways. William Harper inclines to a theory that departs from simple Bayesianism in the third way, a theory that would have one maximize among one's ratifiable or possible stable choices (Harper 1985). I favor a theory that departs from simple Bayesianism in the third way, but in a different manner. The theory I favor departs as well from simple Bayesianism in the first way; see Sobel (1983 [Chapter 10 in this volume]) for both departures. For discussion of the second and third relatively unsimple forms of Bayesianism, see Rabinowicz (1985). For a somewhat different treatment, see Eells (1984b,c).
3. The sense in question is the natural one in which a partition is adequate if it analyzes an action into disjunctive parts such that its expected utility is a weighted average of theirs; "natural" partition theorems for causal decision theory will be analogs of Jeffrey's unrestricted partition theorem. I stress this because, although only restricted theorems of this form hold in causal decision theory, a quite unrestricted partition theorem of *another* form also holds in this theory (though, remarkably, only restricted analogs of it hold in Jeffrey's logic of decision). The following quite unrestricted theorem for expected utilities under logical partitions is proved and discussed in Sobel (1989a [Chapter 9]):

    For any action a and logical partition C, if $\Pr(\oplus a) = 1$ then
    $$U(a) = \Sigma_{c:\, c \in C\ \&\ \Pr(c) > 0} \Pr(c) U(a/c).$$

    For just as only restricted analogs of Jeffrey's unrestricted partition theorem hold for causal expected utility, so only restricted analogs of the just-displayed theorem hold for Desirability – once conditional Desirability has been provided with a natural definition along the lines of the definition for $U(a/c)$.
4. I join Lewis in the suggestion "that we ought to undo a seeming advance in the development of decision theory" according to which "we need not be selective about partitions" (Lewis 1981a, pp. 12-13). But I stress that on reflection this seeming advance should always have seemed too good to be true. It is not an unmixed virtue that "evidential expected utility of an item is the same no

matter how the space of all possibilities is carved up into outcomes and states" (Eells 1985c, p. 102), or that V-maximization "prescriptions are robust under transformations of the way possible outcomes are represented" (Levi 1982, p. 342; see also Levi 1983, p. 541).

5. The point being made is that a partition that is sufficiently exclusive, and thus adequate, need not be sufficiently fine. The further point that a partition can be adequate in the present sense without being either sufficiently exclusive or sufficiently fine is made in Section 4 of Sobel (1989a [Chapter 9]).

9

Partition theorems for causal decision theories

Abstract. Three partition theorems are proved in a causal decision theory for agents who do not believe in objective chances. One is restricted to a very exclusive partition of circumstances and analyzes the utility of an option in terms of its utilities *in conjunction with* circumstances in this partition. This theorem is useful for applications. Another analyzes an option's utility in terms of its utilities *conditional on* circumstances and is quite unrestricted. This theorem is of some theoretical importance. A third theorem of use in applications, one that proceeds in terms of conjunctions of an action and circumstances in very fine partitions, is appended.

0. INTRODUCTION

Causal decision theories need partition theorems. This is demonstrated for theorems of a certain form in Sobel (1985a). In that paper and in Sobel (1986a [Chapter 8 in this volume]) I explain and derive two theorems of this form: for any action a and partition C of kind K,

$$U(a) = \Sigma_{c \in C} [\text{(the probability of } c \text{ given } a)$$
$$\cdot \text{(the expected utility of } a \text{ in } c)],$$

where "the probability of c given a" reflects perceived causal bearings and "the expected utility of a in c" is simply $U(a \& c)$, the expected utility of a in conjunction with c. Specifying kind K, one theorem is based on "sufficiently fine" partitions, where (roughly) a partition is sufficiently fine relative to an action if it analyzes this action into parts all of whose parts are alike in expected utility. The other is based on "sufficiently exclusive" partitions, where (roughly) a partition is sufficiently exclusive if it analyzes an action into versions such that the agent is sure that no two of these are both open.

In this chapter, I explain and prove a partition theorem of *another* form – namely, for any action a and partition C,

Reprinted with revisions by permission from *Philosophy of Science* 56 (1989), pp. 71-93.

$U(a) = \Sigma_{c:\, c \in C\,\&\,\Pr(c)>0}$ [(the probability of c)
· (the expected utility conditional on c of a)].

This theorem differs in form from those I have proved elsewhere, and is not restricted to certain kinds of partitions. Those previous theorems are of the style of a theorem in Richard Jeffrey's theory, according to which, for any positively probable action a and probabilistic partition C,

$$\text{Des}(a) = \Sigma_{c:\, c \in C\,\&\,\Pr(a\&c)>0}\, [\Pr(c/a) \cdot \text{Des}(a\&c)].$$

This theorem is a corollary of Jeffrey's Desirability axiom: If $\Pr XY = 0$ and $\Pr(X \vee Y) \neq 0$ then

$$\text{Des}(X \vee Y) = \frac{(\Pr X\, \text{Des}\, X) + (\Pr Y\, \text{Des}\, Y)}{\Pr X + \Pr Y}$$

(Jeffrey 1983, p. 80). The theorems I have proved in this style of Jeffrey's differ from it not only in that they are restricted to certain partitions, but also in that they feature things rather like probabilities of causal conditionals where Jeffrey's theorem features conditional probabilities.

The new theorem proved here is like Brad Armendt's theorem,

$$U(A) = \Sigma_i \Pr(B_i) U(A, B_i),$$

but may differ in details (Armendt 1986). It is also related to Peter Fishburn's principle,

$$u(xA) = \Sigma_{i=1}^{n} \Pr(A_i/A) u(xA_i),$$

which, when A is the set of all relevant events, presumably simplifies to

$$u(x) = \Sigma_i \Pr(A_i) u(xA_i)$$

(Fishburn 1981, p. 178).

In what follows, I set out the theory within which theorems will be proved. Then a sense is explained in which this theory needs partition theorems. Next, theorems are proved – first a version of the kind I have proved elsewhere, and then the new theorem. After comparing this new theorem with principles of Fishburn and Armendt, I compare the two theorems I prove and assess their merits and significance. Another theorem of a kind I have proved elsewhere is demonstrated in the appendix to this chapter.

1. A CAUSAL DECISION THEORY

We consider a theory for the special case of agents who do not believe in objective chances. Primitives include *complete controlledness* ($\oplus$), *openness* ($\bigcirc$), *and practical conditionality* ($\Box\!\!\rightarrow$). Practical conditionals are strongly centered conditionals constrained by Stalnaker's assumption: There are to be unique "nearest worlds," and $(p \,\square\!\!\rightarrow q)$ is to be true if and only if q is true at the nearest p-world; that is, $(p \,\square\!\!\rightarrow q)$ is to be true if and only if either $(p \,\&\, q)$ or, if it were the case that p, then it would be the case that q either "as a consequence" or "even so." (See Lewis 1973, chap. 1 and p. 78.) It is assumed that if p entails q then $\bigcirc p$ entails $\bigcirc q$: this is principle E. It is also assumed that

$$[\oplus p \,\&\, (p \,\square\!\!\rightarrow q)] \text{ entails } \bigcirc q.$$

This is principle CO. Intuitively, for the agent in a decision problem, a thing is to be *completely under control* if and only if he can, by choice, do it or not, and it can happen only by his doing or choice. For such an agent, a thing is to be *open* if and only if he can bring it about by choice; thus the principle CO. A thing is an *option* in a decision problem if and only if the agent is sure that it is completely under his control. (For a general theory of options for decision problems see Sobel 1983 [Chapter 10].)

Since options are open, the principle that

$$[\bigcirc p \,\&\, (p \,\square\!\!\rightarrow q)] \text{ entails } \bigcirc q$$

would suffice for arguments to come, and it can seem that this principle should be logically valid. But I think that it is not. For it could be that a is open to me even though a is not going to happen and even though, were it to happen, it would happen not by my choice but because I was confused and didn't realize what I was doing. And it could be that *were a* to happen then s would happen, where s is: somebody watching would realize that though a was happening, it was not by my choice that it was happening. In such a case it would be true that $\bigcirc a$ and true that $(a \,\square\!\!\rightarrow s)$, but it would be false that $\bigcirc s$. Recall that a is not going to happen, and that s involves someone believing *truly* that a is happening *and* that a is not happening by my choice; s is not something I could bring about by choice.

We take as fundamental not expected utility but rather conditional expected utility of versions of options, where a is a version of o if and only if a entails o.

Definition of conditional expected utility. For any version a of an option and any condition c, if $\Pr(c) > 0$ and $\Pr(\oplus a/c) = 1$ then

$$U(a/c) = \Sigma_w \, (\Pr[(a \,\square\!\!\rightarrow w)/c] \cdot V(w)).$$

(Throughout, w is used as a variable ranging over worlds or as a stand-in for a world proposition.) The expected utility of a version of an option *on a condition* is, intuitively, its expected utility from the epistemic perspective of that condition – the perspective that would result from conditionalizing on that condition. The expected utility of *a* conditional on *c*, as here explained, is not what would be the agent's expected utility for *a* were it the case that *c*, for *c* is to function as an epistemic condition that determines a certain epistemic perspective and not as a hypothetical fact condition. And, most importantly, the expected utility of *a* conditional on *c* is not the expected utility of *a* in *conjunction* with *c*. Rather, it is the expected utility of *a* from a certain possible epistemic perspective – specifically, that perspective that comes from the agent's actual perspective by conditionalizing on *c*. And though *c* is certain in this possible perspective, not even the expected utility of *a from this perspective* is necessarily the same as the expected utility, from this perspective, of *a* in conjunction with *c*. (Consider that, in a particular case, "condition" *c* might be the action ~*a*.)

Let an *action a* in a decision problem be a version of an option where the agent is sure that if *a* is open then it is completely under his control and can happen only by his doing or choice: $\Pr(\bigcirc a \to \oplus a) = 1$. We take the *unconditional* expected utility of an action an agent thinks may be open to him to be its expected utility conditional on its actually being open to him.

Definition of expected utility. For any action *a*, if $\Pr(\bigcirc a) > 0$ then

$$U(a) = U(a/\bigcirc a),$$

and so (by the definition of conditional expected utility, noting that for any action *a*, $\Pr(\oplus a/\bigcirc a) = 1$)

$$U(a) = \sum_w ([\Pr(a \square\!\!\rightarrow w)/\bigcirc a] \cdot V(w)).$$

Given that options are certainly open, we have the consequence:

For any option *a*, $U(a) = \sum_w \Pr(a \square\!\!\rightarrow w) \cdot V(w)$.

Expected utility could be confined to options, but (as will be explained) at a cost in notational complexity for the first theorem proved hereunder.

The theory is a version of world Bayesianism, according to which rational actions maximize probability-weighted averages of values of total consequences – "worlds" that are complete at least with respect to all things of relevance in the theory. Given their limited completeness it seems a safe, and not merely convenient, assumption that for every world Bayesianism and every agent there are countably many worlds.

2. CAUSAL DECISION THEORIES NEED PARTITION THEOREMS

It will be useful to have before us an argument to show that there is a sense in which causal decision theories require partition theorems. I offer a reprise of an argument I have given elsewhere:

> I am sure ... that I confront ... a machine that delivers one dollar if its left lever is pulled, and a separate jackpot of ten dollars depending on the dispositions of its left and right levers and on how the machine has been set up.... [In one possible set-up] the jackpot is delivered unless both levers are pulled (in which case it is not delivered); [in the other] the jackpot is delivered unless only the left lever is pulled (in which case it is not delivered). Which possibility is realized depends on the actions of a predictor ... in whose predictive capacities I have total confidence. He has set the machine in the first way if and only if he has predicted that I will *not* pull the right lever, and in the *second* way if and only if he predicted that I *will* pull this lever.
>
> Somewhat impetuously I have decided to pull the left lever [and] am sure that I *will* pull this lever.... Let us consider my present expected utilities for the action L and for the ways in which I might do it, with and without R. [We, not the agent "I," consider the expected utility of L. "I" have decided to do it and am presumably not about to question its usefulness.] For definiteness suppose that I consider it *as likely as not* that I will pull the right lever as well as the left.... So I consider [the two possible set-ups] equally likely. Assume finally that my expected utilities [equal] my expected monetary returns. (Sobel 1985a, p. 174)

The machine works with a light beam that activates a release mechanism for the jackpot if it makes contact with a light-sensitive cell, of which there are two. The beam is screened off by a *pair* of screens, one of which blocks half of the spectrum and the other of which blocks the other half. (The screens are depicted by horizontal lines in the boxes: each level is attached to only two screens. See Figure 1.) The dollar is delivered if the left lever is pulled; it is then simply pushed off a shelf. There are two lights and two light-sensitive cells: the left light is on in set-up S1, and the right in S2. A cell, if activated, triggers a mechanism that frees the activated cell's end of a shelf that drops if either end is unattached, and lets the $10 jackpot slide down to be collected. Each level can either be left in place or pulled out exactly one "position." Money freed up falls out of the machine at the bottom. In set-up S1, $10 is delivered if and only if not both levers are pulled: $J \leftrightarrow \sim(L \& R)$. In set-up S2 it is delivered if and only if not only the left lever is pulled: $J \leftrightarrow \sim(L \& \sim R)$.

I am sure that I am going to pull the left lever. So I am sure that either: (1) I will pull only the left lever and not also the right one, the machine is set in the first way (the way the predictor sets it if he predicts that I will not pull the right lever), and I will get both the dollar and the jackpot - ($L \& \sim R \& S1 \& D \& J$); or (2) I will pull both the left lever and the right one, the set-up is S2 (the way the predictor sets it if he predicts that I will

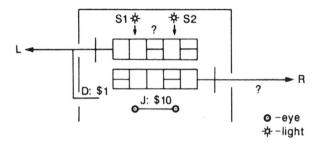

Figure 1

pull the right lever), and again I get both the dollar and the jackpot – $(L \& R \& S2 \& D \& J)$. And so my expected utility for L is 11: I am sure that if I pull the left lever (as I am sure I will) then I will get $11 whether or not I also pull the right lever – $\Pr[L \,\square\!\!\rightarrow (D \& J)] = 1$. (I should feel pretty good about my decision to pull the left lever. It seems to be working out as well as I might have hoped.) In contrast, since I consider the two possible set-ups S1 and S2 equally likely (which follows from my considering R and $\sim R$ equally likely and from my confidence in the predictor, by which $\Pr(S1/\sim R)$ and $\Pr(S2/R)$ are both 1), it follows that my expected utility for $(L \& R)$ is 6: For me, $\Pr[(L \& R) \,\square\!\!\rightarrow (D \& \sim J)]$ and $\Pr[(L \& R) \,\square\!\!\rightarrow (D \& J)]$ are both .5. Similarly, my expected utility for $(L \& \sim R)$ is 6. So $U(L)$ is not any kind of average of $U(L \& R)$ and $U(L \& \sim R)$, the expected utilities of L in conjunction with the members of the partition $\{R, \sim R\}$; in particular,

$$U(L) \neq \Pr(L \,\square\!\!\rightarrow R) U(L \& R) + \Pr(L \,\square\!\!\rightarrow \sim R) U(L \& \sim R).$$

These calculations of expected utilities $U(L \& R)$ and $U(L \& \sim R)$ are intuitive and proceed in terms of a natural principle for analyses of expected utilities under partitions; indeed, the calculations proceed in terms of a principle to be derived shortly. I have allowed myself to look ahead in this way, in order to avoid complications required for calculations that would run explicitly in terms of worlds without implicit dependence on partition principles. Since the partition principle I have used is derivable, it is of course necessary that, given sufficient additional specifications for our case, calculations could have been based directly on the definition of expected utility in terms of worlds. This is done in Sobel (1985a, pp. 176–7), where a formal model corresponding to the case is set out; in this model, values and probabilities are assigned to worlds, "world-nearness" relations are specified, and assumptions concerning openness are made. The preceding calculation of $U(L)$ is already in terms of world values:

What has been brought out, in effect, is that if $\Pr(L \boxdot\!\!\to w) > 0$ then w is a $(D \& J)$-world.

What has been shown is that only *restricted* Jeffrey-style partition principles can be theorems for "expected utility" (as distinct, of course, from Desirability as understood by Jeffrey). It is in this sense, relative to Jeffrey-style partition principles, that *causal* decision theories need partition theorems. Properly restricted theorems will of course not be applicable to L under $\{R, \sim R\}$. Two theorems for special partitions – theorems similar to ones demonstrated in the next section and in the appendix – are derived in Sobel (1985a), and it is shown that the partition $\{R, \sim R\}$ is of neither of these kinds: "$(R, \sim R)$ is not *sufficiently fine* relative to L since not both $(L \& R)$ and $(L \& \sim R)$ are U-elementary. Indeed neither of these conjunctions is U-elementary: though $U(L \& R) = 6$, $U(L \& R \& S1) = 1$; and though $U(L \& \sim R) = 6$, $U(L \& \sim R \& S1) = 11$. The partition $(R, \sim R)$ does *not* make all 'value-relevant' distinctions relative to L" (Sobel 1985a, p. 178). "[T]he partition $(R, \sim R)$ is plainly not *sufficiently exclusive* relative to L.... [I]t is *not* certain that not both $(L \& R)$ and $(L \& \sim R)$ are open; indeed, it is certain that both *are* open" (Sobel 1985a, p. 180, with changes in emphasis).

A final comment adapted nearly verbatim from Sobel (1985a, pp. 175–6): In the case, "I" make a mistake. I should not have decided to pull the left lever and then wondered what to do about the right one. I should instead have wondered from the start what to do about *both* levers, and confronted in the beginning the problem of choosing from the compound actions $(L \& R)$, $(L \& \sim R)$, $(\sim L \& R)$, and $(\sim L \& \sim R)$, all of which (we can assume) were options for me. Of these, $(\sim L \& R)$ and $(\sim L \& \sim R)$ had expected utilities of 10, as did their disjunction $\sim L$, since these actions would have guaranteed my getting the jackpot while I passed up the sure-thing-if-L dollar. In contrast, $(L \& R)$ and $(L \& \sim R)$ had expected utilities of 6, as did their disjunction L in the beginning: I assume that in the beginning the two set-ups were equally probable. In the beginning the maximum expected utility for my options was evidently 10, and I should have decided *not* to pull the left lever, and to pull the right lever or not as the spirit moved me. (I am assuming not only that all disjunctions of the four compound actions were certainly open, but also that they were all certainly open in distinctively disjunctive ways so that, for example, I was certain that I could have opted for $\sim L$ without opting for either $(\sim L \& R)$ or $(\sim L \& \sim R)$. In terms of Sobel 1983 [Chapter 10, Section 4], I am assuming that the disjunctions were the subjects of my certainly possible "precise choices." According to principle R5 of that paper, my rational precise choice was for $\sim L$, which choice had – amongst certainly possible precise choices – maximum expected utility, and a decision for which choice would have been ideally stable. And according to that principle, given that my rational precise choice was for $\sim L$, the *action* $\sim L$ was rational, as were its disjuncts $(\sim L \& R)$ and $(\sim L \& \sim R)$. They get counted as rational even though decisions for *them* would not have been ideally stable.)

3. Two partition theorems

3.1

The first theorem to be proved here analyzes the utility of an option in terms of its utilities *in conjunction with circumstances* in a certain partition of possible circumstances. The second analyzes its utility in terms of its utilities *conditional on circumstances* in any partition of possible circumstances. We begin with several definitions and lemmas for the first theorem.

Definition. C is a *practical partition* for an option a if and only if the agent is sure that there is exactly one c in C such that $(a \,\square\!\!\rightarrow c)$. That is, at each positively probable world w, there is exactly one c in C such that $(a \,\square\!\!\rightarrow c)$.

Lemma 1. *If C is a practical partition for an option a, then, if*
$$\Pr(a \,\square\!\!\rightarrow w) > 0,$$
there is exactly one c in C such that w is an $(a \& c)$-world.

Proof. Suppose w were not only an $(a \& c)$-world but also, for some c' in C distinct from c, and $(a \& c')$-world. Then, since at some positively probable world it is true that $(a \,\square\!\!\rightarrow w)$, at this positively probable world it would also be true both that $(a \,\square\!\!\rightarrow c)$ and $(a \,\square\!\!\rightarrow c')$. So C would not be a practical partition for a. But, by hypothesis, C *is* a practical partition.

Lemma 2. *For any option a, if $\Pr[\bigcirc(a \& c)] = 0$ then, for every $(a \& c)$-world w, $\Pr(a \,\square\!\!\rightarrow w) = 0$.*

Proof. At every positively probable world, we have by hypothesis
$$\sim\!\bigcirc(a \& c),$$
since it is given that $\Pr[\bigcirc(a \& c)] = 0$. Thus at every positively probable world we have, for every $(a \& c)$-world w, the negation $\sim(a \,\square\!\!\rightarrow w)$. (Assume, for indirect argument, that $(a \,\square\!\!\rightarrow w)$. It then follows that $[a \,\square\!\!\rightarrow (a \& c)]$, and, given that a is an option, it follows by our special principle CO that $\bigcirc(a \& c)$, which contradicts the hypothesis $\sim\!\bigcirc(a \& c)$.)

Definition. C is a *sufficiently exclusive partition* for an option a if and only if (i) for each c in C, $(a \& c)$ is an "action," and (ii) for any distinct c and c' in C, the agent is sure that not both $(a \& c)$ and $(a \& c')$ are open. That is, $\Pr(\sim[\bigcirc(a \& c) \& \bigcirc(a \& c')]) = 1$ or, equivalently, $\Pr[\bigcirc(a \& c) \& \bigcirc(a \& c')] = 0$.

Lemma 3. *For option a, practical and sufficiently exclusive for a partition C, c in C, and $(a \& c)$-world w:*

$$\Pr(a \,\square\!\!\rightarrow w) = \Pr(\bigcirc(a \& c) \& [(a \& c) \,\square\!\!\rightarrow w]).$$

Proof. At any positively probable world:

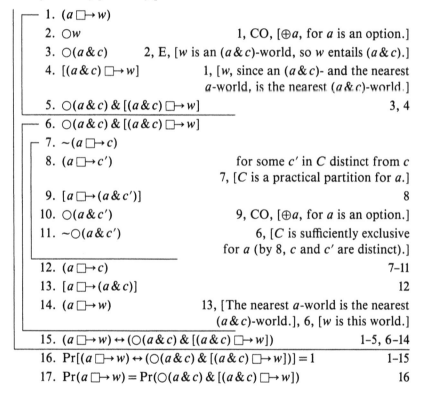

Lemma 4. *For option a, practical and sufficiently exclusive for a partition C, and c in C:*

$$\Pr[\bigcirc(a \& c)] = \Pr(a \,\square\!\!\rightarrow c).$$

A proof of this lemma, in the style of the proof of Lemma 3 but simpler than that proof, is possible. We proceed now to our first partition theorem.

Theorem 1 (Utilities of options under sufficiently exclusive practical partitions). *For option a and for C that is a practical and sufficiently exclusive partition for a,*

$$U(a) = \Sigma_{c: c \in C \,\&\, \Pr[O(a\&c)] > 0} [\Pr(a \,\square\!\!\rightarrow c) \cdot U(a\&c)].$$

Proof.

1. $U(a) = \Sigma_w [\Pr(a \,\square\!\!\rightarrow w) \cdot V(w)]$ ExpUt, [a is an option.]

2. $= \Sigma_{c \in C} \Sigma_{w \in (a\&c)} [\Pr(a \,\square\!\!\rightarrow w) \cdot V(w)]$
   Lemma 1, [C is a practical partition for a.]

3. $= \Sigma_{c: c \in C \,\&\, \Pr[O(a\&c)] > 0} \Pr[O(a\&c)]$
   $\cdot \Sigma_{w \in (a\&c)} [(\Pr(a \,\square\!\!\rightarrow w)/\Pr[O(a\&c)]) \cdot V(w)]$ Lemma 2

4. $= \Sigma_{c: c \in C \,\&\, \Pr[O(a\&c)] > 0} \Pr(a \,\square\!\!\rightarrow c)$
   $\cdot \Sigma_{w \in (a\&c)} \left[\dfrac{\Pr(O(a\&c) \,\&\, [(a\&c) \,\square\!\!\rightarrow w])}{\Pr[O(a\&c)]) \cdot V(w)} \right]$

   Lemmas 3 and 4, [C is a practical and sufficiently exclusive partition for option a, and w is an $(a\&c)$-world.]

5. $= \Sigma_{c: c \in C \,\&\, \Pr[O(a\&c)] > 0} \Pr(a \,\square\!\!\rightarrow c)$
   $\cdot \Sigma_w \left[\dfrac{\Pr(O(a\&c) \,\&\, [(a\&c) \,\square\!\!\rightarrow w])}{\Pr[O(a\&c)]) \cdot V(w)} \right]$

   [For $w \notin (a\&c)$, $\Pr[(a\&c) \,\square\!\!\rightarrow w] = 0$ and so $\Pr(O(a\&c) \,\&\, [(a\&c) \,\square\!\!\rightarrow w]) = 0$.

6. $U(a) = \Sigma_{c: c \in C \,\&\, \Pr[O(a\&c)] > 0} [\Pr(a \,\square\!\!\rightarrow c) \cdot U(a\&c)]$
   CondPr, CondExpUt, [Since C is sufficiently exclusive, $(a\&c)$ is an action, and $\Pr[\oplus(a\&c)/O(a\&c)] = 1$.], ExpUt.

A theorem for a theory suited to agents who believe in objective chances is proved in Sobel (1986a [Chapter 8]), which theory and theorem I would now change in line with some of the preceding developments.

Were expected utility confined to options, this theorem would need to take a somewhat more complex form that provided analyses of expected utilities of options – not in terms of the expected utilities of their conjunctions with circumstances in partitions, but rather in terms of expected utilities of these conjunctions *on condition of their being open*. In Theorem 1, $U(a\&c)$ would be replaced by $U[(a\&c)/O(a\&c)]$.

Note on adequate partitions. Let a partition C be *adequate* for an option a if and only if

$$U(a) = \Sigma_{c: c \in C \,\&\, \Pr[O(a\&c)] > 0} [\Pr(a \,\square\!\!\rightarrow c) \cdot U(a\&c)].$$

Adequacy, as here defined, relates to Jeffrey-style analyses of expected utilities of options. It is clear that not every adequate partition is sufficiently exclusive, because sufficiently fine partitions are adequate (Sobel 1985a, 1986a [Chapter 8]) and not every sufficiently fine partition is also sufficiently exclusive. Is every adequate partition, then, either sufficiently exclusive or sufficiently fine? No. For this point, consider the partition {Left, Not left} in a Newcomb problem (that is, the partition into worlds in which I take with only my left hand, and all other worlds – suppose the boxes are small enough so that I can take both boxes with only my left hand, as well as take only box 2 with this hand). Given natural stipulations, this partition is neither sufficiently fine nor sufficiently exclusive, but it *is* adequate relative to both options. It suffices for its adequacy relative to Both, for example, that $U(\text{Both \& Left}) = U(\text{Both \& Not left}) = U(\text{Both})$; that is, it is sufficient for its adequacy that it does not matter whether I use only my left hand. For then it will be that

$$U(\text{Both}) = \Pr(\text{Both} \,\square\!\!\rightarrow \text{Left})U(\text{Both \& Left})$$
$$+ \Pr(\text{Both} \,\square\!\!\rightarrow \text{Not left})U(\text{Both \& Not left});$$

clearly, $\Pr(\text{Both} \,\square\!\!\rightarrow \text{Left}) + \Pr(\text{Both} \,\square\!\!\rightarrow \text{Not left}) = 1$, which is sufficient for the adequacy (for what it's worth) of {Left, Not left} relative to Both.

I know of no simple characterization of adequate partitions that, like {Left, Not left} under suitable stipulations, are neither sufficiently exclusive nor sufficiently fine, other than this very relatively simple characterization. I know of no simple characterization of adequate partitions that are neither sufficiently fine nor sufficiently exclusive which does not say right out that they are adequate. But if, as I think, all natural adequate partitions are sufficiently exclusive, then simple characterizations of these other adequate partitions (that do not say right out that they are adequate) would be of only limited, theoretical interest.

3.2

For our second partition theorem, we start with a definition and a lemma.

Definition. C is a *probability partition* if and only if, for every positively probable world w, exactly one c in C is true at w.

Lemma 5. *For any probability partition C and proposition a,*

$$\Pr(a) = \Sigma_{c \in C} \Pr(c \,\&\, a).$$

We come now to the theorem and its proof.

Theorem 2 (Utilities of options under probability partitions). *For any option a and probability partition C,*

$$U(a) = \sum_{c:\, c \in C\ \&\ \Pr(c) > 0} [(\Pr(c) \cdot U(a/c)].$$

Proof.

1. $U(a) = \sum_w [\Pr(a \,\square\!\!\rightarrow w) \cdot V(w)]$ ExpUt, [*a* is an option.]
2. $= \sum_w \sum_{c \in C} [\Pr[c\,\&\,(a \,\square\!\!\rightarrow w)] \cdot V(w)]$
   Lemma 5, [*C* is a probability partition.]
3. $= \sum_w \sum_{c:\, c \in C\ \&\ \Pr(c) > 0} [\Pr[c\,\&\,(a \,\square\!\!\rightarrow w)] V(w)]$
4. $= \sum_{c:\, c \in C\ \&\ \Pr(c) > 0} \Pr(c) \cdot \sum_w [(\Pr[c\,\&\,(a \,\square\!\!\rightarrow w)]/\Pr(c)) \cdot V(w)]$
5. $U(a) = \sum_{c:\, c \in C\ \&\ \Pr(c) > 0} [\Pr(c) \cdot U(a/c)]$
   CondExpUt, [Since *a* is an option, $\Pr(\oplus a/c) = 1$.]

A stronger theorem for a theory suited to agents who believe in objective chances can be proved. In what follows I refer to Theorem 2 as *Partitions for U* (PFU) and to Theorem 1 as *Exclusive Partitions* (EP).

It is noteworthy that, for the ultimate partition into worlds, Theorem 2 (PFU) has the corollary

$$U(a) = \sum_{w \in \Pr(w) > 0} [\Pr(w) \cdot U(a/w)].$$

Furthermore, since $U(a/w)$ is (intuitively) $u_w(a)$, the actual utility of *a* at world *w*, we have the corollary

$$U(a) = \sum_{w:\, \Pr(w) > 0} [\Pr(w) \cdot u_w(a)].$$

It is then a short step to

$$U(a) = \sum_x (\Pr[u(a) = x] \cdot x),$$

where

$$\Pr[u(a) = x] = \sum_{w:\, u_w(a) = x} \Pr(w).$$

That is, it is an easy corollary of Theorem 2 that *the expected utility of an option is a probability-weighted average of its possible actual utilities,* which is, of course, exactly as it should be.

4. Fishburn and conditional acts

Fishburn's principle,

$$u(xA) = \sum_i \Pr(A_i/A) u(xA_i),$$

becomes, I assume,

$$u(x) = \sum_i \Pr(A_i) u(xA_i)$$

when event A is the universal set in $\mathcal{E}$, an algebra of relevant events. However, the resemblance of this principle to PFU is largely superficial, since u does not measure conditional preferences but rather *preferences for conditional acts,* which, in the idiom of logic, are acts conjoined with conditions. Whereas U measures preferences *on* conditions for acts, u measures preferences for acts *in* conditions.

In substantiation, I note that Fishburn takes for his framework sets S, $\mathcal{E}$, and F of states, consequences, and certain functions from states to consequences. The elements of F "would ... be the acts of real interest to the decision maker" (Fishburn 1973, p. 5), and "the state that obtains ... does not depend on the selected act" (p. 1). In a Prisoners' Dilemma, F would presumably include confessing, not confessing, and nothing else; "f_A denotes the restriction of f to A ... a function from A to $\mathcal{C}$ with $f_A(s) = f(s)$ for all $s \in A$" (p. 3). *Conditional acts* are "restrictions f_A of acts $f \in F$ on non-empty events $A \in \mathcal{E}$" (p. 5), $\mathcal{E}$ being an algebra of subsets of states. Conditional acts in a Prisoners' Dilemma will be confessing alone. The formula "$(xA) > (xB)$" could, Fishburn indicates, say "that you would rather eat harmless mushrooms (xA) then eat poisonous mushrooms (xB)" (pp. 9-10). He tells us that, in general, "We interpret xA ... as whatever might happen if you do x ... *and A obtains*" (Fishburn 1981, p. 177, emphasis added).

For further evidence that conditional acts are, for Fishburn, acts conjoined with conditions, we consider his axiom,

A4: $A \& B = \emptyset \& xA \geqslant xB \rightarrow xA \geqslant x(A \vee B) \geqslant xB$.

Fishburn (1973, p. 9) characterizes A4 as "a generalized version of Bolker's averaging condition," which is

if A and B are disjoint then $A \Pr B$ implies $A \Pr (A \vee B) \Pr B$ and $A \sim B$ implies $A \sim (A \vee B) \sim B$. (Bolker 1967, p. 336)

Taking xA to be (in effect) the conjunction $(x \& A)$, A4 is in fact nearly an instance of Bolker's condition: in particular, $x(A \vee B)$ becomes the conjunction $[x \& (A \vee B)]$, which is logically equivalent to the disjunction $[(x \& A) \vee (x \& B)]$.

There is, however, a problem for Fishburn's A4 on this interpretation. While it is correct for preferences for *news,* it seems wrong for preferences for *facts,* and one supposes that Fishburn would have said that he was axiomatizing the latter kind of preference. For a prima facie case against A4 as a condition on preferences for facts, we have the lever-case set out in Section 2. In this case, $U(L) = 11$, $U(L \& R) = 6$, and $U(L \& \sim R) = 6$. Given that $U(L) = U[L \& (R \vee \sim R)]$, we have

$$(L\&R) \geqslant (L\&\sim R),$$

but, contrary to A4, *not* that

$$(L\&R) \geqslant [L\&(R \vee \sim R)] \geqslant (L\&\sim R).$$

What follows from this confrontation of A4 with the lever case is not, however, that A4 is wrong in the context of Fishburn's theory, but rather that (probably unnoticed by Fishburn) A4 implies restrictions on $\mathcal{E}$, the set of conditioning events, in relation to acts – for example, that partitions constituted of events from $\mathcal{E}$ should, in my terms, be sufficiently exclusive relative to acts in F. Fishburn (1973, p. 2) describes $\mathcal{E}$ as a set of "relevant events." He does not elaborate on this idea in that paper, though he writes in Fishburn (1974, p. 27) that states, and thus events, are "not subject to the agent's control." That should ensure that partitions of events are sufficiently exclusive relative to acts.

To make plain the distance between Fishburn's

$$u(x) = \sum_i \Pr(A_i) u(x \& A_i)$$

and PFU, and to relate the problems of axiom A4 to this principle of Fishburn's, I note that (assuming no restrictions on conditioning events A) the U-analog of Fishburn's principle,

$$U(x) = \sum_i \Pr(A_i) U(x \& A_i),$$

is not valid. For valid principles, one can (i) replace $U(x \& A_i)$ by $U(x/A_i)$, or (ii) restrict the displayed principle to partitions that are *both* independent of x in the sense that $\Pr(A_i) = \Pr(x \square \!\!\rightarrow A_i)$ as well as sufficiently exclusive relative to x.

For completeness, we observe that the Des analog of this principle of Fishburn's,

$$\text{Des}(x) = \sum_i \Pr(A_i) \text{Des}(x \& A_i),$$

is also not valid, though in its case it is sufficient, for a valid principle, to impose a restriction on it to partitions that are independent of x in the sense that $\Pr(A_i) = \Pr(A_i/x)$. I note that while Fishburn says that states, and thus events, are to be *causally* independent of acts (Fishburn 1974, p. 25), he nowhere says that they are also to be probabilistically or epistemically independent of acts.

5. Armendt on conditional preferences

5.1

Partitions for U is similar to Armendt's principle for logical partitions B:

$$U(A) = \sum_i \Pr(B_i) U(A, B_i),$$

where $U(-,-)$ measures conditional preferences – that is, preferences on conditions for acts. The relation of this principle to PFU depends on the nature of the conditional preferences to which his principle is addressed. Armendt, however, provides "less than a complete account of conditional preference" (1986, p. 11). I set out things he says about conditional preferences in the paper just cited, indicate four possible explications of his concept, and then comment on these in relation to his theoretical purposes. One seems to suit his purposes, and another, even if not equivalent to this one, *may* suit them.

Preferences for propositions may be thought of as ... preferences for news. (p. 8)

[An agent's] conditional preferences for P, Q and for m, Q (where m is a mixture) are understood as preferences the agent has for P and for m, under his hypothesis that Q is true. [Conditional preferences are not preferences for conditional propositions.] And P, Q is understood as the agent's *present* preference for P, under his hypothesis that Q, not (necessarily) as the preference he *would have* for P if P were true, or if he were to *believe* Q is true. [A conditional preference need not have the same ranking as] an unconditional preference for a corresponding conjunction.... We may think of unconditional preferences for P ... as preferences for the actual world's being a P-world.... [R]egarding preference as preference-for-news, may sometimes be useful in thinking about conditional preference.... But for many conditional preferences this metaphor will not be useful. (p. 9)

Since to hypothesize is not to acquire news, preference is not perturbed by way of alterations in the *degree of belief* in the information contained in P. So a preference for P, P need not be equal to the preferences [for] T, P or T (where T is the valid proposition).... But sometimes the hypothesis that the actual world is a P-world may perturb the value P has in the actual world, since the hypothesis may carry information about states (the agent believes are) correlated with P. (p. 10)

5.2

Alternative explications of Armendt's remarks on conditional preference can take the form of definitions of possible *measures* of such preferences. As a preliminary, we distinguish ways in which a condition can operate to determine an epistemic perspective. Let Pr be a probability function. Then $\Pr_c$ comes from Pr by *conditionalizing on c* if and only if, for every proposition p,

$$\Pr_c(p) = \sum_{w \in (p)} \Pr(w/c).$$

By this rule, $\Pr_c$ is the probability function such that $\Pr_c(p) = \Pr(p/c)$. "Conditionalizing Pr on c gives a minimal revision in this ... sense: unlike all other revisions of Pr to make c certain, it does not distort the profile fo probability ratios ... among sentences that imply c" (Lewis 1976, p. 311; I have put "c" for "A" and "Pr" for "P". For a probability function Pr, $\Pr_{I(c)}$ comes from Pr by *imaging on c* if and only if, for every proposition p,

$$\Pr\nolimits_{I(c)}(p) = \sum_{w \in (p)} \Pr(c \mathbin{\Box\!\!\rightarrow} w),$$

where $(c \mathbin{\Box\!\!\rightarrow} w)$ is a Stalnaker conditional. By this rule, $\Pr_{I(c)}$ is a probability function such that $\Pr_{I(c)}(p) = \Pr(c \mathbin{\Box\!\!\rightarrow} p)$. "Imaging Pr on c gives a minimal revision in this sense: unlike all other revisions of Pr to make c certain, it involves no gratuitous movement of probability from world to dissimilar world.... [Every share stays as close as it can to the world where it was originally located]" (Lewis 1976, p. 311). Different ways of resolving the vagueness of counterfactuals and the parallel vagueness of the relations of nearness or similarity of worlds (see Lewis 1979b) determine different modes of imaging. We take $\Pr_{BI(c)}$ to result from imaging under a certain manner of backtracking, acausal correlating resolution of vagueness.

The following measures are to be universally defined, but restricted in the manners indicated only for certain propositions. For example, assuming standard conditional probabilities, U_1 is restricted in the manner indicated only for a and c such that $\Pr(a \& c) > 0$; for simplicity, the restrictions are not made explicit.

$$U_1(a, c) = \sum_w [\Pr_c(w/a) \cdot V(w)];$$
$$U_2(a, c) = \sum_w [\Pr_{BI(c)}(w/a) V(w)];$$
$$U_3(a, c) = \sum_w [\Pr_{BI(c)}(a \mathbin{\Box\!\!\rightarrow} w) \cdot V(w)];$$
$$U_4(a, c) = \sum_w [\Pr_c(a \mathbin{\Box\!\!\rightarrow} w) \cdot V(w)].$$

Of these measures, the first two, though suggested by some of Armendt's remarks, are inconsistent with others and are clearly not suited to his main purposes. The third fits many of his remarks, and may suit his purposes. The fourth seems suited to his main purposes, though it does not fit his remarks as well as the third does.

5.3

Before considering the first two measures, note that Armendt derives a measure for unconditional preferences by taking T, a valid proposition, as a condition: $U(a)$ is to equal $U(a, T)$. But then neither U_1 nor U_2 are suited to Armendt's purposes, for he seeks foundations for causal decision theory, thinks Jeffrey's evidential decision theory that would have an agent maximize Desirability gives wrong answers in some cases, and presumably favors a theory that would have one maximize Utility instead.

The problem with the first two measures (U_1 and U_2) is that, for every a, both $U_1(a/T)$ and $U_2(a/T)$ equal $\text{Des}(a)$. Regarding the second identity, note that $(T \mathbin{\Box\!\!\rightarrow} w)$ is true at world w and at no other world, and that $\Pr_{I(T)}(p) = \Pr(p)$ for every proposition under every resolution of vagueness.

Measure U_4 seems suited to Armendt's purposes, even if not to all of his remarks. It is, modulo restrictions not explicit in this section, U as defined in Section 1 and used in PFU. Central to Armendt's system is an averaging principle that figures prominently in his representation theorem: For incompatible A and B,

$$U(x, A \vee B) = \Pr_{A \vee B}(A) U(x, A) + \Pr_{A \vee B}(B) U(x, B).$$

I think it thus all but settles the credentials of U_4 for Armendt's purposes that (equating $\Pr_{A \vee B}(-)$ with $\Pr(-/A \vee B)$)

$$U_4(x, A \vee B) = \Pr_{A \vee B}(A) U_4(x, A) + \Pr_{A \vee B}(B) U_4(x, B)$$

can be proved from definitions by an argument very similar to the one I have used to prove PFU. Given that A and B are incompatible, it follows that

$$\Pr[(A \vee B) \& (x \,\square\!\!\rightarrow w)] = \Pr[A \& (x \,\square\!\!\rightarrow w)] + \Pr[B \& (x \,\square\!\!\rightarrow w)],$$

and from this, by the definition of U_4 and the equality $\Pr_{A \vee B}(x \,\square\!\!\rightarrow w) = \Pr[(A \vee B) \& (x \,\square\!\!\rightarrow w)] / \Pr(A \vee B)$, it follows that

$$U_4(x, A \vee B) = \sum_w [(\Pr[A \& (x \,\square\!\!\rightarrow w)] / \Pr(A \vee B)) \cdot V(w)]$$
$$+ \sum_w [(\Pr[B \& (x \,\square\!\!\rightarrow w)] / \Pr(A \vee B)) \cdot V(w)].$$

Armendt's averaging principle follows from this by elementary mathematics, probability theory, and the definition of U_4.

U_3 *may* be suited to Armendt's purposes. Whether or not it actually *is* suited depends, I think, on whether or not his averaging principle is valid for it – that is, on whether or not

$$U_3(x, A \vee B) = \Pr_{A \vee B}(A) U_3(x, A) + \Pr_{A \vee B}(B) U_3(x, B)$$

is provable from definitions. This principle is, of course, provable *if* for "relevant" p and q – for example, in connection with $U(x, A \vee B)$, for $A \vee B$ as p and $(x \,\square\!\!\rightarrow w)$ as q – the vagueness is resolved in such a manner that $\Pr_{\text{BI}(p)}(q)$ equals $\Pr_p(q)$ (i.e., $\Pr(p \,\square_{\text{BI}}\!\!\rightarrow q)$ equals $\Pr(q/p)$), so that $U_3 = U_4$. The question left open may thus be divided into two. First, is vagueness here to be resolved in a way that reduces U_3 to U_4? (See Lewis 1976 and 1986 for the perils of assuming $\Pr(p \,\square\!\!\rightarrow q) = \Pr(q/p)$ without restriction on p and q.) And second, if vagueness is not to be resolved in a manner that reduces U_3 to U_4, is Armendt's averaging principle, even so, provable for U_3?

5.4

For completeness, I recall here that Paul Weirich has proposed that, for any logical partition S,

$$U(a) = \sum_i \Pr(s_i/a) U(a/s_i \text{ if } a)$$

(Weirich 1980, p. 712). He explains that "$\Pr(s_i/a)$ is just the probability of" s_i given a, where "there are good reasons to suppose that [this] equals not $\Pr(a \& s_i)/\Pr(a)$ but rather the probability that if a were performed [then] s_i would obtain. See Allan Gibbard and William Harper..." (p. 707, n. 6). We are also told that "s_i if a" in the displayed formula is a Stalnaker conditional under an unspecified resolution of vagueness for which an analysis would "pursue the leading idea of Robert Stalnaker... and Gibbard and Harper" (p. 712, n. 9). And, from what he has written earlier in the paper, we gather that the "conditional utility" (or, we might say, the "doubly conditional utility") $U(a/s_i \text{ if } a)$ is not to be "what the... utility of the action would be if it were learned only that the [conditional] condition [that s_i if a] obtains" (pp. 704-5), but rather "the rational degree of desire, supposing that the condition obtains" (p. 705), where this is not "subjunctive supposing" but "indicative supposing." Weirich writes: "I will assume that, in conditional utilities, indicative supposition is appropriate for the condition and that subjunctive supposition is appropriate for the action" (p. 707). I recall that Weirich says that there are good reasons to think that $\Pr(s_i/a)$ (which occurs in the preceding formula) is not a conditional probability, but rather the probability of a conditional that involves (he plainly implies) subjunctive supposition after the manner of conditionals in the theory of Gibbard and Harper.

Weirich's principle is very different in form from both EP and PFU. How different it is in substance from each of these principles, and whether or not, like them, it is valid, are matters that depend on (among other things) how "indicative supposing" is to work, and how it is related to imaging and conditionalizing. Given explanations of the intended logics and semantics of supposing, indicative and subjunctive, it should be possible either to demonstrate Weirich's principle for utilities under partitions or to invalidate it. However, the explanations Weirich has so far provided are not sufficient for such exercises.

6. Partitions for U and Exclusive Partitions compared

Partitions for U is different in three ways from *Exclusive Partitions*. First, different *values* are averaged. In EP, the expected utility of an action is an average of its expected utilities *in* various circumstances, whereas in PFU what is averaged are expected utilities *on* epistemic conditions that various circumstances obtain. Second, different *weights* are employed. In EP, the weight assigned to the value of a, *in* circumstance c, is the unconditional probability of this circumstance on the causal condition that this action obtain – that is, the unconditional probability of the causal conditional ($a \square\!\!\rightarrow c$). In PFU, the weight assigned to the value of a, *on* circumstance c, is simply $\Pr(c)$ – the unconditional probability of this circumstance.

The third contrast is of course that, while EP does, PFU does *not* place a restriction on partitions to which its averaging principle is applicable. (The extensions of these theories to practical and probability partitions are not restrictions on, but licenses to go variously beyond, strictly logical partitions.) For example, PFU can be applied even to partition $\{R, \sim R\}$ in our lever case. Important for its application to $U(L)$ under this partition is the fact that, in this case, R and $\sim R$ are independent of L in a *strong* sense not relevant until now: both $\Pr_R[L \square\!\!\rightarrow (L \& R)] = 1$ and $\Pr_{\sim R}[L \square\!\!\rightarrow (L \& \sim R)] = 1$. (*Proof:* We have $\Pr(L) = 1$. From this it follows that $\Pr([L \square\!\!\rightarrow (L \& R)] \leftrightarrow R] = 1$. So $\Pr(R \& [L \square\!\!\rightarrow (L \& R)]) = \Pr(R)$, and so $\Pr_R[L \square\!\!\rightarrow (L \& R)] = 1$; similarly for $\Pr_{\sim R}[L \square\!\!\rightarrow (L \& \sim R)] = 1$.) While the expected utilities $U(L \& R)$ and $U(L \& \sim R)$ are both 6, given the just-noted independence, the *conditional* expected utilities $U(L/R)$ and $U(L/\sim R)$ are both 11. Hence $U(L)$, while it is *no* kind of average of $U(L \& R)$ and $U(L \& \sim R)$, is *every* kind of average of $U(L/R)$ and $U(L/\sim R)$, including the PFU average

$$\Pr(R) U(L/R) + \Pr(\sim R) U(L/\sim R).$$

(*Proof that* $U(L/R) = 11$: It is given that $\Pr(S2/R) = 1$, and it follows from this that $\Pr_R[(L \& R) \square\!\!\rightarrow (D \& J)] = 1$. But, by the independence of R from L in the way just noted, $\Pr_R[L \square\!\!\rightarrow (L \& R)] = 1$. And so we have $\Pr_R[L \square\!\!\rightarrow (D \& J)] = 1$, and thus $U(L/R) = 11$. For the next-to-last inference I rely on the principle that, for any p, q, and r, from $[p \square\!\!\rightarrow (p \& q)]$ and $[(p \& q) \square\!\!\rightarrow r]$ it follows that $(p \square\!\!\rightarrow r)$. By a similar argument, $U(L/\sim R) = 11$.)

7. Uses of partition theorems

7.1

Causal decision theories need partition theorems. But why is this – what do they need them for? Mainly, I think, in order that these theories should be humanly applicable in real and imagined (for purposes of theoretical discussion) cases. Averaging values of all of an action's possible worlds is not a humanly manageable task. But then, in many cases, neither is averaging the values of an action's "parts" under a partition that is sufficiently fine.

The need for partition theorems is, however, not equally satisfied by all partition theorems. What one seeks are theorems that underwrite familiar and natural matrix procedures for decision problems, theorems that legitimate small matrices for such simple decision problems as Newcomb's Problem and the Prisoners' Dilemma; PFU and EP are candidates for that work. Sufficiently fine partition theorems are not. Partitions that are really sufficiently fine, while possibly smaller than ultimate partitions into individual worlds, are in most cases like ultimate partitions in that they are still impractically large. As for the relative practical merits of our two candidates, I think EP has a certain clear advantage.

It is relatively natural and easy to consider and assess the expected utilities of actions *in* various circumstances, that is, in conjunction with various circumstances. In contrast, fixing the expected utilities of actions *on* various epistemic conditions can be a difficult task demanding considerable concentration. Consider, for example, Newcomb's Problem under the partition {C: the being has correctly predicted my action; $\sim C$: the being has not correctly predicted my action}. The problem's matrix for PFU is

	$\Pr(C)$	$\Pr(\sim C)$
Take both boxes	$U(\text{Both}/C)$	$U(\text{Both}/\sim C)$
Take just one box	$U(\text{One}/C)$	$U(\text{One}/\sim C)$

The probabilities are simple unconditional ones, and their values are obvious: nearly 1 and 0 respectively in a standard Newcomb Problem. But what about the conditional expected utilities in this matrix? What, for example, is the expected utility on the epistemic condition (i.e., from the epistemic perspective) that the being has correctly predicted one's action of taking both boxes? It can seem that this conditional expected utility must be the same as the simple utility of $1000, for that is what one gets by taking both boxes when the being *has* correctly predicted your action (and so has predicted that you will take both boxes). But that rationale is suited specifically to evaluating the expected utility of taking both *in* the circumstance in which the being correctly predicts your action, and so it should be viewed with extreme caution in the present context in which the issue is the expected utility of taking Both *on* the epistemic condition that the being correctly predicts your action. Regarding this issue, consider first that {$C, \sim C$} is not independent of the action 'Take both' in the sense that $\Pr_C[\text{Both} \,\square\!\!\rightarrow (\text{Both} \,\&\, C)] = 1$ and $\Pr_{\sim C}[\text{Both} \,\square\!\!\rightarrow (\text{Both} \,\&\, \sim C)] = 1$. Even if I were to become convinced that the predictor had correctly predicted what I was going to do (in a standard problem I am already nearly convinced), I should be sure that – were I to act other than I in fact will act (however I will act; I need not yet have made up my mind) – the being would not have correctly predicted my action. One consequence of this dependence of $\{C, \sim C\}$ on the action 'Take both' is that it can by no means be assumed that, for example, $U(\text{Both}/C) = U[(\text{Both}\,\&\,C)/C]$.

Consider next that if, as could be, I were quite sure that C ($\Pr(C) = 1$) then, by definitions, both

$$U(\text{Both}) = U(\text{Both}/C) \quad \text{and} \quad U(\text{One}) = U(\text{One}/C).$$

In this limit case, PFU, though applicable under $\{C, \sim C\}$, is of no help at all in determinations of $U(\text{Both})$ and $U(\text{One})$. Furthermore, since in standard Newcomb Problems I am nearly sure that C, one supposes that PFU under $\{C, \sim C\}$ can be of at most little help in them, for presumably $U(\text{Both}/C)$ and $U(\text{One}/C)$ are in these standard cases nearly $U(\text{Both})$ and $U(\text{One})$, respectively. Far from being of help, PFU would if anything be positively unhelpful in these cases, given the confusions – between $U(\text{Both}\,\&\,C)$ and $U(\text{Both}/C)$ and between $U(\text{One}/C)$ and $U(\text{One}\,\&\,C)$ – courted by applications of PFU under $\{C, \sim C\}$.

7.2

I conjecture that PFU is helpful and easy to apply only under partitions C such that, for each c in C, both (1) $U(a/c) = U(a\,\&\,c)$ and (2) $\Pr(a \,\square\!\!\rightarrow c) = \Pr(c)$; and that such partitions will always satisfy conditions for the kind of averaging conducted in EP. I conjecture more narrowly that, for an option a, PFU is helpful and easy to apply only under strongly independent partitions such that, for each c in C, (3) $\Pr[c \leftrightarrow \bigcirc(a\,\&\,c)] = 1$ and $\Pr[c \rightarrow (a\,\square\!\!\rightarrow c)] = 1$, from which condition (1) and (2) follow. Such partitions are, by condition $\Pr[c \leftrightarrow \bigcirc(a\,\&\,c)] = 1$, sufficiently exclusive.

But what is helpful and easy for one theorist may not be for another, and, of course, relative practical merits of partition theorems as instruments for applications of theories to cases are distinct from the relative theoretical merits of these theorems. Though of little practical value, PFU, given its unrestricted character, has certain advantages for theoretical purposes – for example, in connection with foundations for and axiomatizations of causal decision theories.

7.3

Of some interest are relations between PFU and Jeffrey's Desirability axiom, and I will close with comments on these. In general form, Jeffrey's axiom may be stated as follows.

Theorem (*Partitions for Des*). *For any a such that $\Pr(a) > 0$ and for any logical partition C,*

$$\text{Des}(a) = \sum\nolimits_{c:\, c \in C\,\&\,\Pr(c) > 0} [\Pr(c/a) \cdot \text{Des}(a\,\&\,c)].$$

What I wish to bring out is the way in which this theorem and PFU, and their analogs for U and Des respectively, are "symmetrical opposites." To explain, I set a definition.

Definition of conditional Desirability. For a and c such that $\Pr(c \& a) > 0$,
$$\mathrm{Des}(a/c) = \sum_w [\Pr_c(w/a) \cdot V(w)],$$
where $\Pr_c$ comes from Pr by conditionalization on c.

Given this definition, we have, for such a and c, the identity $\mathrm{Des}(a/c) = \mathrm{Des}(a \& c)$. In contrast with expected utility, Desirability *on* a condition is the same as Desirability *in conjunction with* that condition. Thus the Des analog of PFU,

for any a such that $\Pr(a) > 0$ and for any logical partition C,
$$\mathrm{Des}(a) = \sum_{c:\, c \in C \,\&\, \Pr(c \& a) > 0} [\Pr(c) \cdot \mathrm{Des}(a/c)],$$
is *not* a theorem, though a theorem results if a restriction is imposed to partitions that are independent in the sense that $\Pr(c) = \Pr(c/a)$ for every c in C.

Partitions for Des is a quite unrestricted partition theorem for evidential Desirability, though only restricted analogs of it are valid for causal utility. This, it has been suggested, is an important advantage for evidential decision theory; it is (so far) a structurally simpler theory. (See, e.g., Horwich 1985, pp. 444–5.) Neutralizing this structural advantage, however, is the fact that while PFU is a quite unrestricted partition theorem for causal utility, only restricted analogs of *it* are valid for evidential Desirability.

APPENDIX: A THEOREM FOR SUFFICIENTLY FINE PARTITIONS

Definition. For any proposition a, a is *U-elementary* if and only if if $\Pr(\bigcirc a) > 0$ then, for every proposition a' that entails a, if $\Pr(\bigcirc a') > 0$ then $U(a/\bigcirc a) = U(a'/\bigcirc a')$.

Definition. C is a *sufficiently fine partition* for an option a if and only if, for every c in C, $a \& c$ is U-elementary.

Theorem 3 (Utilities of options under sufficiently exclusive practical partitions). *For option a with C a practical and sufficiently exclusive partition for a,*
$$U(a) = \sum_{c:\, c \in C \,\&\, \Pr[\bigcirc(a \& c)] > 0} [\Pr(a \,\square\!\!\rightarrow c) \cdot U(a \& c)].$$

Proof.

1. a is an option
2. C is a practical partition for a
3. C is sufficiently fine for a
4. $U(a) = \sum_w \Pr(a \,\square\!\!\rightarrow w) \cdot V(w)$ ExpUt, 1
5. $\sum_{c \in C} \sum_{w \in c} \Pr(a \,\square\!\!\rightarrow w) \cdot V(w)$ 2, 4
6. $\sum_{c \in C} \sum_{w \in (a \& c)} \Pr(a \,\square\!\!\rightarrow w) \cdot V(w)$ 5
7. $\sum_{c \in C} \sum_{w:\, w \in (a \& c)\,\&\, \Pr(\bigcirc w) > 0} \Pr(a \,\square\!\!\rightarrow w) \cdot V(w)$
   1, 6, [When $\Pr(\oplus a) = 1$, if $\Pr(a \,\square\!\!\rightarrow w) > 0$ then $\Pr(\bigcirc w) > 0$.]
8. $\sum_{c \in C} \sum_{w:\, w \in (a \& c)\,\&\, \Pr(\bigcirc w) > 0} \Pr(a \,\square\!\!\rightarrow w) \cdot U(w/\bigcirc w)$
   7, [Since $w \in (a \& c)$, w is a version of an option; since w is a world, $\square(\oplus w \leftrightarrow \bigcirc w)$ and $\Pr(\oplus w / \bigcirc w) = 1$.]
9. $\sum_{c:\, c \in C\,\&\, \Pr[\bigcirc(a \& c)] > 0} \sum_{w:\, w \in (a \& c)\,\&\, \Pr(\bigcirc w) > 0} \Pr(a \,\square\!\!\rightarrow w) \cdot U(w/\bigcirc w)$
   1, 8, [For $w \in (a \& c)$ and $\Pr(\oplus a) = 1$, if $\Pr(a \,\square\!\!\rightarrow w) > 0$ then $\Pr[\bigcirc(a \& c)] > 0$.]
10. $\sum_{c:\, c \in C\,\&\, \Pr[\bigcirc(a \& c)] > 0} \sum_{w \in (a \& c)} \Pr(a \,\square\!\!\rightarrow w) \cdot U[(a \& c)/\bigcirc(a \& c)]$
                                                                                                   3, 9
11. $\sum_{c:\, c \in C\,\&\, \Pr[\bigcirc(a \& c)] > 0} \Pr(a \,\square\!\!\rightarrow c) \cdot U[(a \& c)/\bigcirc(a \& c)]$ 10
12. $\sum_{c:\, c \in C\,\&\, \Pr[\bigcirc(a \& c)] > 0} \Pr(a \,\square\!\!\rightarrow c) \cdot U(a \& c)$ 11, ExpUt

10

Expected utilities and rational actions and choices

Abstract. According to a simple Bayesian rule, an action is rational if and only if its expected utility is at least as great as that of each alternative action. Two refinements of this rule are made in this chapter. First, it is maintained that the rationality of an option requires not only that its expected utility be maximum but also that a decision for it be ideally stable. And second, problems with treating actions themselves as basic options and applying tests of utility directly to them are held to recommend treating as basic not the actions themselves but, rather, *choices* in which actions are made certain or given specific chances.

0. INTRODUCTION

Bayesian decision theories say that whether or not actions and choices are rational are issues decided by expected utilities. But we want to know just *which* expected utilities are decisive in these ways, and exactly *how* these expected utilities are decisive. As a first simple approximation, one might meet these demands by saying that rational actions and choices maximize expected utilities. This chapter is aimed at a more explicit and more exact response.

Section 1 concerns certain preliminaries. The concept of expected utility assumed for this study is explained and several simplifying assumptions and restrictions are stated. Section 2 deals with a part of the "how" question. It is maintained that to be rational it is necessary not only that an option should maximize expected utility, but that a decision for this option should be ideally stable or (roughly stated) after-the-fact maximizing. Subsequent sections deal mainly with the "which" question. Section 3 starts with the natural proposal that an action's rationality should depend on how its expected utility compares to the expected utilities of its alternatives. Difficulties with this proposal lead to the idea that tests of expected utility should be confined to fully specific actions, and then to the idea that they should be confined to fully specific minimal actions.

Reprinted with revisions by permission from *Theoria* 49 (1983), pp. 159–83.

Difficulties with all proposals that feature the expected utilities of actions lead, in Sections 4 and 5, to the idea that it is the expected utilities of choices that should be fundamental in determinations of rationality, where we understand a choice to be "an organization of will in which actions are made certain or given specific objective chances." Throughout this chapter I make a distinction between choices and actions, both overt actions such as depressing a lever and other actions such as silently repeating certain words of resolution; I avoid such phrases as "acts of choice."

1. Definitions, Assumptions, and Restrictions

Principles examined in Sections 3 through 5 are addressed only to agents who are possessed, at least "just before" times of action, of quite extensive and definite probabilities and preferences. Confidence of our agents in propositions is to be measured by a probability function defined for all propositions, a function based on a function defined for all worlds; and preferences are to extend to every pair of possible worlds, and to be represented by a bounded value function.

The *expected utility* of a proposition for an agent at a given time is to depend on his probability and value functions at this time, and to be a weighted average of the values of the worlds at which this proposition is true, the weight assigned to the value of a world being proportional to the probability of the proposition's coming true in that way or at that world. More explicitly, we assume for the present study the following analysis of expected utility.

For any logically possible proposition p,

$$U(p) = \sum_{w \in W(p)} [\Pr(p \,\square\!\!\rightarrow w) \cdot V(w)],$$

where $W(p)$ is the set of worlds at which p is true.

In this analysis, "$p \,\square\!\!\rightarrow w$" can be read "if it were the case that p, then it would be the case that q" and is to be understood as a causal conditional. This is like the theory of expected utility favored in Gibbard and Harper (1978), and is, I think, a superior analysis to that in Jeffrey (1965b) which runs in terms of conditional probabilities rather than probabilities of conditionals. Though, because it is in a certain sense less general, it is inferior to some other causal decision theories; it is the simplest such theory and it serves all present purposes. One consequence of this analysis of expected utility is that the expected utility of a proposition is a weighted average of the values of its "parts" under a logical partition that is sufficiently fine in the sense that any further distinctions would not resolve any part into parts that had different expected utilities. More precisely, if a proposition

p is *U-elementary* if and only if, for every possible proposition p' that entails p, $U(p) = U(p')$, then:

> For any possible proposition p and logical partition Q such that, for each q in Q, $p \& q$ is U-elementary,
>
> $U(p) = \Sigma_{q:\, q \in Q \text{ and } (p \& q) \text{ is possible}} [\Pr(p \,\square\!\!\rightarrow q) \cdot U(p \& q)]$.

Of particular relevance to the rationality of choices and actions that would take place or commence at a certain time would be the agent's beliefs and preferences just before this time. But if time is dense then there are no times "just before" given times. To meet this difficulty, we make issues of rationality depend on beliefs and preferences that are maintained throughout an interval bounded above by what would be the time of action or choice, during which period there are no changes in beliefs and preferences. This interval can be very short. To avoid circumlocutions we will speak of the beliefs and preferences that an agent maintains throughout such a period as his beliefs and preferences just before the period's least upper bound, and issues of rationality at t will be said to depend upon the agent's probability and world-value functions just before t.

Our principles make rationality and irrationality of actions and choices functions of expected utilities just before times of actions and choices. This means that, according to our principles, possession of probabilities and preferences that are settled and constant for a period leading up to a time is a necessary condition to any of an agent's actions' at this time being either rational or irrational. Our principles all imply that if an agent cannot in this sense "make up his mind," none of his actions can be rational or irrational; these categories apply only to agents who have quite extensive, determinate, and (at least for a short time) settled credences and preferences. This implication of our principles is of interest if there are, as I think there are, possible situations in which ideally rational agents – owing to their perfections – would not be *able* to make up their minds. Prominent among such situations are ones in which very rational agents would, if they could, interact. Related matters are taken up in Section 2.

Principles examined herein are restricted in another way. They are addressed only to agents who, just before a time of action or choice, are sure of what they can then do. It is not clear whether or not categories of rationality and irrationality extend beyond such agents, but in any case no attempt is made to extend them here, and the principles considered are designed only for such agents. Put in general terms, our principles apply to an agent only if there are things he is sure he can do that "cover" the things he thinks he might do. We now make precise a form of this restriction that suffices for the principles of Section 3. Let an action be *open* at a time if and only if its agent can then at will make its occurrence objectively certain. An action can be open at times before what would be its first moment. Let an action be at a time or during a period *certainly open* for an agent if and only if he is then subjectively certain that this action is or will be open to him (that is, that he is or will before or at this action's first moment be able to make it objectively certain). The precise requirement we now impose is that there be actions our agent is sure that he can do that cover all things he thinks he might do and make a set from which he is sure that he must choose. More exactly, for any time t at which an action is open to him, there is to be a set of actions with first moment t such that: each action in this set is certainly open to him just before t; every action with first moment t that he thinks just before t that he might do is a disjunct of an action in this set; and he is sure just before t that he must do some action from this set, perhaps because there is an action in this set that is such that he is sure just before t that not doing any other action in the set would be tantamount to doing it.

One more restriction recommends itself. Ordinary agents are sometimes certain of things that are not true. But to simplify matters, the principles examined here are addressed exclusively to agents who are unerring in their certainties. One important consequence of this restriction is that any action that is certainly open for one of our agents will actually be open.

2. The ideal stability of rational decisions

Maximizing expected utility may be necessary, but is not sufficient, to rationality. For it can happen that, though an action would maximize expected utility, no decision for it could be rationally sustained given effects that a decision for it would have on expected utilities.

Consider a fanciful case. Let me be sure just before t that I can drink water or not, and that I must do one of these or the other. I am sure that I cannot effectively randomize, and that I must make one or the other action certain. Assume further that I am sure that I have either disease A or disease B, but not both, and that I am sure that whether or not I drink water will (of course) have absolutely no effect on which disease I have. Suppose further that just before t I am nearly certain that I have disease A, not disease B. (Never mind what makes me nearly certain of this or whether this, or any, conviction of mine is reasonable.) Other things of which I am to be nearly certain include: that having disease A would, possibly for what I took to be good reasons, make me drink water; that having disease B would make me not drink water; that if I have disease A I will recover if and only if I do not drink; and that if I have disease B I will recover if and only if I do drink.

The case has this structure:

	Possible disease A	Possible disease B
Drink	Death	Life
Do not drink	Life	Death

Possible actions

Relevant probabilities just before t include

$$\Pr(D \square\!\!\rightarrow A) \simeq 1, \qquad \Pr(D \square\!\!\rightarrow B) \simeq 0,$$
$$\Pr(\sim\!D \square\!\!\rightarrow A) \simeq 1, \qquad \Pr(\sim\!D \square\!\!\rightarrow B) \simeq 0.$$

Given that my interest is almost exclusively in my health – that I place little value in drinking or not drinking just as such – it is clear that *not* drinking would maximize expected utility just before t, the time of action. This is mainly because I am nearly sure that I have disease A; as a consequence of this, near-certainty values of possible outcomes in the B-column get near-zero probability weights in expected utility calculations and so count for almost nothing.

But would my not drinking be *rational*? Might I decide rationally not to drink? I think not. For suppose I were to make this decision, reflect on my decision and firm intention not to drink, and be thereby nearly certain that I was not going to drink. Then I should be nearly certain that I had disease B, for my probabilities just before t would include

$$\Pr(\sim\!D\,\&\sim\!B) = \Pr(\sim\!D\,\&\,A), \qquad \Pr(\sim\!D\,\&\,A) \simeq 0,$$

and thus

$$\Pr(B, \sim\!D) \simeq 1$$

as well as

$$\Pr(A, \sim\!D) \simeq 0.$$

That is, if I were to revise my probabilities in a way that would be reasonable given my probabilities just before t, I would be nearly certain that I had disease B: given these probabilities, my not drinking at t would be for me a near-certain sign that I had disease B. But this means that were I nearly certain that I was not going to drink and to revise my probabilities accordingly by "approximately conditionalizing" on this near certainty, then not drinking would not remain maximizing; relevant probabilities would then include:

$$\Pr(D \square\!\!\rightarrow A) \simeq 0, \qquad \Pr(D \square\!\!\rightarrow B) \simeq 1,$$
$$\Pr(\sim\!D \square\!\!\rightarrow A) \simeq 0, \qquad \Pr(\sim\!D \square\!\!\rightarrow B) \simeq 1.$$

Supposing that I am a moderately reflective and rational agent, a decision not to drink would not be *stable*. And I think that it follows from this that a decision not to drink would not be *rational:* It would not be rational even if it would in fact be stable for me because I am rather unreflective, or because I would not have time to reflect, or . . . .

I hold that not drinking would not be rational in this case. For similar reasons I think that drinking would not be rational either. I think that in this case *no* action certainly open to me would be rational, because no decision for any such action would be ideally stable. The general conclusion illustrated is that an action or choice would be rational only if a decision for it would be ideally stable. This stability condition is incorporated in principles examined in coming sections by a clause framed in terms of *conditional expected utilities,* which I now define.

For any logically possible proposition q and proposition p such that $\Pr(p) > 0$,

$$U(q/p) = \sum_{w \in W(q)} \Pr[(q \,\square\!\!\rightarrow w)/p] \cdot V(w).$$

What is rational for a possibly imperfect agent cannot, I think, in general be identified with what would be rational for him, and what he would decide to do, were he perfectly rational. But I think that something like this is true, and that what is rational for an agent must be such that a decision for it would be ideally stable, and so must be something he *could* decide to do even if he were perfectly rational. (Consider Gibbard and Harper 1978, sec. 11; Jeffrey 1981b, sec. VI; and Sobel 1982.)

3. Principles that apply tests of expected utility to actions

I have argued that maximizing expected utility is not sufficient to rationality, but it can seem that for a Bayesian it should be at least necessary. An action is rational on such a theory, one supposes, only if its expected utility would be at least as great as that of each of its alternatives, and an action is rational if it would in this way maximize expected utility and a decision for it would be ideally stable. We begin our investigation of just which expected utilities are decisive for rational actions with this simple and natural idea. After spelling it out, I argue against it. It can give wrong results for some complex, disjunctive actions. Difficulties with this initial idea lead to others which we examine in turn.

3.1

Let an action x' be an *alternative* to x if and only if x and x' would have the same agent and first moment and cannot both take place. Our simple idea can be spelled out as follows.

R1 *An action x that is certainly open just before its first moment t is rational* if and only if, for each action y that is certainly open just before t, if the agent of x is sure just before t that y is an alternative to x then

$$U_t(x) \geq U_t(y),$$

where U_t assigns to propositions the expected utilities they have for the agent just before t, and

$$U_t(x/x) \geq U_t(y/x),$$

where $U_t(-/x)$ assigns to propositions conditional-on-x expected utilities they have for the agent just before t.

(As predicted in Section 1, principle R1 is suited only to agents for whom there are actions they are sure they can make certain, which actions constitute sets from which they are sure they must choose. Suppose that an agent does not satisfy this condition, and that – though actions of his are open for t – no actions are certainly open just before t for him. Then, regardless of the merits of actions this agent thinks (but is not sure) are open to him, principle R1 makes no action with first moment t rational, and makes no such action not rational. Or suppose that, while a given agent is sure that several actions are open to him, the actions he is sure are open to him do not make a set from which he is sure he must choose. Then, according to principle R1, the best of the actions he is sure he can choose is rational, regardless of the merits of actions he thinks but is not sure he can choose instead. Principle R1 is suited only to the agents described in Section 1. It is specific to these agents and can have strange, and worse than strange, consequences for other agents. Similar points hold for principles R2-R5, but will not be made explicit.)

Principle R1 encounters difficulties in some cases where expected utilities of disjunctions exceed those of their disjuncts.[1] We proceed to a case of this kind.

Case 1. The agent is sure the situation is as follows. There are three levers, *a*, *b*, and *c*. He can grasp and press any one but at most one of these levers, and he must grasp and press at least one. (Perhaps he is falling toward the levers.) His pressing lever *c* would depress *c* and only *c*. In contrast, levers *a* and *b* are released in one of the following three ways, though he is not sure which: (i) pressing either of these levers would depress both; (ii) pressing *a* would depress *a* and only *a*, though pressing *b* would depress both *a* and *b*; (iii) pressing *b* would depress *b* and only *b*, though depressing *a* would depress both *a* and *b*. In any case, were either *a* or *b* pressed, *c* would be neither pressed nor depressed. Using primed letters to stand for acts of grasping and pressing levers and overbars to signify negation, the acts that are certainly open to the agent are precisely $(a'\&\bar{b}'\&\bar{c}')$, $(\bar{a}'\&b'\&\bar{c}')$, $(\bar{a}'\&\bar{b}'\&c')$, and every act that is entailed by any one of these.[2] Regarding the disjunctive act $(a'\vee b')$, we assume that while the agent is sure that this act is open, he is also sure that it is *not* open in a "distinctively disjunctive" way: We assume that he is sure that he can choose $(a'\vee b')$ only by choosing, or at will making certain, one or the other of its disjuncts.

Monetary payoffs are functions of what levers are depressed, regardless of which lever is actually grasped and pressed. Using uppercase letters to stand for depressions of levers, payoffs for combinations of lever depressions that the agent considers possible are

$$A\&B\&\bar{C}: \$1$$
$$A\&\bar{B}\&\bar{C}: \$7$$
$$\bar{A}\&B\&\bar{C}: \$7$$
$$\bar{A}\&\bar{B}\&C: \$4.50$$

We assume that utilities equal monetary payoffs and that, for example,

$$U[a'\&(A\&B\&\bar{C})] = U(A\&B\&\bar{C})$$

and

$$U[(a'\vee b')\&(A\&B\&\bar{C})] = U(A\&B\&\bar{C}).$$

We suppose that the eightfold partition

$$(A\&B\&C), (A\&B\&\bar{C}), (A\&\bar{B}\&C), (A\&\bar{B}\&\bar{C}),$$
$$(\bar{A}\&B\&C), (\bar{A}\&B\&\bar{C}), (\bar{A}\&\bar{B}\&C), (\bar{A}\&\bar{B}\&\bar{C})$$

is "sufficiently fine" relative to each action that is certainly open in the case.

Another specification for the case is that in the agent's view the three possible ways in which levers a and b may be related are equally likely, so that it is twice as likely that A would take place with B as without B. More precisely, we assume that, for the agent,

$$\Pr[a'\,\square\!\!\rightarrow(A\&B\&\bar{C})] = \tfrac{2}{3},$$
$$\Pr[a'\,\square\!\!\rightarrow(A\&\bar{B}\&\bar{C})] = \tfrac{1}{3},$$

and the probability for each of the other six conditionals with antecedent a' is 0; and that, similarly,

$$\Pr[b'\,\square\!\!\rightarrow(A\&B\&\bar{C})] = \tfrac{2}{3}$$

and

$$\Pr[b'\,\square\!\!\rightarrow(\bar{A}\&B\&\bar{C})] = \tfrac{1}{3}.$$

We assume also that
$$\Pr[c' \square\!\!\rightarrow (\bar{A} \& \bar{B} \& C)] = 1$$
and that
$$\Pr[(a' \vee b') \square\!\!\rightarrow (A \& B \& \bar{C})] = \tfrac{1}{3},$$
$$\Pr[(a' \vee b') \square\!\!\rightarrow (A \& \bar{B} \& \bar{C})] = \tfrac{1}{3},$$
and
$$\Pr[(a' \vee b') \square\!\!\rightarrow (\bar{A} \& B \& \bar{C})] = \tfrac{1}{3}.$$

These probabilities are consistent. They are realized in the following partially specified model,[3] wherein it is assumed that the three actions a', b', and c' are equally probable. We adopt the following abbreviations.

$L(ab)$: Levers a and b are linked so that pressing either depresses both.
$L(a)$: Levers a and b are linked so that pressing a depresses both, but pressing b depresses only b.
$L(b)$: Levers a and b are linked so that pressing b depresses both, but pressing a depresses only a.

World probabilities

	a'	A	b'	B	c'	C	$L(ab)$	$L(a)$	$L(b)$	
W1	T	T	F	T	F	F	T	F	F	1/9
W2	F	T	T	T	F	F	T	F	F	1/9
W3	F	F	F	F	T	T	T	F	F	1/9
W4	T	T	F	T	F	F	F	T	F	0
W5	F	F	T	T	F	F	F	T	F	2/9
W6	F	F	F	F	T	T	F	T	F	1/9
W7	T	T	F	F	F	F	F	F	T	2/9
W8	F	T	T	T	F	F	F	F	T	0
W9	F	F	F	F	T	T	F	F	T	1/9

Nearness orderings

W1	W2	W3	·	·	·
W2	W1	W3	·	·	·
W3	W1	W2	·	·	·
W4	·	·	·	·	·
W5	W4	W6	·	·	·
W6	W5	W4	·	·	·
W7	W8	W9	·	·	·
W8	·	·	·	·	·
W9	W7	W8	·	·	·

To illustrate computations: $[a' \square\!\!\rightarrow (A \& B \& \bar{C})]$ is true at W1. It is also true at W2, for its consequent is true at W1, the nearest-to-W2 world at which its antecedent is true. (See the second nearness ordering.) For similar reasons it is true at W3, W5, and W6, and false at W7 and W9. Thus

$$\Pr[a' \,\square\!\!\rightarrow (A\,\&\,B\,\&\,\bar{C})] = \tfrac{1}{9} + \tfrac{1}{9} + \tfrac{1}{9} + \tfrac{2}{9} + \tfrac{1}{9} = \tfrac{2}{3}.$$

In contrast, $[a' \,\square\!\!\rightarrow (A\,\&\,\bar{B}\,\&\,\bar{C})]$, while true at both W7 and W9, is false at the other positively probable worlds, so that

$$\Pr[a' \,\square\!\!\rightarrow (A\,\&\,\bar{B}\,\&\,\bar{C})] = \tfrac{2}{9} + \tfrac{1}{9} = \tfrac{1}{3}.$$

It is noteworthy that in this model not all probabilities of conditionals equal corresponding conditional probabilities. For example,

$$\Pr[(A\,\&\,B\,\&\,\bar{C})/a'] = \tfrac{1}{3},$$

so

$$\Pr[a' \,\square\!\!\rightarrow (A\,\&\,B\,\&\,\bar{C})] \neq \Pr[(A\,\&\,B\,\&\,\bar{C})/a'].$$

A related point is that $[a' \,\square\!\!\rightarrow (A\,\&\,B\,\&\,\bar{C})]$ is not probabilistically independent of a':

$$\Pr([a' \,\square\!\!\rightarrow (A\,\&\,B\,\&\,\bar{C})]/a') = \tfrac{1}{3},$$

so

$$\Pr([a' \,\square\!\!\rightarrow (A\,\&\,B\,\&\,\bar{C})]/a') \neq \Pr[a' \,\square\!\!\rightarrow (A\,\&\,B\,\&\,\bar{C})].$$

What lies behind the probabilistic dependence of $[a' \,\square\!\!\rightarrow (A\,\&\,B\,\&\,\bar{C})]$ on a' is that, in the case being modeled, learning or becoming certain that a' would make $L(a)$ absolutely improbable for me while enhancing the probability for me of $L(b)$: $\Pr[L(a)/a'] = 0$ and $\Pr[L(b)/a'] = \tfrac{2}{3}$. Learning that a' would have these effect on probabilities, even though it is certain a' would not affect how levers a and b were linked. The case modeled is "Newcomblike." It is possible that *any* case in which the expected utility of an exclusive or non-exclusive disjunction exceeded that of its disjuncts would be Newcomblike. Whether or not this is so I leave as an open question.

The key feature of our model is that, at c'-worlds that are not $L(ab)$-worlds (worlds W6 and W9), the nearest $(a' \vee b')$-world is either a $[b'\,\&\,L(a)]$-world or an $[a'\,\&\,L(b)]$-world. It is mainly this feature that makes possible the equality of the probabilities of

$$[(a' \vee b') \,\square\!\!\rightarrow (A\,\&\,B\,\&\,\bar{C})],$$
$$[(a' \vee b') \,\square\!\!\rightarrow (A\,\&\,\bar{B}\,\&\,\bar{C})], \text{ and}$$
$$[(a' \vee b') \,\square\!\!\rightarrow (\bar{A}\,\&\,B\,\&\,\bar{C})],$$

despite the equal probabilities of the patterns of linkage and the causal independence from my actions of these patterns. One way to "explain" this feature of our model would be in terms of a predictor and manipulator, someone who I am sure has predicted what I will do and, if I will do c', how I *would* do $(a' \vee b')$ – by way of a' or by way of b'. I could then be assumed to think that if he has predicted c' (which in this case is for me only ⅓ probable) then it is equally likely that he has established that $L(ab)$, predicted that a' is the way I would do $(a' \vee b')$ and established that $L(b)$, or predicted that b' is the way I would do $(a' \vee b')$ and established that $L(a)$.

Specifications for our case are consistent; in particular, specifications for probabilities in it are consistent, as our formal model shows. Given these specifications, we have the following expected utilities:

$$U(a') = \Pr[a' \square\!\!\rightarrow (A \& B \& \bar{C})] \cdot U(A \& B \& \bar{C})$$
$$+ \Pr[a' \square\!\!\rightarrow (A \& \bar{B} \& \bar{C})] \cdot U(A \& \bar{B} \& \bar{C})$$
$$= \tfrac{2}{3}(1) + \tfrac{1}{3}(7)$$
$$= 3,$$

$$U(b') = 3,$$
$$U(c') = 4.5,$$
$$U(a' \vee b') = \tfrac{1}{3}(1) + \tfrac{1}{3}(7) + \tfrac{1}{3}(7)$$
$$= 5.$$

The certainly open alternatives to $(a' \vee b')$ are

$$(\bar{a}' \& \bar{b}' \& c'),$$
$$(\bar{a}' \& c'),$$
$$(\bar{b}' \& c'),$$
$$(\bar{a}' \& \bar{b}'),$$

and c'. We are assuming that $U(c')$ equals $U(\bar{a}' \& \bar{b}' \& c')$, $U(\bar{a}' \& c')$, $U(\bar{b}' \& c')$, and $U(\bar{a}' \& \bar{b}')$. Therefore, since

$$U(a' \vee b') > U(c'),$$

it follows by principle R1 that $(a' \vee b')$ is rational. *But this conclusion is unacceptable.* The agent is sure that $(a' \vee b')$ is *not* open to him in a distinctively disjunctive way: He is sure that making $(a' \vee b')$ certain would consist either in making a' certain or in making b' certain. And the expected utility of c', which he is sure is an open alternative both to a' and to b', exceeds the expected utility of each of these.[4] Hence it would be irrational for him to do a', b', or $(a' \vee b')$.[5] (A similar case is used to make a similar point in Sobel 1971, pp. 378–80. But there the point is made in relation to Jeffrey's "news value" logic of decision, and in terms of a disjunction whose disjuncts are not probabilistically incompatible.)

3.2

A natural reaction to problems posed by Case 1 is to confine tests of expected utility to fully specific actions. Since not only $(a' \vee b')$, but also a' and b', are certainly open in the case, it can seem that its status should depend upon theirs or, if not on theirs, then on yet more specific certainly open actions. For a principle that gives expression to this reaction, let an action x be a *fully specific* action for time t that is certainly open at time t' if and only if (i) x is an action for t that is certainly open at t', and (ii) no agent-identical action for t that is certainly open at t' entails but is not entailed by x. Our new principle is then:

R2   *An action x that is fully specific among actions for its first moment t that are certainly open just before t is rational* if and only if, for every agent-identical action y that is fully specific among actions for t that are certainly open just before t,

$$U_t(x) \geq U_t(y) \quad \text{and} \quad U_t(x/x) \geq U_t(y/x);$$

an action x with first moment t is irrational if and only if there is a certainly open agent-identical action with first moment t and, for each agent-identical rational action y that is fully specific among actions for t that are certainly open just before t, the agent of x is sure just before t that y is an alternative to x; *an action x with first moment t is rational* if and only if there is a certainly open agent-identical action with first moment t, and x is not irrational.[6]

The basic issue, according to this principle, concerns fully specific actions; other actions are rational and irrational only derivatively. In Case 1, the set of fully specific certainly open actions is $\{(a' \& \bar{b}' \& \bar{c}'), (\bar{a}' \& b' \& \bar{c}'), (\bar{a}' \& \bar{b}' \& c')\}$, and the expected utility of $(\bar{a}' \& \bar{b}' \& c')$ exceeds that of the other two actions in this set. So principle R2 makes $(\bar{a}' \& \bar{b}' \& c')$ the rational fully specific certainly open action, c' a rational action, and a', b', and $(a' \vee b')$ irrational actions.

Principle R2 copes with Case 1, but this principle encounters difficulties in other cases. It encounters difficulties in some cases in which the agent is capable of settling or leaving open options somewhat *removed* from a given moment of choice. We proceed with a case to this point.

Case 2. Let there be two levers, a and b. Assume that the agent is sure that he can at no time pull both a and b, and at each of $t1$ and $t2$ must pull one of a and b. Let $A1$, his pulling a at $t1$, be "now" (i.e., just before $t1$) certainly open for him. And assume that he is sure that by $t1$ he can at will not only make certain $A1$, but also either $A2$ (his pulling a again at $t2$) or $B2$ (his pulling b at $t2$). Assume that not only the "immediate action" $A1$ but also the "extended actions" $(A1 \& A2)$ and $(A1 \& B2)$ are certainly open just before $t1$. Our agent is sure that he has the capacity (the self-control) not only to settle by $t1$ what he will do then, but also to commit himself in relation to $t2$ and settle by $t1$ what actions with first moment $t2$ he will do.[7] However, let him be sure that, though he can settle what he does at $t2$, if he decides now for $A1$ he can leave *open* what he will do at $t2$. But matters are not the same if he decides now for $B1$; we assume that he is sure that pulling b would render a inoperable, so that in making $B1$ certain he would make $B2$ certain as well. Although both $B1$ and $(B1 \& B2)$ are certainly open just before $t1$, $(B1 \& A2)$ is not certainly open then nor at any time. We take the set of fully specific actions with first moment $t1$ that are certainly open just before $t1$ to be $\{(A1 \& A2), (A1 \& B2), (B1 \& B2)\}$; it is clear that $A1$ is not in this set, since there is an action in the set that entails but is not entailed by $A1$ – indeed, there are two such actions. For similar reasons $B1$ is not in this set.

We assume that the agent thinks that only two possible patterns of payoffs are at all likely, and that he thinks these patterns are equally likely. These patterns are: $0 each time a is pulled, and $3 each time b is pulled; and $2 each time a is pulled, and $0 each time b is pulled. Utilities for actions are to equal expected monetary returns: If M is the set of possible monetary returns on an action x then, in harmony with the analysis we have adopted for expected utility, we take the *expected monetary return* of x to be

$$\sum_{m \in M} \Pr(x \,\square\!\!\!\rightarrow m) \cdot m.$$

Thus,

$$U(A1 \& A2) = .5(0) + .5(4)$$
$$= 2,$$
$$U(A1 \& B2) = .5(3) + .5(2)$$
$$= 2.5,$$
$$U(B1 \& B2) = .5(6) + .5(0)$$
$$= 3.$$

According to principle R2, the rational fully specific certainly open action is $(B1 \& B2)$, action $B1$ is rational, and action $A1$ is irrational. *But this conclusion is unacceptable;* it would be irrational in this case to do

B1. The rational course for our agent would be to do A1 and leave open his options for $t2$, these options to be settled only after information regarding the payoff for A1 is in hand. The expected monetary return of *that* course, assuming that having left his options open he would do at $t2$ what he would then be sure was most lucrative, is

$$.5(0+3)+.5(2+2),$$

which is 3.5.

3.3

The problem with principle R2 is that, in some cases where it would be rational for an agent to leave future options open, this principle selects actions as rational that would close them. Principle R2 has this effect because it makes fully specific issues basic, and because an agent's fully specific certainly open options will extend as far into the future as he is sure his will can reach. One solution to this problem refines the basic issue, and makes it what very specific action it would be rational to do "now." The principle envisioned would call first for the selection of actions fully specific among certainly open minimal actions. Let an action be *minimal* if and only if it is open and, were the agent to begin it, he would not be able to stop it short of completion. Count an action x as *fully specific among minimal actions certainly open at t* if and only if (i) the agent of x is sure at t that x is a minimal action open to him, and (ii) there is no action x' with the same first moment and agent as x such that he is sure that x' is a minimal action open to him, and such that x' entails but is not entailed by x.

R3   *An action x that is fully specific among minimal actions certainly open just before its first moment t is rational* if and only if, for every agent-identical action y that is fully specific among minimal actions for t that are certainly open just before t,

$$U_t(x) \geq U_t(y) \quad \text{and} \quad U_t(x/x) \geq U_t(y/x);$$

an action x with first moment t is irrational if and only if there is a certainly open agent-identical action with first moment t and, for each agent-identical rational action y that is fully specific among minimal actions for t that are certainly open just before t, the agent of x is sure just before t that y is an alternative to x; *an action x with first moment t is rational* if and only if there is a certainly open agent-identical action with first moment t, and x is not irrational.

In Case 2, the extended action ($A1 \& A2$) is not a certainly minimal action, for the agent is not sure that were he to begin this action there would be no turning back. Indeed, the agent realizes that he can begin this action without being already committed to $A2$; he can be sure (we could add this to the case) that even if he *were* already committed to $A2$, though this would mean that he would not turn back he still *could* turn back. Similarly, the extended action ($A1 \& B2$) is not a certainly minimal action. Presumably, the set of the agent's fully specific minimal actions with first moment $t1$ that are certainly open 'just before' $t1$ is in this case $\{A1, (B1 \& B2)\}$. If this is right, principle R3 selects as rational $A1$ and (derivatively) other actions – for example, actions with first moment $t1$ that are entailed by $A1$. (I assume that the agent is sure that if he were to do A at $t1$ he would leave open his options for $t2$ and would, when the time came, make good use of them.) Principle R3 does not have the problems that embarrassed principle R2. Principle R3 does not advocate actions that would close off options that ought to be left open. But this principle can encounter difficulties of an opposite kind: It can select actions that would leave open options that ought to be closed.

Case 3. The agent is sure that the set-up is as follows. There are two levers. They cannot at any time both be depressed. Each can be depressed at $t1$, but each can be depressed at $t2$ if and only if one or the other was depressed at $t1$. Depressing a at $t1$ delivers \$5. Depressing b at $t1$ delivers \$1. After $t1$ and before $t2$, \$10 is added either to the \$5 that a delivers in any event at $t2$ or to the \$1 that b delivers in any event at $t2$. The position of the \$10 bonus is determined by a random device, so that the chance of its being associated with a is the same as that of its being associated with b (this chance is thus .5). If b is depressed at $t1$, then shortly thereafter the position of the \$10 bonus is announced by a clear signal that the agent would interpret correctly and be able successfully to act upon at $t2$. But if a is depressed at $t1$ then – though the position of the bonus is announced by a signal – the signal is made so late, is so unclear, and is so brief that it is only as likely as not that the agent correctly interprets it and successfully acts upon it.

So much for the set-up. We now proceed to several of the agent's beliefs concerning actions. $A1$ and $B1$ are to be fully specific among minimal actions that are certainly open just before $t1$. Extended actions ($A1 \& A2$) and ($A1 \& B2$) are certainly open just before $t1$, but he is sure they are not minimal; and ($B1 \& A2$) and ($B1 \& B2$) are not certainly open just before $t1$. Furthermore, the agent is sure that if he were to opt for $A1$ then, though he *could* incorporate its choice in the choice of a more extended action, he would not. He is sure that if he were to opt for $A1$ he would gamble (unwisely) on the unclear signal and try to act on it. In contrast, he is sure that if he were to opt for $B1$ then he not only would not but could not incorporate its choice in the choice of a more extended action: He is sure that, were he to do $B1$, then whatever his state of mind at $t1$ – whatever his "resolve" – he would at $t2$ be completely under the sway of the clear signal that would have announced the position of the $10 bonus.

Utilities for actions are to equal expected monetary returns. Pertinent expected monetary returns include, for $A1$,

$$5 + (.25 \cdot 15 + .25 \cdot 1 + .25 \cdot 5 + .25 \cdot 11) = 13$$

and, for $B1$,

$$1 + (.5 \cdot 15 + .5 \cdot 11) = 14.$$

Assuming that the agent's fully specific minimal actions with first moment $t1$ that are certainly open just before t are exactly $A1$, $B1$, and $(A1\&B1)$, and assuming that his expected monetary return for $(A1\&B1)$ is 0, principle R3 selects as rational $B1$ and makes $(A1\&A2)$ irrational. *But this conclusion is unacceptable.* The extended action $(A1\&A2)$ is certainly open just before $t1$: The agent can not only adopt for $A1$, but can do this *and* settle his options for $t2$ in favor of $A2$. And the expected monetary return (expected just before $t1$) of this extended action exceeds that of $B1$: The expected monetary return of this extended action is

$$5 + 5 + .5 \cdot 10 = 15.$$

It is clear that this is what he should do, that this is what he should "now" make certain. He should not do $B1$, as principle R3 would have him do, leaving open his options for $t2$; he should settle his options for both $t1$ and $t2$ and opt "now" (by $t1$) for $(A1\&A2)$.

4. A principle that confines tests of expected utility to choices

Reflection on difficulties posed in the previous section for principles that make expected utilities of *actions* fundamental can suggest that what is of first importance in decision problems is exactly which action options are to be settled and exactly which are to be left open. What seems wanted is a principle that makes expected utilities of *choices,* not of actions, fundamental. (This suggestion is made in Sobel 1971, p. 383.)

For present purposes, *choices* are organizations of an agent's will in which acts of this agent are made certain: for the present, the choice of an act is possible at a time if and only if this act is then open in the sense of Section 1. Let a *precise choice* of an action x be a choice of this action that is not accompanied by a choice of any action y such that y entails but is not entailed by x. Let the precise choice of an action x be *certainly possible* at a time if and only if the agent of x is at this time sure that he can by the first moment of x make a precise choice of x. An assumption set out in Section 1 implies that there are choices our agent is sure that he can make, that cover all of the things he thinks he might do, and that are such that he is sure that he cannot avoid them all. We now require that there be *precise* choices that meet these conditions. More exactly, we require that, for a time t at which an action is open to an agent, there be a set that contains all and only precise choices for t that are certainly possible just before t, such that every action with first moment t that he thinks just before t he might do is a disjunct of an action in a choice in this set; and he is sure just before t that he must make a choice from this set, perhaps because there is a choice in this set such that he is sure just before t that his not making any other choice in the set would be his making that choice by default. We now state a principle of rational choice, and derivatively of action, and test it on Cases 1 through 3.

R4 *A precise choice c of an action with first moment t, which precise choice is certainly possible just before t, is rational* if and only if, for every agent-identical precise choice c' for t that is certainly possible just before t,

$$U_t(c) \geq U_t(c') \quad \text{and} \quad U_t(c/c) \geq U_t(c'/c);$$

an action x with first moment t is irrational if and only if there is a certainly open agent-identical action with first moment t and, for each action y that is the subject of a rational precise choice for t that is certainly possible just before t, the agent of x is sure just before t that y is an alternative to x; *an action x with first moment t is rational* if and only if there is a certainly open agent-identical action with first moment t, and x is not irrational.

Principle R4 is free of difficulties posed for principles in Section 2. The problem in Case 1 is made by a disjunctive action $(a' \vee b')$, which action while certainly open is certainly not open in a distinctively disjunctive way: The expected utility of $(a' \vee b')$ exceeds that of its certainly open alternatives, but the expected utility of one of these, c', exceeds that of the certainly open ways, a' and b', in which $(a' \vee b')$ can be made certain. Despite its relatively excellent expected utility, action $(a' \vee b')$ should not be selected as rational in this case. And, according to principle R4, this action is not rational: although the choice of $(a' \vee b')$ is, the *precise* choice of $(a' \vee b')$ is *not* certainly possible in this case. The agent is sure that he can choose $(a' \vee b')$ only by choosing one or the other of a' and b'. Certainly possible precise choices are presumably limited to $(a' \& \bar{b}' \& \bar{c}')$, $(\bar{a}' \& b' \& c')$, and $(\bar{a}' \& \bar{b}' \& c')$. Of the three, $(\bar{a}' \& \bar{b}' \& c')$ excels in expected utility. So in Case 1, according to principle R4: The rational, certainly possible precise choice is of $(\bar{a}' \& \bar{b}' \& c')$; rational actions include c'; and actions a', b', and $(a' \vee b')$ – which the agent is sure are alternatives to $(\bar{a}' \& \bar{b}' \& c')$ – are irrational. All this is as it should be in Case 1.

The problem in Case 2 is that the fully specific, certainly open action with greatest utility, $(B1 \& B2)$, entails an immediate action, $B1$, that would settle options for $t2$ best left open at $t1$. It would be best not to do $B1$ in this case, since doing $B1$ precludes choice after $t1$ for $t2$; it would be best to do $A1$ and leave open the issue for $t2$ so that its resolution can be based on information concerning the payoff for depressing lever a at $t1$. Principle R4 endorses this best strategy. Precise choices that are certainly possible just before $t1$ are limited to $A1$, $(A1 \& A2)$, $(A1 \& B2)$, and $(B1 \& B2)$; presumably, the precise choice of $B1$ – making certain $B1$ without making certain $(B1 \& B2)$ – is not certainly possible just before $t1$. Of these certainly possible choices, that of $A1$ excels in expected utility. So in Case 2, according to R4: The rational precise choice that is certainly possible just before $t1$, is of $A1$ – the agent should make certain $A1$ without making certain either $(A1 \& A2)$ or $(A1 \& B2)$; $A1$ is a rational action; and $B1$ is an irrational action.

The problem in Case 3 is that action $B1$ (the action that, among fully specific minimal actions for $t1$ certainly open just before $t1$, excels in expected utility) would leave open an option for $t2$ that it would be best to close – an option which, furthermore, the agent is sure can be closed only if he does not do $B1$. He is sure that were he to do $B1$ he would, regardless of his resolve at $t1$, attend to the clear signal that would announce the position of the $10 bonus. In this case, it would be best not to choose $B1$ and then decide for $t2$ on the basis of this clear signal, but instead to choose by $t1$ the course $(A1 \& A2)$ and ignore what, given $A1$, would be an unclear and useless signal for $t2$. Principle R4 endorses this strategy. Precise choices for actions with first moment $t1$ that are certainly possible just before $t1$ are limited to $A1$, $(A1 \& A2)$, $(A1 \& B2)$, and $B1$. Of these certainly possible precise choices, that of $(A1 \& A2)$ excels in expected utility. So in Case 3, according to R4: The rational precise choice certainly possible just before $t1$ is of $(A1 \& A2)$ – the agent should by $t1$ not only make certain $A1$ but also, as he is sure he can, settle his option for $t2$ in favor of $A2$; $A1$ is a rational action; and $B1$ is an irrational action.

5. A PRINCIPLE FOR AGENTS WHO ARE SURE THEY CAN MAKE MIXED CHOICES

We now extend the notion of a choice to organizations of will in which objective chances are established for actions without these actions being made certain. More exactly, let a choice for a moment t be either a *pure* choice, in which some action with first moment t is made certain (that is, in which this action is assigned the chance of 1), or a *mixed* choice, in which exclusive chances are distributed to several actions with first moment t: chances in such a choice sum to 1, and the chance of an action in a choice is established for it alone, to the exclusion of every other action in the choice. Let a choice c' be a *refined version of a choice* c if and only if each action in c is a disjunction of actions in c', and the chance of an action in c is the sum of the chances of its disjuncts in c'. We say that a choice for t is at a time a *certain choice* if its agent is then sure that he can make it by t, and we say that a choice is at a time a *certain precise choice* if its agent is then sure he can make it without making any refined version of it. We still require that there be precise choices that our agent is sure that he can make, which cover all of the things he thinks he might do and are such that he is sure that he cannot avoid them all. But this requirement is now to be understood in terms of notions of choice and precise choice that extend to mixed choices. Our final principle is suited to agents that satisfy this requirement.

R5 *A precise choice c for moment t, which precise choice is certainly possible just before t, is rational if and only if, for every agent-identical precise choice c' for t that is certainly possible just before t,*

$$U_t(c) \geq U_t(c') \quad \text{and} \quad U_t(c/c) \geq U_t(c'/c);$$

an action x with first moment t is irrational if and only if there is a certainly open agent-identical action with first moment t, and – for each agent-identical action y that is the disjunction of the actions of a rational precise choice for t that is certainly possible just before t – the agent is sure just before t that y is an alternative to x; an action x with first moment t is rational if and only if there is a certainly open agent-identical action with first moment t, and x is not irrational.

6. CONCLUSION

The problems taken up have been exactly *which* expected utilities are decisive for the rationality of actions and choices, and exactly *how* these expected utilities are decisive. Solutions suited to agents who are sure, just before times of action of their powers of choice, are contained in principle R5. The tests of expected utility are to be applied directly only to precise choices. Expected utilities and conditional expected utilities of precise choices determine directly which precise choices would be rational. Rational actions are determined derivatively by their relations to rational precise choices.[8]

NOTES

1. That the value of a disjunction is not bounded by the values of its disjuncts is true of "news values" based on conditional probabilities (consider Jeffrey 1965b, p. 79, exercise 8), and also of expected utilities based on probabilities of conditionals. In partial contrast, while expected utilities of disjunctions with probabilistically incompatible disjuncts remain not necessarily bounded by the expected utilities of their disjuncts, the case is otherwise for news values. (Consider Jeffrey 1965b, p. 70, principle 5-2.) This contrast explains why the rule for values of propositions under logical partitions can be quite unrestricted for news values but not for expected utilities. In Section 1 the rule for expected utilities is restricted to sufficiently fine logical partitions. That such partitions are adequate is explained and demonstrated in Sobel (1985a), and in Chapters 8 and 9 of this volume, under more complicated and general definitions for expected utilities than that of the present chapter.
2. Strictly, basic certainly open acts should be $[(A \vee B) \& (a' \& \bar{b}' \& \bar{c})]$, $[(A \vee B) \& (\bar{a}' \& b' \& \bar{c}')]$, and $[C \& (\bar{a}' \& \bar{b}' \& c')]$, where uppercase letters stand for depressions of levers; however, given our assumptions, we need not for the argument to follow attend to these more complex acts. In particular, given our assumptions it will be clear that $U(C \& c')$ and $U(C \vee c')$ equal $U(c')$.
3. I owe this model to Włodek Rabinowicz. It is better than the model I used originally, and I am grateful for it and for several other valuable suggestions.
4. The expected utility of the disjunctive action $(a' \vee b')$ exceeds that of each of its disjuncts. The argument against principle R1 turns on this. For a simpler case that involves such a disjunctive action, see Sobel (1985a, sec. II) and Chapter 9, Section 2, of this volume. The present argument could have been based on that simpler case, in which $U(L) = U[(L \& {\sim}R) \vee (L \& R)] = 11$, $U(L \& {\sim}R) = U(L \& R) = 6$, and $U({\sim}L) = 10$.
5. The R1-rational action $(a' \vee b')$ is a disjunction of actions that are not R1-rational. I do not say that they are R1-*ir*rational, R1-irrationality, as distinct from R1-*non*rationality, has not been explicitly defined, and might best be *distinguished* from R1-nonrationality. But it does seem that a' and b' would be R1-irrational on any natural definition of this term, which is very strange given that their disjunction $(a' \vee b')$ is R1-rational.

   Another anomaly related to the expected utilities of disjunctions attaches to principle R1, but I do not offer it as a criticism of this principle. This anomaly is that it is not necessary for an action entailed by an R1-rational action to itself be R1-rational. Let an agent be certain that x, y, and z are alternatives, and that z is an alternative to $(x \vee y)$. Assume that x is rational, that the expected utility of z exceeds that of y, and that it is certain that y is the way in which $(x \vee y)$ would be done. Then the expected utility of $(x \vee y)$ would equal

that of y, and $(x \vee y)$, though entailed by an R1-rational action, would not itself be R1-rational. It is not clear that this anomaly is a defect, for it is not clear that every action that is entailed by a rational action must itself be rational.
6. A fully specific action is R2-rational only if it is certainly open. But other actions can be R2-rational (and R2-irrational) without being certainly open.

   Principle R2 is concerned with rationality in the sense of *consonance* with reason, as distinct from *exclusive* consonance. If a fully specific action is R2-rational then it is "R2-alright" that it take place; but perhaps it is not the case that it "R2-ought" to take place, for there may be other fully specific R2-rational actions with the same first moment. All principles studied in this chapter are concerned with rationality in the sense of consonance (not exclusive consonance) with reason.
7. Our agent is sure that he can make commitments and resolutions that are fully effective for times separated by temporal intervals from the times of commitment and resolution. It has been said that such resolutions are not possible – that an agent who thinks that they are possible and that he can thereby reduce his future freedom, or can thereby have reduced his present freedom, is deceiving himself (Sartre 1956, pp. 32–3).
8. An application of principle R5 to a case involving only pure choices is adumbrated in Sobel (1985a, sec. II). Arguments against principles R1, R2, and R3 that are rather different from the arguments presented here could be based on that case. For another treatment of issues taken up in this chapter, see Pollock (1983).

11

Maximization, stability of decision, and actions in accordance with reason

Abstract. I maintain that whether or not an action is rational depends on beliefs and preferences just before its time, and that an action is in accordance with reason given an agent's beliefs, preferences, and evidential commitments if and only if, in view of these, not only would that action maximize expected utility but a decision for it would be ideally stable or ratifiable. Arguments against this view are taken up, and alternative positions are considered.

0. INTRODUCTION

Rational actions reflect beliefs and preferences in certain orderly ways. The problem of theory is to explain which beliefs and preferences are relevant to the rationality of particular actions, and exactly how they are relevant. One distinction of interest here is between an agent's beliefs and preferences *just before* an action's time, and his beliefs and preferences *at* its time. Theorists do not agree about the times of beliefs and desires that are relevant to the rationality of action. Another distinction is between actions that would, in one sense or another, *maximize* expected utilities given relevant beliefs and desires, and actions that decisions for which would, in one sense or another, be *stable*. There is disagreement about the relevance of these factors to the rationality of actions. In Section 1, several possible positions on these issues are explained and compared. In Section 2, I contrast perspectives on these issues, and comment on arguments that might be brought against the position I favor. In Section 3, I restate and elaborate on this position.

1. POSITIONS

1.1

The simplest theory says that an action is rational if and only if it would *maximize actual expected utility just before what would be this action's*

Reprinted with revisions by permission from *Philosophy of Science* 57 (1990), pp. 60–77.

time. For this rule we assume that the actual expected utility at the time of an action *a* is

$$U(a) = \Sigma_w [\Pr(a \,\square\!\!\rightarrow w) \cdot V(w)],$$

with ($a \,\square\!\!\rightarrow w$) a causal conditional, Pr a probability function that measures the agent's credences at this time, and V a measure of his preferences for worlds at this time. According to our first pure maximization theory, an option *a* in an appropriate partition of options is rational if and only if, just before what would be the time of actions in this partition, $U(a) \geq U(a')$ for each option a' in this partition. Appropriate partitions of options are discussed in Sobel (1983 [Chapter 10 in this volume]). Options in such partitions would have the same times. More complicated theories of causal expected utility are set out in Sobel (1986a, [Chapter 8]) and Sobel (1989a [Chapter 9]).

A pure maximization theory can be embarrassed by a case or state of mind, stipulations for which are completed at the beginning of the next paragraph and begun here with the supposition of the following structure:

Case I

	c_1	c_2
a_1	2	1
a_2	2	0

Several conditions pertain to our structures. First, (a_1, a_2) is an appropriate partition of options. Note that this partition does not include mixtures of its two options. We assume for this case, and for all other cases below, that no mixtures of options are themselves options. Second, (c_1, c_2) is an adequate partition of possible circumstances. Such partitions of possible circumstances are discussed in Sobel (1985a, 1986a [Chapter 8], 1989a [Chapter 9]). Third, for each *i*, option a_i, and possible circumstance c_i, $\Pr(c_i/a_i) = 1$, where the conditional probability part of Pr measures what are for the agent possible evidential bearings just before what would be the time of *a*. Similar simplifying stipulations should be understood to hold for all cases considered here. (Option *a* could be pressing a button of some color, and circumstance *c* that the agent's exact twin will press a button of this color, or that a perfect predictor will predict that the agent will press a button of this color.) We employ, as measures of possible evidential bearings, extended conditional probability functions that are defined even for conditions of zero probability. We do this not only for simplicity, but because zero-probability conditions can have possible evidential bearings, and these should be covered by a measure of such bearings. These extended conditional probability measures are explained in Sobel (1987). Fourth, we assume for this and for all our cases that circumstances are certainly causally independent of options, so that for each option a and circumstance c, $\Pr(a \,\square\!\!\rightarrow c) = \Pr(c)$.

A further assumption completes Case I – namely, that $\Pr(c_1) = 1$. Given this assumption, according to our pure maximization theory of rationality both a_1 and a_2 are rational. Nothing relevant distinguishes these options. In particular, it is not relevant from the standpoint of rational choice that while a decision for a_1 would be ideally stable, a decision for a_2 would not be, since proper reflection on a decision for a_2 would lead to a belief in c_2 (recall that $\Pr(c_2/a_2) = 1$) and thus to a preference for a_1 over a_2. Our first theory makes such considerations not relevant to the rationality of actions, but this seems wrong. It seems that the instability of a decision for a_2 *is* relevant to what (if anything) would be rational in Case I; in this case, while a_1 would be rational, a_2 – because a decision for it would not be ideally stable – would not be rational.

For a formal explication of the idea of decision instability, we let the expected utility at a time of an option a conditional on a proposition b be

$$U(a/b) = \sum_w \Pr[(a \,\square\!\!\rightarrow w)/b] \cdot V(w).$$

We then say that a decision for an option a in an appropriate partition of options would be ideally *stable* if and only if, just before what would be its time, a would maximize expected utility conditional on a; that is, if and only if, just before what would be the time of a, $U(a/a) \geq U(a'/a)$ for each option a' in this partition. (A slightly different definition, closer to that of "ratifiable" in Jeffrey 1983, would equate ideal stability of a decision for an option a with $U(a/a^*) \geq U(a'/a^*)$ just before what would be the time of a, where a^* is a decision for a.)

1.2

The rule to maximize actual expected utility leaves out considerations of decision stability, and seems for this reason (if no other) to be inadequate as a rule for rational actions. One possible extreme reaction would be to exchange maximization for stability, and to identify rational actions with actions decisions for which would be stable. However, a pure stability theory encounters in its turn several difficulties. For one thing, it seems an overreaction that itself leaves out something relevant to rational action – namely, maximization. For this point we have a case or state of mind of the following structure.

Case II
$\Pr(c_1) = 1$

	c_1	c_2
a_1	1	0
a_2	0	1

We assume for this and all our cases that certain features already noted *persist through all conditionalizations on options*. We assume for any sequence of options (in which options can recur) that if U' is the agent's expected utility function conditional on the proposition recording this sequence, and if $\Pr'$ is the extended probability function that (i) comes from $\Pr$ by conditionalization on this proposition and (ii) has a conditional part that would reflect evidential bearings for the agent as revised in the light of this sequence of shifting decisions, then the U'-structure of the case is identical with the U-structure just displayed; for each option a_i and corresponding circumstance c_i, $\Pr'(c_i/a_i) = 1$; and for each option a and circumstance c, $\Pr'(a \,\square\!\!\rightarrow c) = \Pr'(c)$. (We say that $\Pr^*$ comes from an extended probability function $\Pr$ by conditionalization on p if and only if, for each q, $\Pr^*(q) = \Pr(q/p)$ and, for each r and s, if $\Pr^*(r) > 0$ then $\Pr^*(s/r) = \Pr^*(r \& s)/\Pr^*(r)$. Note that the conditionial part of $\Pr^*$ is not regimented for q such that $\Pr^*(q) = 0$.)

Decisions for both a_1 and a_2 would be stable in this case, and so – according to the present theory of rational actions – both of these actions would be rational. According to the present theory, there are from the standpoint of rationality no relevant differences between them, and in particular it is from this standpoint not relevant that of the two only a_1 would maximize actual expected utility. But this seems wrong.

1.3

Our first difficulty with a pure stability theory of rational action was that it seems to leave out something relevant. A second possible difficulty is that it is in a certain other sense incomplete. The same probabilities as in previous cases, together with the following structure,

Case III
$\Pr(c_1) = 1$

	c_1	c_2
a_1	0	1
a_2	2	0

make a case in which no decision would be stable, and thus a case in which (according to the present theory) no action would be rational. A *complete* theory would make selections of rational actions in every case; our pure stability theory is not in this sense complete.

This second possible difficulty with a pure stability theory suggests that the demand for stable decisions may be too stringent. Perhaps decisions for rational actions need not be stable, so long as they are not ideally defeatable or bound to be irrevocably withdrawn. Neither decision in Case III would be ideally defeatable. Each decision would in fact lead eventually, and again and again, back to itself. A pure undefeatability theory would hold that an action is rational if and only if it is not defeatable, or, equivalently, that an action is not rational if and only if it is defeatable. No decision in either Case II or Case III is defeatable. And, more importantly, it can be shown in every case of the sort studied in this chapter (I allude here to independence and persistence conditions) in which there are finitely many options, or infinitely many mixed options based on finitely many pure options, that at least one decision would not be defeatable and that, according to our pure undefeatability theory, at least one action would be rational. Exchanging undefeatability for stability would, at least in finite cases, meet the second supposed difficulty with a pure stability theory of rationality.

For a formal explication of defeatability, based on ideas derived with modifications from Weirich (1986b, pp. 444–5), we use the idea of a "deliberational path" (a "maximizing path" or "path of strong improvement") between options. Let there be a *path* from option a to option a' in an appropriate partition of options if and only if there is a sequence S of options in this partition with first member a and last member a' such that, for the ith member of S, b, and its immediate successor b', and for each option o in the partition of options, $U_i(b'/b) \geq U_i(o/b)$, where

$$U_i(a/c) = \sum_w \Pr_i[(a \,\square\!\!\!\rightarrow w)/c] \cdot V(w).$$

$\Pr_i$ is the extended probability function that comes from $\Pr$ by conditionalization on the proposition recording the subsequence through b_i of S whose conditional part would reflect evidential bearings for the agent as revised in the light of this subsequence of shifting decisions.

Using this idea of a path, we let a decision for an option a in an appropriate partition be ideally *defeatable* if and only if

(1) there is an option a' in the partition such that (i) $U(a'/a) > U(a/a)$, (ii) for every option o in the partition, $U(a'/a) \geq U(o/a)$, and (iii) there is no path from a' to a; and

(2) for each a'' in the partition, if (i) and (ii) obtain with a'' in place of a' then (iii) holds with a'' in place of a'.

1.4

The first difficulty with our pure stability theory of rationality was that it leaves out something that seems relevant to rational action – namely, maximization of actual expected utility. One proposal would meet this difficulty part way, and meet the second supposed difficulty of incompleteness, by combining maximization (though not of *actual* expected utility) with undefeatability. A mixed, hierarchical theory somewhat like that proposed by Weirich ("hierarchical" is his term – see Weirich 1986b, p. 445; 1988) would make an action rational if and only if it is both undefeatable and, among actions that are undefeatable, it would maximize projected expected utility, where the *projected expected utility* of an action a is its expected utility self-conditional $U(a/a)$. It is noteworthy that, since $[a \& (a \square\!\!\rightarrow w)]$ and $(a \& w)$ are logically equivalent, $U(a/a)$ equals $\text{Des}(a) = \Sigma \Pr[(w/a) \cdot V(w)]$; a requirement simply to maximize projected expected causal utility is equivalent to a requirement to maximize evidential Desirability. The present theory would have one maximize evidential Desirability *among undefeatable options.*

Let me stress that the present hierarchical theory is only somewhat like Weirich's. The main difference is that where I use probability functions that come by conditionalization from given probability functions, he uses functions that (if I understand him rightly) come by "imaging" on conditionals that are in a certain manner backtracking. The manner is to be somehow "evaluational" rather than "evidential" (Weirich 1986b, p. 445 and p. 448, n. 19). For the contrast between conditionalizing and imaging, see Lewis (1976, p. 311) and Sobel (1989a [Chapter 9]). Further comments on Weirich's own hierarchical theory are made in Section 2.5 and in Sobel (1988a).

Our Weirich-style hierarchical theory selects a_2 in Case III, and is in general complete for cases of the kind we study, selecting at least one action in each. And this theory, though mixed, is in a certain sense all of a piece: in it, all relevant probabilities are alike in that none are actual probabilities just before an action, all being instead probabilities that would come from these by various conditionalizations. But this Weirich-style mixed theory would leave standing and undiminished the first difficulty with our pure stability theory – that in some cases, and in particular in Case II, *actual* expected utilities seem to be relevant. According to the present mixed theory both a_1 and a_2 are rational in Case II even though, given that $\Pr(a_1) = 1$, only a_1 would maximize actual expected utility.

1.5

To respond to just the first difficulty with our pure stability theory, it is necessary and sufficient to combine actual expected utility maximization with ideal stability. There are, however, several ways of doing this. The easiest way, and the way I favor, is simply to combine such maximization with ideal stability, and to rule that an action in an appropriate partition is rational if and only if a decision for it would be stable *and* this action would maximize actual expected utility among all options in this partition. William Harper integrates the two conditions and endorses a hierarchical theory, according to which an action in an appropriate partition is rational if and only if this action would maximize actual expected utility among options for which decisions would be stable. Our theories diverge in the following case, where I assume the same probabilities as in previous cases.

Case IV
$\Pr(c_1) = 1$

	c_1	c_2	c_3
a_1	0	0	1
a_2	0	0	0
a_3	1	0	0

According to my theory, *no* action is rational in this case: only a_3 would maximize, and a decision for it would not be ideally stable. According to Harper's theory, a_2 would be rational: only this option satisfies the ideal stability condition, and so of course it satisfies the maximization condition *restricted* to options that satisfy the ideal stability condition. It can be shown that Harper's theory and mine diverge only in cases in which, according to my theory, no action is rational, though our theories do not diverge in all such cases. They do not diverge in Case III.

1.6

For a third way of combining stability and maximization, we have the rule that an action a in an appropriate partition is rational if and only if (i) there is an option o in this partition such that o would maximize actual expected utility among members of this partition, (ii) there is a path from o to a, and (iii) if reached by this path, a decision for a would be stable. This rule would identify as rational those actions the decisions for which could result from ideally persistent deliberation that involved repeated applications of the rule – decide for some action that would maximize actual expected utility – in which deliberation each application of this rule is made after updating probabilities in the light of the option just previously decided upon. (Comments on "in light of" updating were made in Section 1.3.) Throughout episodes of ideally persistent deliberation in our cases, probabilities of circumstances conditional on actions are to remain constant, and circumstances are to remain certainly causally independent of options.

This ideally persistent maximization theory is suggested by ideas in Eells (1984b,c). Ellery Eells, however, would endorse only a theory that was more realistic in its modeling of deliberation. In a more realistic model of deliberation – in a model of "deliberation with metatickles" (Skyrms 1986) – updatings would be on tentative decisions and tendencies to decision, and would take account of such things as degrees of tentativeness, strengths of tendencies, the amount of time left to deliberate, competing demands on capacities for thought and computation, and so forth. It would be arbitrary to require, in this more realistic model, that probabilities of circumstances conditional on actions should remain constant throughout episodes of ideally persistent deliberation. In this more realistic model, it would be plausible to maintain (as Eells himself has) on the contrary that, during episodes of deliberation, changes in conditional probabilities of circumstances on actions will, as deliberations stabilize, leave circumstances probabilistically independent of actions. (Ideas – "a few qualitative points" – for yet more realistic models can be found in Skyrms 1990, p. 102.) Ideal Bayesian deliberations could include *metadeliberations* on strategies for first-order deliberation in problems at hand and, pursuant to these metadeliberations, implementations of strategies favored. One important idea is that, in cases where ideally persistent deliberation could not terminate, metadeliberations could recommend strategies that would have one not persist.

For a case in which our Eells-style theory diverges from both my theory and Harper's, and in which Harper's theory and mine diverge again from each other, we assume the same probabilities as for previous cases, except that instead of $\Pr(c_1)$ we let $\Pr(c_4)$ equal 1. We assume the following structure:

Case V
$\Pr(c_4) = 1$

	c_1	c_2	c_3	c_4
a_1	0	0	0	2
a_2	1	1	0	0
a_3	0	0	2	1
a_4	0	0	0	0

According to my theory, no action would be rational in this case, since only a_1 would maximize and a decision for it would not be stable. Harper's theory selects a_3 for this case: Decisions for a_2 and a_3 would be ideally stable, and of these a_3 has the greater actual expected utility ($1 > 0$). Our ideally persistent maximization theory selects a_2: Only a_1 would maximize actual expected utility; there is a path from a_1 to a_2; a decision for a_2 would be ideally stable; there is no path from a_1 to any other action. For another case, we assume the same probabilities as for Cases I–IV, and the following structure:

Case VI
$\Pr(c_1) = 1$

	c_1	c_2	c_3
a_1	1	0	0
a_2	1	0	0
a_3	0	1	2

In this case, Harper's theory and mine select only a_1 – it would maximize and a decision for it would be stable – whereas the ideally persistent maximization theory selects *both* a_1 and a_3. Here, of course, a_1 is selected because it would maximize, there is a path from it to itself, and a decision for it would be stable; a_3 is selected because a_2 would maximize, there is a path from it to a_3, and a decision for a_3 would be ideally stable.

These three theories – Harper's, the Eellslike one, and mine – combine considerations of stability and actual expected utility maximization. It is clear that similar theories could combine considerations of undefeatability with actual expected utility maximization, though no one has seen fit to endorse theories in the spirit of such combinations. (Our Weirich-style theory combines undefeatability with *projected* expected utility maximization.) I note that in contrast with possible undefeatability theories, none of our stability and maximization theories is complete in the sense of selecting at least one rational action in every possible case; none of these theories selects an action in Case III. If mixed strategies were open, then – under certain assumptions concerning probabilities for possible circumstances, probabilities of possible circumstances conditional on these strategies, and utilities of mixed strategies in circumstances – each of these theories could be made to select the mixed strategy ($\frac{1}{2}(a_1)$, $\frac{1}{2}(a_2)$). But mixed strategies are not open in this case. (Problems concerning utilities of lotteries are taken up briefly in Sobel 1988c.)

2. Arguments and perspectives

2.1

I maintain that maximization of actual expected utility and ideal stability of decision are singly necessary and jointly sufficient for rational action. A consideration of arguments against aspects of this theory, and of perspectives from which it is somewhat implausible, can make plainer my intentions and perspective. We begin with arguments by Rabinowicz, Eells, and Weirich against the relevance of stability to rational action, consider next an argument of Weirich's against the relevance of maximization of actual expected utility, and conclude with arguments by Harper against simply combining the two conditions in a rule for rational actions.

2.2

Włodzimierz Rabinowicz argues against the practical relevance of how things would look given a decision for an action. He would argue similarly against the relevance of how things would look given the action itself, and so against the relevance of stability to rational action. I adapt his remarks to my Case I:

> One might ask: shouldn't this foreseeable rational belief-change [consequent to a decision to a_2] be taken into consideration . . . when he makes his decision? Shouldn't it give him a good reason against deciding [to a_2]? This seems to be the position taken by Sobel. . . . [He] would claim that the decision [to a_2] is not to be recommended due to its foreseeable effect on the rational agent's probability assignments.
>
> I disagree [because] in terms of [the agent's] pre-choice probabilities, deciding [to a_2] should lead him to adopt . . . beliefs about [circumstances c_1 and c_2] that most probably would be mistaken! . . . To put it shortly, [he] might very well admit, before the choice, that, upon deciding [to a_2], it would be rational of him to change his beliefs about [circumstances]. . . . But, at same time, he deems it highly probable that these new beliefs, while rational, would be false. Therefore, he has no good reason to take such most probably mistaken hypothetical beliefs of his into consideration when he makes his decision. (Rabinowicz 1985, pp. 188-9)

In defense of my theory of rational action, and in particular of the relevance it accords to stability of decision, I note first that this theory is offered neither as a guide to making rational decisions nor in a prescriptive spirit as a system for generating practical directives and recommendations. Instead, it is intended in a descriptive-explanatory spirit as a statement concerning how rational actions are related to beliefs and preferences.

Turning to specific points in or suggested by Rabinowicz's argument, I do *not* think that the instability of a decision for a_2 should be taken into consideration by the agent, or that it gives him a reason against deciding for a_2. I claim only that if an agent is rational then he will not make a decision he cannot maintain on reflection: that rational processes for translating beliefs and desires into actions cannot, if allowed to run their course, issue in ideally unstable final decisions. I do not say that hypothetical beliefs that from an agent's point of view would be false (were they to come to pass) sometimes figure in reasons and grounds for decisions. My claim is rather that such beliefs can, without ever figuring in reasons for decisions, imply constraints and limitations on possibilities for rational decisions.

Can such beliefs figure in reasons for actions? I think not, but stress again that I am not concerned now with such questions. Maximization *cum* stability is proposed as a theory of how rational actions will stand to beliefs, preferences, and evidential commitments; not as part of a theory of reasons for actions, and of what and how things figure in such reasons. This theory is proposed as a theory of actions in accordance with *reason*, and not as a theory of actions in accordance with *reasons*. I think that every action that is – given the agent's beliefs, preferences, and evidential commitments – in accordance with reason will be in accordance with reasons which could be operating. But I think that the converse is not necessary. For I think that an agent's reasons can tell for decisions that would not be sustainable on reflection and ideally persistent deployment of reason's practical resources, given his beliefs, preferences, *and* evidential commitments.

2.3

Ellery Eells objects to ratifiability (Jeffrey 1983) on the ground that "one cannot really assess the ratifiability of a choice until and unless the choice has been made, at least tentatively" (Eells 1984c, p. 185). I assume that he would object to stability, as a criterion of choice, on the same ground. But I doubt that prechoice assessments relevant to stability are problematic in the way Eells suggests, though the point I wish to stress here is that it would not be important to my theory if they were. I endorse stability of decision not as a criterion or partial basis for choice nor as an *end* of reason but rather, though unintended and unpursued, as a constraint or condition on rational choice – that is, as a *feature* of rational choice. The idea is not that a thoroughly rational agent would see to it that he made only ideally stable decisions but rather that this is how things would work out for him, given the nature of rational processes for the generation of decisions. Indeed, maximization itself is endorsed in my theory not as a criterion of choice but as a constraint on rational choice. Supposed problems with its implementation (e.g., the availability of data concerning quantitative probabilities and values) are not problems for my theory, which is - as regards even this condition - an "as if" theory. There is, however, this difference: Though I think that an agent (insofar as he is reflective and deliberate) *is* trying to maximize - or, more colloquially, to do what would probably work out best - I do not suppose that in ideal deliberative ratiocinations an agent has as a distinct goal - let alone, I add with Harper's theory in mind, a prior goal - to make a stable decision. In this, I think Eells and I agree.

2.4

Weirich objects to the incompleteness of stability theories of rational action. He observes (I adjust his remarks to the present context) that

> we can construct cases of decision instablity where not reaching a decision is out of the question. We can, for example, suppose that indecision is ruled out by a penalty attached to failure to reach a decision....
> Perhaps at this point someone might try to [argue] that in cases of decision instability... there is no option whose realization is not irrational.... Although it is not his fault, whatever option he realizes is irrational.... This argument is flawed, however. We have a strong intuition that a decision maker can always choose rationally, even in cases of decision instability. (Weirich 1985, p. 467)

A "drawback" of stability theories, Weirich writes, is that they do

> not yield satisfactory results in cases where mixed strategies are not available. For instance, [they make] no recommendation in [Case III. Some say that] such cases are pathological.... But I hold that there are choices that are intuitively rational in these cases, and that a comprehensive rule for decisions in dynamic settings ought to yield these choices. (Weirich 1986b, p. 442)

I agree that Case III need not be in any way pathological. In particular, it need not involve any kind of pathology on the part of agent. (See Rabinowicz 1985, pp. 182-5, for a nonpathological elaboration of a very similar case.) I agree that neither possible action in this case would be *irrational*. But I maintain that each would be *not rational,* and that in this case no rational choice is possible. It is I think one thing for a choice to be not rational and, given an agent's beliefs and desires, not in accordance with reason and a choice that could not result from ideally complete and thorough rational processes of deliberation. It is another thing for a choice to be irrational and contrary to reason, and a decision that would be *ruled out* by - set finally and irretrievably aside in - any ideally persistent effort to make a reasoned decision. Consider Case III changed by the presence of a "no decision" option added (which, however, has been made "out of the question") and by the presence of a circumstance for which that option would be evidence:

Case III'
$\Pr(c_1) = 1$

	c_1	c_3	c_2
a_1	1	1	2
a_3	0	0	0	(a_3: neither a_1 nor a_2)
a_2	3	1	1

I say that every action would be not rational in Case III', since no decision for an action would be ideally stable, but that a_3 would in addition be *irrational,* since a decision for a_3 would be ideally defeatable. The extension of my theory assumed here consists in part of the rule that an action in an appropriate partition would be (not only not rational but) positively irrational if a decision for it would be ideally defeatable. In Case III', the "no decision" option a_3 is irrational – it would be irrational not to make a decision either for a_1 or a_2. But even so, each of the actions a_1 and a_2, while not irrational, remains not rational, since a decision for either action remains ideally unstable. Though the option of not deciding between the original options is ruled out of the question, decisions for these original options have not been made ideally stable or (I contend) possible for a thoroughly rational and persistent deliberator.

For an irrationality condition that is not only sufficient but also necessary, I propose that an action in an appropriate partition be irrational if and only if either it would be not rational and some action in this partition would be rational, or it would be ideally defeatable. It has been noticed by Rabinowicz (in correspondence and conversation) that – in contrast with my nonhierarchical treatment of rationality, where maximization and stability are fully coordinate conditions – my treatment of irrationality accords a preeminent role to defeatability in the case in which no action is rational. In that case defeatability is necessary and sufficient to irrationality, and nonmaximization is neither necessary nor sufficient.

One consequence of my explication of irrationality is that, in contrast with nonrationality, it cannot happen in a finite case that every action is irrational. This is, I think, as it should be; for even if it can happen that there is no decision in a case that would be consonant with a full and never-ending deployment of the agent's practical rational faculties, it seems that it should not be possible for *all* decisions to be definitely and irrevocably ruled out in the course of a finite and finished well-conducted process of practical ratiocination.

I note that an explication of irrationality that added nonmaximization as an independent sufficient condition would have as a consequence that in some cases every action would be irrational. In the following case, given standing assumptions concerning strong perceived causal independence and probabilistic dependence, only a_2 would maximize, and it is defeatable.

Case VII
$\Pr(c_1) = 1$

	c_1	c_2	c_3
a_1	0	0	0
a_2	1	0	0
a_3	0	1	1

2.5

Weirich maintains that maximization of actual expected utility is "not a goal of rationality," since to maximize actual expected utility would be to maximize "with respect to past information" (Weirich 1985, p. 470). Reason, he maintains, requires an action "that has maximal expected utility with respect to current information, i.e., the information one has at the moment of decision." Reason requires an action "that has maximal expected utility at the moment it is realized" (p. 470).

> The general problem with expected utilities at the last moment for deliberation is that they are out of date when a decision is reached. They do not take account of the information provided by the decision itself. A method of deciding among . . . options should be based upon the information one anticipates having if those options are realized. (Weirich 1986b, p. 442)

I disagree with Weirich's position here, but am inclined to accept some of his words. Reason, I think, does demand reliance on one's opinions – on one's latest and then-current opinions at the time of decision. This is the information on which the decision should be based. Being based on this information, information concerning the decision that will be made cannot itself be included in this "latest and then-current" information. To base a decision in part on information one would have if it is made, and on information one would have if other decisions were made, would be to base it in part on information one does not *have* when it must be made. It seems obvious that such looking ahead to hypothetically anticipated information would be wrong. This is the point on which Rabinowicz – I think, rightly – insists.

Furthermore, it seems that looking ahead in anything like the way Weirich's words suggest must in some cases lead to patently wrong and irrational decisions. Weirich himself allows that unrestricted maximization of anticipated expected utilities would give wrong answers in Newcomb problems:

> the exhortation is not to realise an option o such that, for every option o', the expected utility of o at the moment of decision, if o were realised, is at least as great as the expected utility of o' at the moment of decision if o' were realised. That exhortation would urge one-boxing in Newcomb problems. (Weirich 1985, p. 470)

Weirich's solution is to restrict his forward-looking maximization test to something like options the decisions for which are undefeatable: Only two-boxing decisions are undefeatable in Newcomb problems. However, I think that, even if restricted along such lines, maximization of anticipated expected utility – however exactly it is explicated – must give wrong answers in other cases. For a structure for cases to this point, retain all probabilities and conditional probabilities in Case II and modify its structure as follows:

Case II'
$\Pr(c_1) = 1$

	c_1	c_2
a_1	1	0
a_2	0	2

Decisions in such cases for *both* options would be stable and undefeatable (in this it is like the "nice demon" – Weirich 1986b, p. 440), and suitable stories should entail that the second option (a_2) would maximize anticipated expected utility, however exactly this notion is explicated. And yet it seems clear that – regardless of details of these stories – since the agent is certain that c_1, the rational choice in any such case must be for a_1. It seems clear that, notwithstanding the promise a_2 would have immediately upon its execution, it would be irrational for the agent actually to do a_2.

A case of a similar structure in which the rational action is neither dominant nor optimal – I call it "the popcorn problem" – is, in Sobel (1986a [Chapter 8]), brought against Jeffrey's "logic of decision." I use versions of this case against Eellsian metatickle defenses of evidential decision theory, and use a version of it against ratificationism – a revision of the logic of decision contemplated in Jeffrey (1983) – in Sobel (1986b; 1988a; 1990b [Chapter 2, Section 5 and Appendix V]). It seems that versions of the popcorn problem should tell against final reliance, however it is explained in detail, on "information one anticipates having if . . . options are realized" (Weirich 1986b, p. 442). I argue in Sobel (1988a) that a version does tell against Weirich's own worked-out hierarchical theory.

2.6

Harper holds that maximizing should be restricted to ratifiable alternatives for which decisions would be ideally stable. He realizes that this rule can clash with unrestricted maximization, but claims that

> it is quite correct to have ratifiability override unrestricted maximization in this way, because choosing an unratifiable option cannot be a non-pathological application of causal decision theory. If A is unratifiable then as soon as you commit yourself to A you will give yourself evidence which shows that this commitment is irrational. (Harper 1986, p. 33)

This argument seems insufficient to support its conclusion. It is true, I think, that the choice of an unratifiable option cannot be rational, since it could not be sustained on reflection. But all that follows from this is that ratifiability is a condition of rationality, not that it is a condition lexically prior to maximization or a condition that sometimes overrides maximization. Harper observes that if ratifiability is merely another condition then no action will be rational in a case where no ratifiable action would unrestrictedly maximize, but he does not explain why this is wrong as a comment about rational actions. Harper does not think that there always is a rational action; in his view, there is no rational action if no decision for an option would be ratifiable or ideally stable. I hold that there is no rational action even if a decision for some option would be ideally stable – if no such option would, in terms of the agent's actual beliefs and desires, be recommended at least as highly as any other option. Harper has not given a reason, and I think there is no good reason, for curtailing the maximization condition to actions for which decisions would be ideally stable, or, equivalently, for thinking that there is a rational action whenever there is an action for which a decision would be ideally stable.

3. THE MAXIMIZATION AND STABILITY THEORY: RESTATEMENT AND ELABORATION

The theory I favor is for ideal agents whose beliefs and evidential commitments are represented by unique probability functions, and whose preferences for worlds are represented by functions that are unique to positive linear transformations. It is a theory of actions that are in accordance with reason and subjectively appropriate given an agent's beliefs, evidential commitments, and desires. Such actions may not be *fully* in accordance with reason: Actions can be appropriate given certain beliefs, evidential commitments, and desires whether or not these beliefs and desires are themselves reasonable, and even if they would be changed if subjected to ideally searching rational criticism. And such actions may not be "out of" or "due to" reason, and *actually informed by* reflection on the beliefs and desires with which they are in accordance and for which they are appropriate. Furthermore, even if fully in accordance with and out of reason, actions may not be displays of complete rational control and wills so good that these actions would have been chosen no matter what pressures, distractions, or persuasion had been present – provided only that the agent's preferences for worlds, his probabilities for worlds as upshots of actions, and his evidential commitments were, in all practically relevant ways, what they in fact are (or, less demandingly, provided only that his expected utilities for actions in an appropriate partition, and his expected utilities conditional on these actions, remained what they in fact are).

There are *preconditions* for actions in accordance with reason. An agent is ideally ready to act rationally only when his credences, evidential commitments, and preferences are for all practical purposes settled. An agent is ideally ready to act rationally at a time only if there is an interval bounded by (but not including) this time, during which interval the agent's preferences for worlds, probabilities for worlds as upshots of actions, and evidential commitments do not change in relevant ways; that is, during which interval his expected utilities for actions in an appropriate partition, and his expected utilities conditional on actions in this partition, are constant. A vacillating agent is not ready to act rationally. And an agent is not ideally ready to act rationally if he *thinks* that he would think differently about upshots of actions if he thought more about things – including, in particular, his decisions and apparently pending actions. An agent who thinks he should be vacillating is not ideally ready to act rationally, even if he is in fact not vacillating.

There are also *conditions* for actions in accordance with reason. Suppose that an agent is ready to act rationally and has for some time had settled beliefs and preferences. He acts rationally only if his action is in accordance with these beliefs and preferences, his latest and (at the time for decision) current beliefs and desires. The agent's then-current beliefs and desires just before he acts will inform and determine his action, insofar as it *is* informed and determined by his beliefs and desires. And it is these beliefs and desires – and not, for example, beliefs and desires that he would have were he to deliberate, or had he been deliberating, with ideal persistence – that his action reflects, insofar as it is reasonable. One condition on rational actions, I maintain, is that an action in an appropriate partition is rational only if it fully reflects these beliefs and desires – that is, only if it would maximize actual expected utility unrestrictedly, and its actual expected utility would be at least as great as that of any action in this partition.

However, maximization is not, I think, *sufficient* for action in accordance with reason, for an agent acts rationally and in a subjectively appropriate manner only if a decision for his action is one that could have survived an ideally persistent process of deliberation, given his preferences for worlds, his credences for worlds as upshots of actions, and his evidential commitments, and given no information inputs other than those that could come from reflection on the course of this deliberation. An agent's evidential commitments, while (in contrast with his beliefs and desires) not constituting part of the possible basis for his decision, do imply limitations on possible rational decisions and on actions that would be fully consonant with his state of mind just before their times. If an action is rational, then it seems that it should have been possible for a decision for it to have resulted from an ideally full deployment just before this action's time. My explication of this second condition is that an action in an appropriate partition is rational only if a decision for it would have been ideally stable – that is, only if this action maximizes expected utility, conditional on itself.

According to the theory I favor, these two necessary conditions are jointly sufficient, and an action in an appropriate partition would be in accordance with reason (given its agent's beliefs, desires, and evidential commitments) if and only if (i) it would, in this partition, maximize actual expected utility, and (ii) a decision for it would be ideally stable. I offer this as an explication of the idea that an action would be rational for an agent, given his beliefs, desires, and evidential commitments, if and only if it is an action he could do ("could," not "would," because of the possibility of ties) were he to have these very beliefs, desires, and commitments, were he to make full use of his capacities for affecting his actions by thought – full use, that is, of his capacities for vivid and effective reflection, and of his capacities for bringing his beliefs and desires to bear on his actions in an ideally balanced manner in which their influences were impartially proportioned to their intensities[1] and his choice and action constrained thereby ("constrained," not "determined," again to allow for ties).

NOTE

1. Full uses of an agent's capacity for vivid and variously ordered reviews of (what are in his view) facts might be unending. Against this possibility, references to "full uses" can be replaced by references to "searching uses" that have attained steady states in their practical bearings.

12

Useful intentions

Abstract. It can be useful to intend actions that one will have reasons on balance for not doing, and this suggests that it can be rational to intend irrational actions. I maintain that there is in this suggestion only an air of paradox. Distinctions are made, and ways are explained in which ideally rational Bayesian agents can be capable of intentions whose adoptions make actions rational that would otherwise be irrational.

0. Introduction

My point of departure is a paper of Daniel Farrell's in which he defends

C_1: An ideally rational individual cannot, consistent with her rationality, intend to do what she grants it will be irrational for her to do.

With more hesitation he defends also

C_2: An ideally rational individual could not, consistent with her rationality, *adopt,* at least in the normal way, an intention to do what she grants it will be irrational for her to do. (Farrell 1989, p. 283)

Regarding C_2, he concedes that perhaps "a fully rational person [*can,*] consistent with her rationality, attempt to *bring about* that she intends to do what she believes it would be irrational for her to do" (p. 285, cf. p. 291). He thinks that: "All that follows [for C_2, from his argument for C_1] is that in knowingly attempting to bring this about, she would be attempting to bring it about that she is less than fully rational" (p. 285), and that even when it would be rational she cannot do so "by simply *adopting the intention* to perform the relevant act" (p. 285).

I endorse a version of C_1 and, opposing C_2, say that ideally rational people can have normal ways of adopting intentions to do things that they believe would be irrational were these means not taken, and to do so without inter alia bringing it about that they are less than fully rational. My perspective on rationality is Bayesian, as implicitly is Farrell's. I assume

---

I am grateful to Daniel Farrell, Willa Freeman-Sobel, and anonymous readers for generous comments on previous versions of this chapter.

throughout that an option is rational if and only if it both maximizes expected value and is "ratifiable" in that it maximizes expected value epistemically conditional on its performance. Fine points, including especially some to do with ratifiability (for which see Sobel 1983 [Chapter 10 in this volume] and 1989 [Chapter 11]), are not relevant and will not be retailed here. To have a particular case in mind, here is the best-known one that Farrell discusses, modified somewhat to isolate presently relevant issues.

"The Toxin Puzzle" . . . [S]uppose an eccentric billionaire has offered you a million dollars to agree, today, to drink a certain toxic fluid tomorrow – a fluid guaranteed to be toxic enough to make you quite sick for a day or two, but guaranteed as well to have no ill effects beyond these two days of misery. Having made the offer, he tells you he will return later today, at a prearranged time, to [find out without asking as a Newcomb impresario might do] whether you intend to drink the toxin tomorrow. [If you do, then] he will give you the million dollars, no strings attached. You will not have in fact to *drink* the toxin tomorrow, that is to say, in order to keep the money. The money will have become yours by then, simply because [at that prearranged time] you had the intention to drink the toxin. (Farrell 1989, p. 285)

A "spin doctor" might modify the last sentences thus:

If you do, he will give you the million dollars *after the time for drinking the toxin,* no strings attached. You will not have in fact to *drink* the toxin to get the money. It will be yours simply because, at that prearranged time, you had the intention to drink the toxin.

1. Senses in which the rational cannot intend irrational actions

1.1

It is possible for an individual who is ideally rational to expect to become less than ideally rational and to do things that would be irrational by her now present lights (her credences and preferences), and/or that will be irrational by her future lights. Furthermore, an ideally rational individual can have good reasons for bringing it about that she will be less than ideally rational. However,

(i) an ideally rational person cannot intend while remaining ideally rational to do things that will be irrational by her time-of-action lights, and
(ii) an ideally rational person cannot while remaining ideally rational intend to do things that are irrational by her time-of-intention lights.

All that follows "intend" in (i) and (ii) lies in its scope and characterizes the content and circumstances of a contemplated intention. Principles (i) and (ii), and principles (iii) and (iv) to come, are of the form "an A cannot F," which is to be understood not in the sense of "an A is incapable of F" but rather in the sense of "an A necessarily does not F" or, more perspicuously, "it is necessary that for any x, if x is an A then x does not F." An honest man, though capable of lying, necessarily does not tell lies, and would not be honest if he did, for necessarily an honest man does not lie.

Certainly an ideally rational person cannot do irrational things. Even we know this. So it is clear that an ideally rational person cannot intend while remaining ideally rational to act irrationally by her future, time-of-action lights. No one can intend what they believe to be impossible. That is sufficient for (i). But I note as a matter of interest that given that an ideally rational person knows that ideally rational persons do not do irrational things, (i) follows also from

(iii) a person cannot intend an action she is sure that she will not do,

which principle "derives appreciable support from ordinary usage" (Mele 1989, p. 20 – Alfred Mele is commenting on a principle equivalent to one that comes from (iii) by putting "thinks she probably" for "is sure that she"). Intending to *try to* leap an awful chasm is consistent with believing one will not succeed, but intending to leap it is not (cf. Brand 1984, p. 148). Evidence for (iii) can be found in words of Hector-Neri Castañeda that are meant to deny it:

Many a time an agent proposes to carry out a plan of activities that is fraught with obstacles. . . . In such cases the agent can have both a very *strong* belief that she will not succeed and a very *firm* intent to pursue her plan. (Castañeda 1983, p. 402)

Castañeda writes "pursue," not "execute." Lacking confidence, I can, while working on a theorem with the object (intent?) of proving it, hope to prove it even if I think that I almost certainly will not succeed; but, if I lack all confidence, I cannot plan or intend to prove it. Although I can pursue a plan that I think I will not carry out, I cannot, thinking so about it, propose to carry it out. Again, if – with a view to my absentmindedness – I know that even if I set my mind to stop for milk on my way home I will probably not stop, then though I can (for what it is worth) set my mind to stopping, I cannot decide or intend or plan to stop (cf. Bratman 1992, pp. 11-12).[1]

So much for (i). The truth of (ii) is nearly as clear. "Intending," even if it does not involve being expectant, certainly does involve being willing:

(iv) A person cannot intend an action she is not willing to do.

And an ideally rational person is willing to do, wants to do, and hopes to do only what, by her time-of-willing lights, will maximize expected value. Thus she cannot while remaining ideally rational intend to do things that are irrational by her time-of-intention lights.

For the record, I reject the converse of (iii) according to which a person must intend what she is sure she will do. A person can be sure she will do what she is not yet willing to do and (contrary to Ginet 1962) what she has not yet decided to do. She can be sure she will do something without yet knowing why she will do it or what will make it her best bet and something she will want to do. I also reject the converse of (iv). A person can be willing to do and wish to do what she is sure she will not do either for lack of opportunity or ability.

1.2

From (i) and (ii) combined, which is the version of C_1 I endorse, it follows that no ideally rational individual can bring it about that she intends while remaining ideally rational to do what will, notwithstanding her intention to do it, be irrational by her time-of-action lights. And it follows that no ideally rational individual can bring it about that while remaining ideally rational she intends to do what will be irrational by her time-of-intention lights. Regarding the toxin puzzle, the optimum course of intending to drink and then declining the cup, which would yield the million without days of misery, is not an option for a rational agent who knows she will remain rational.

This much follows from (i) and (ii) combined, and from the logics of intention and rationality. But it does not also follow that an ideally rational person, in knowingly attempting to intend to do what she believes it would be irrational for her to do, would necessarily be attempting to bring about that she is less than ideally rational. Nor does it follow that it is not possible for her to adopt in the normal way an intention to do what she grants would, as things stand, be irrational. These things do not follow, because it may be that she can in adopting intentions in normal ways *change* how things stand.

2. FORMING AND ADOPTING INTENTIONS

2.1

Even ideally rational people can sometimes – by exploiting external features of their situations – bring about intentions to do things that they believe are (as things stand) irrational, and bring about such intentions while remaining ideally rational. In the toxin case there could be the possibility of a legally binding transfer of $10,000 to her dearest enemy conditional on her failing to drink the toxin. That might give her a sufficient reason not only to drink it but to intend to drink it. However, the question, is not whether, in Farrell's words, "a rational person [may be able] to take steps of *some* sort to bring it about that she has [a] desired intention," but whether she can do this "in the simplest . . . way: namely, by straightforwardly adopting the intentions her interests require her to have" (Farrell 1989, p. 291).

2.2

Farrell (1989, p. 294, n. 8) maintains in conscious opposition to David Lewis, that adopting an intention to do an action is strictly speaking not itself any kind of possibly intentional action. He says that

even if it is an "act" of *some* sort, the act of adopting a future-directed intention is the sort of act a person "performs" when, in the light of certain considerations about the state of the world, she comes to have one psychological attitude rather than another. (Farrell 1989, p. 288)

"[T]o form or adopt a future-directed intention" (p. 285), rather than being something that people *do* is, properly speaking according to Farrell, something that when circumstances are right *happens* to them, and consists in "com[ing] to have a certain attitude" (p. 285). "[A]dopting an intention . . . is very much like 'adopting' or coming to have a belief: it is something that happens to one, in light of certain of one's circumstances, rather than something that one actually does" (Farrell 1992, p. 273).

2.3

According to a theory he describes as "recent and very powerful" (Farrell 1989, p. 289), an intention is

a "pro-attitude" of a certain sort: one that is best represented, semantically, as an all-out evaluative judgment according to which, all things considered, it will be best for one to act in the relevant way when the time for action arrives. (p. 289)

It may be noted that this "pro-attitude" includes both an evaluative component concerning the attractiveness of an action, as well as a factual component concerning its possibility and that, when what will be its time comes, it will be an option.

I think that there is more to intending than having such pro-attitudes; I think that being decided and settled in one's mind is a further factor. But I agree that making up one's mind and forming intentions can happen in Farrell's way, that it can be the willing aspect of a finding that, given how things stand, some action would be one's best bet. This may indeed as a matter of statistics be how minds are *usually* made up, and this may as a matter of ordinary language semantics be *the* way in which intentions are *formed,* and with reference to which they can be more or less well informed. But Farrell implies that this is *the only normal way* in which minds are ever made up and intentions established, and that, I think, is an excessively intellectual conception that pushes the model of belief formation much too far. Acts of commitment and resolution are also normal ways in which minds are made up and intentions established, and may as a matter or ordinary semantics be *the* ways in which intentions are *adopted.* "*[A]cts* of intention-*formation*" (Mele 1993a, p. 335, emphasis added) has, I think, an odd ring. The substantive point, however, is that there are two ways in which intentions can come about. There may not, by way of contrast, be two analogous ways by which beliefs come about. Relevant here is that it is somewhat odd to speak of adopting beliefs, as one gathers from Farrell's use of scare quotes that he realizes: see the quotation at the end of Section 2.2 (from Farrell 1992, p. 273). I return in Section 6 to the business of would-be adoptions of beliefs in acts of faith.

2.4

It has been said that "intending to do A is, contrary to what most modern theorists maintain, a goal-directed, actionlike state" (Pink 1991, p. 344). I say that intending, even if not itself actionlike, can be the result of bona fide acts. Commitments and resolutions are ways in which minds can be made up and wills determined. They are ways of attaining intentions to do things that are not antecedently recommended by reasons. Before proceeding to descriptions of mechanisms by which acts of commitment and resolution can work, here are things to think about that can make plausible that rational intentions are sometimes realizable by such acts, voluntarily and "just like that" (contrary to Gibbard 1990, pp. 38-9).

Suppose that I ask you please to hand me a pencil, and that there are two pencils between which there is, as far as you can see, "nothing to choose." No problem. If you choose to respond to my please, you'll hand me a pencil, maybe the red one.

Now suppose that, before you grab one, I say, "I'll give you a nickel if you hand me the red one." Done. For now you have a reason to hand me the red one.

Finally, suppose that I say, "Wait. Don't touch a pencil for five minutes. I'll give you a nickel if you now *intend* to hand me the red one in five minutes. I'll give you the nickel now. I don't care if you do hand me the red one when the five minutes are up. The nickel will be yours whether or not you do that then, *if* you manage now to intend to do it then."

Can you realize this intention? A consideration of facts does not reveal that handing over the red pencil would be best, and the offered nickel is yours whether or not you hand over the red pencil. Can you even so decide now to hand it over; can you resolve without reservations to hand it over, and so intend to hand it over, no fingers crossed? Can you, for example, say to yourself, "I'll do it. I'll hand over the red pencil" (cf. Farrell 1989, p. 285), and by saying this to yourself make up your mind and reach a firm and completely sincere intention to hand over the pencil? Can you, even though as things stand there is no *point* in doing the thing to be intended rather than the alternative, no particular point in doing the thing as distinct from *intending* to do it?

3. MAGICAL BOOTSTRAPPING, AND RATIONAL INTENTIONS AND PREFERENCES

3.1

Can people "at will" and "just like that" make up their minds to do things, even when they see that they have no reasons for doing them rather than other things? That is an easy question, for people are not Buridan's asses. Can ideally rational people sometimes do that? We hope so, and Bayesianism says so, for it says that their actions are maximizing, not that they are uniquely maximizing. But can ideally rational people sometimes make up their minds to do *future* things, even when they see that they lack reasons for doing them rather than other things, and – to come at last to our main question – even when they see that they have, as things stand, reasons for *not* doing them?

Yes they can, for (i) ideally rational people can be capable of acts of commitment that either embody or lead directly to intentions to do actions that are not rationally prescribed – that are even proscribed – by things as they stand before these commitments are made, and (ii) when such acts of commitment are possible, they can be rationally permitted and even rationally prescribed. They therefore provide ways in which useful intentions for antecedently nonrational and even irrational actions can be established by rational people who remain rational, and established "in the simplest imaginable way: namely, by straightforwardly adopting" them (Farrell 1989, p. 291). Acts of commitment are, for some rational people, "bootstrapping" ways of establishing useful intentions given their psychologies – their capacities, propensities, and values.

3.2

Farrell writes that

> someone might hold that actions that *would* be irrational, absent a prior (and rational) intention to perform them, can be rendered rational *given* a prior (and rational) intention to perform them. (Farrell 1989, p. 291)

He has in mind philosophers who maintain that it is necessary that if it would be rational for a person to bring it about that he has an intention to do something then, if he does bring this about, his doing the thing will be rational. Farrell (1989, p. 295, n. 23) has in mind such philosophers as David Gauthier (1984) who garner from this supposed logical connection – between intentions it is rational to bring about and the rationality of actions intended in them – the idea that adopting intentions that it would be rational to adopt is a way of "just like that" (presto!) making acts intended in them rational. Actions are made rational on Gauthier's view by the logic of rationality as it relates to intentions and actions – or by the magic of it, an orthodox Bayesian detractor might say.[2]

Farrell is skeptical of such revisionary post-Bayesian views and of the kind of bootstrapping they say is possible for useful intentions. So am I. But in his preoccupation with views that would have those intentions it is rational to adopt render, as a matter of logic, the acts intended in them rational, Farrell does not consider ways in which this might work for a person as a matter of fact. He does not consider ways of nonmagical psychological bootstrapping that I will describe in Section 4.

3.3

It may be noticed that I have written not of rational intentions, but of intentions it would be rational to bring about or establish. Regarding rational intention, two kinds may be distinguished. An intention i is rational$_1$ at time t if and only if it is for an action that is rational by the intender's time-t lights; an intention i is rational$_2$ at time t if and only if, in comparison with having no intention and having different intentions, having intention i maximizes expected value in terms of the intender's time-t lights. Also (this could be a third sense) there is the matter of intentions it would be rational to bring about. An intention it would be rational to bring about need not, if brought about, be rational in either of those two senses. In particular, what justifies our bringing about intentions is not precisely what justifies "our holding the intentions we do" (Pink 1993, p. 332). It would be rational to bring about an intention if doing so would maximize expected value in terms of the would-be intender's lights "just before" what would be the time of bringing it about, and so before what would be its time.

The clause "it would be rational to intend to do A" can thus be seen to be at least three ways equivocal. It can mean "it would be rational to bring about an intention to do A"; "an intention to do A would be rational$_1$"; or "an intention to do A would be rational$_2$." Connectedly, there is a sense in which "if it would be rational to intend to do A, then it would be rational to do A" is a tautology. It is a tautology in the sense of "if it now would be rational$_1$ to intend to do A then, assuming no changes in relevant beliefs to the time for A, it would at the time for A be rational to do A." There are many other senses in which "if it would be rational to intend to do an action, then it would be rational to do it" is not a tautology. In particular, intentions that are always rational$_2$ can be for actions it would always be irrational to do. Reasons for having an intention can differ from reasons for doing the action intended.

Alfred Mele takes up the question of which psychological attitudes explain and justify intentions, and observes that Pink (1991) opposes an identity thesis according to which they are the same psychological attitudes that, were it time for the acts intended, would explain and rationalize them. (See Mele 1992, p. 326.) For the issues of concern to Mele and Pink, I distinguish among the establishment of, the having of, and the execution of intentions. I distinguish psychological attitudes (i.e. credences and preferences) that would rationalize establishments of intentions from psychological attitudes that would rationalize having established intentions, and then distinguish both from psychological attitudes that, were it time for the act intended, would rationalize it.

3.4

Turning briefly to rational preferences, Duncan MacIntosh conflates what could be termed rational$_1$ preferences with preferences it would be rational to have (rational$_2$ preferences), and does not notice that the preferences it would be rational to have may not, owing to the *costs* of bringing them about, be preferences that it would be rational to bring about (MacIntosh 1993, p. 167). In this connection, he claims that "one should revise one's preferences – and . . . they would automatically so change – if one sees that this would cause conditions targeted in one's original preferences . . . much as my beliefs automatically change to fit new evidence" (p. 165). Preferences do tend automatically to change in the light of, or on deeper appreciation of, considerations relevant to the goodness of their objects. As beliefs seek the true and tend automatically to be rational$_1$, so preferences seek the good and tend automatically to be rational$_1$. But preferences do not tend automatically to be rational$_2$.

Wishful thinking does not come easy for everyone, and is not always a good idea. This holds similarly for wishful preferring, though as explained in Section 4 there are possible means to wishful preferring that do not have analogs for wishful thinking. In Section 4, ways are explained by which an agent may be able at will to change his overall preferences for states, while leaving unchanged his intrinsic preferences for possible worlds. These ways are not, however, relevant to MacIntosh's purposes. He wants ways in which *intrinsic* preferences might be changed, if not automatically upon reflection on the utility of changes, then "by taking a pill, undergoing . . . conditioning, [or] hanging out with the wrong or the right people" (MacIntosh 1993, p. 165).

3.5

The next section is about ways of rendering actions rational by making decisions and commitments that lead to intentions, and not just or especially by making decisions and commitments that lead to intentions that are rational$_1$, or rational$_2$, or that it would be rational to establish. That is one contrast between the processes to be detailed and what would be a bootstrapping of intentions that is grounded in the logic of rationality. Also, it is not intentions however and by whomever attained that will be said to make intended actions rational. Processes to be detailed feature acts of commitment that for their agents at once generate or involve intentions to do actions, and effect changes that make intended actions rational. It will not be intentions themselves that render intended actions rational, but things to do with how they are established and by what kind of agent.

4. Rational adoption of intentions to do things that would otherwise be irrational

4.1

Complementing the concentration of many theorists on intellectual intention *formation* in the manner of reflective credence and preference formation, I attend to ways of willful intention *adoption*. Three ways that can be possible for ideally rational Bayesian agents will be described. Extended forms of the first, while possible for ideally rational agents, are not plausible for real agents. The second and third ways are plausible, and combined are I think the main ones actually in play in the world.

4.2. First way

A person can have a capacity to set her mind adamantly, and to make decisions for the future that are not only firm but irrevocable. By dint of sheer will, a person may be able to "tie her hands" and make necessary and inevitable some action, while making impossible every action that would otherwise have been a possible alternative to it. In order to be sure that she does not answer the door, a person could, instead of handcuffing herself to the stove, do that sort of thing "all in her head." Such acts of commitment could change things relevant to the comparative assessment of an option for which she has decided, by changing the options with which it is compared and, more specifically, by removing every alternative option.

Does this make sense? Is this a capacity that ordinary people possess in any measure, and that a person might have in full measure? I answer "yes" to both questions. It is plausible that everyone has a capacity for irrevocable decisions relative to possible imminent actions that would follow without appreciable passage of time upon decisions to do them. It is plausible that choices, at least in these cases, generally do initiate processes over which deciders have no further control - processes that, in the absence of external obstacles, will issue in the decided-upon actions without possibility of interventions by their agents. It is derivatively plausible that acts of commitment and decision that initiate processes aimed at nonimminent (more or less distant) actions can be similarly irreversible.

I think that ideally rational agents who are well equipped for life would have capacities for irreversible commitments, and that completely ideal agents would always have unlimited control over not only their present possible actions but also their future ones. A capacity for irrevocable resolutions would have potential value for imperfect agents, as the case of Ulysses and the Sirens shows - had he possessed this capacity he would not have needed to be lashed to the mast. And it could be useful for both perfect and imperfect agents in toxin puzzles, for deterrent threats when credible bluffs are not possible, and more generally, given agents' "basic needs for coordination, both social and intra-personal" (Bratman 1992, p. 1). Having a capacity for irrevocable decisions is consistent with ideal rationality, as would be the exercise of such a capacity. While it may usually be wise to keep one's options open so that final decisions when made can be better informed, absolutely irrevocable advance decisions - as well as decisions revocable only given unexpectedly untoward circumstances - can be rationally required even for perfect agents. (For a case to this point see Sobel 1983, pp. 177-8 [Chapter 10, Section 3.3].) But suppose that an ideally rational person exercised such a capacity, and had quite settled that she was going to do some future thing. Could she be ideally rational still? Yes. The rationality of her coming action would be assured because, assuming an exercise of the capacity, at the time of action done would be the best thing she could do by being the only thing she could do. She would have made it best and rational by default. But she would do it not for reasons, but out of necessity; is this consistent with her being ideally rational? Yes. Doing the only thing one can do, acting out of necessity, is not in general a problem for ideal rationality, for the ideally rational can be governed by necessities and constrained in their choices. And acting out of an inner necessity is consistent with being ideally rational, when this inner necessity has been self-generated for good reasons. Inner necessities manufactured by reasoned irrevocable decisions would be deliberately controlled and contrived conditions not of driven and compulsive minds, but of in-charge and self-controlled "hyperautonomous" minds. (Cf. Christman 1993, and Mele 1993b.)

4.3

Were we to lack completely capacities for irrevocable decisions (and for decisions revocable only given unexpectedly untoward circumstances), one might conjecture that evolution would invent some for us. In Friedrich Nietzsche's words, which have the force of a Darwinian prediction,

an animal [with such capacities would have] the right [the power] to make promises.... [T]his animal... [would have] created for itself [a] power... of... continuing to will what has once been willed, a veritable "memory of the will"; so that, between the original determination and the actual performance of the thing willed, a whole world of new things ... can be interposed without snapping the long chain of the will. [It would have made of itself one who can] stand pledge for his own future as a guarantor does. (Nietzsche 1956, pp. 189–90: Section 1 of the Second Essay in *The Genealogy of Morals*)

David Gauthier is

concerned ... with the use of intentional structures to enable me to "stand as my own guarantor" – the phrase is adapted from Nietzsche. An adequate theory of rational deliberation, did we have one, would articulate ... the interplay of intension and reason that makes possible the realization of this ideal. (Gauthier, "Assure and Threaten: An Essay on Rationality," manuscript of June 30, 1989, privately communicated)

I too want to understand these structures. But my speculation, in contrast with Gauthier's, accepts Bayesian maximization as an adequate theory of ideally rational deliberation and studies the play of psychologies – various capacities, propensities, and values – that are consistent with perfect reason. Having "a power of continuing to will what has once been willed" –

having a power of irrevocable willing – is, I say, one way in which ideally rational agents could stand as their own guarantors and so generate useful intentions.

4.4. Second way

An ideally rational agent can have another way of making up her mind to do things even when there are, antecedent to her decision, reasons on balance for not doing them. She can do this if she puts a premium on steadfastness and on being a person of her word to herself, the premium varying perhaps with the firmness with which she makes a decision. For, given such a premium, she could change her reasons for acts by committing herself to them. That she had committed herself to doing something, or resolved to do it, would be for her a new reason for doing it, much in the way in which promises to others provide most people with new reasons for doing what they have promised to do. This is at least part of the way in which advance decisions work for most people. Most people value being steadfast, and perhaps even more would do so if they thought about it and appreciated its place in the economy of an integrated psyche.

Even if one is skeptical about the mechanisms of the first way – of irreversible wills – there is nothing dubious about the value of resoluteness, or about how it can work to make effective commitments and useful intentions possible.[3] There is nothing strange about this common caring for and valuing of steadfastness and resoluteness, any more than there is about the common valuing of being trustworthy and a person of one's word to others. A Bayesian maximizer who values resoluteness will take his plans seriously, even if not as seriously as would what McClennen (1990, p. 230) terms a "resolute chooser." McClennen's resolute choosers, having resolved to act on maximizing plans, do planned acts without regard to their values and whether they are maximizing; they do them simply for the reason that they plan to do them. In contrast, Bayesian maximizers who value resoluteness do not necessarily do planned acts, not even when they are pursuant to maximizing plans. For no matter how highly a maximizer values resoluteness, acts pursuant to maximizing plans – though of course collectively maximizing, and though they will have the value of resoluteness in their favor – need not all be individually maximizing. McClennen's resolute choosers and Bayesian choosers who value resoluteness highly, even when they agree (as they often do) in actions performed pursuant to plans and intentions, differ in their reasons and in their explicit or implicit cogitations. Bayesians, but not McClennen's resolute choosers, are always concerned with what they stand to lose in acting according to their plans. I have made similar suggestions concerning relations between Gauthier's "constrained maximizers" and Bayesian straight maximizers who value fidelity to fair agreements. (See Sobel 1975, p. 684. Also relevant are Sobel 1976, pp. 47-9, 53 [Chapter 15, Section 1]; 1988a; 1988e, pp. 259-60 [Chapter 7, Appendix].)

4.5. Third way

Two ways in which acts of simple commitment can work have been described. For a third way in which a Bayesian agent might, in Nietzsche's words, "stand pledge for [and indeed be answerable for] his own future as a guarantor," consider a person who has the capacity by acts of commitment to make alternatives not impossible, as in our first way, but difficult and costly according to the firmness of her commitment and consequent intention. Having this capacity, she can by resolutions change things that are relevant to assessments of actions for which intentions are adopted in or consequent to resolutions. In contrast with our second way, rather than attaching bonuses that make new reasons for doing intended things given the attractiveness to her of steadfastness and her appreciation of its importance, this third way fixes penalties that make new reasons against failing to do intended actions. In Reinhard Selten's words,

One could assume that human beings have some kind of "internal commitment power." Once somebody has made a plan, a negative utility will be attached to any change of the plan. This idea is in agreement with the theory of cognitive dissonance (Festinger 1957). (Selten 1978, p. 146)

That is another way in which resoluteness could work for ideally rational Bayesian agents.

4.6

There are possible objections to the first way of reaching useful intentions (by irreversible decisions), which can seem less a way of generating intentions to do future actions then a way of long in advance actually starting to do them. One may also question (though I would not) the metaphysical possibility of this way. Michael Bacharach, echoing Jean-Paul Sartre (for whose words see Section 5), would say that an agent's belief in the first way would involve

the fallacy of that Rule of the Present [whereby] we are lulled into a sort of current self-deception, into forgetting the unpalatable thought that the plans we have for ourselves ... are out of our present hands, that we-later will be unconstrained by our present wisdom, we-later will be of our present creation but not of our present mind, like disrespectful offspring. (Bacharach 1992, p. 269)

The second way of reaching useful intentions, by valuing steadfastness, while not subject to theoretical objections can still seem to be of limited utility: Although most persons do value personal resoluteness and would, I think, on reflection value it highly, many would not value it highly enough to make reasonable the drinking of toxin that they know will make them "quite sick," even given the guarantee that there will be no "ill effects beyond . . . two days of misery" (Farrell 1989, p. 285).

The third way of reaching useful intentions – by generating psychic resistance to, and penalties for, backsliding – is thus a welcome complement. Leaving aside as of little relevance to the psychologies of ordinary reasonable people the first devices (of absolute irrevocability and of revocability only given unexpectedly untoward circumstances), many decisions of ordinary people can in the second and third ways combined be made practically irrevocable in that they are resistible only with difficulty and subject to psychic penalties, including prominently "the loss of a good character with oneself." In this combination way, ordinary reasonable people can reach useful intentions to do things for which they antecedently lack sufficient reasons, and to do things against which they antecedently have sufficient reasons.

I think, however, that there are some useful intentions that few if any ordinary reasonable people could reach in these or any ways – useful intentions to carry out threats of massive nuclear retaliation coming first to mind, given the hell to which these intentions (even if good and useful) could pave the way. A propos a recent critique of nuclear deterrent intentions (Pink 1991, pp. 346–8), the problem with such intentions is not that, when it came time to retaliate, a would-be deterrent intender would no longer have what had been an essential reason for adopting the intention. The problem, I think, is that she might not be capable of generating, in adopting a deterrent intention, sufficient reasons for maintaining and executing it. In that case – that is, if she cannot in adopting it generate sufficient reasons for maintaining and executing it – she cannot without error adopt it.

5. CONCLUSION

Whether a Bayesian maximizer has the capacity to establish useful intentions, and (if so) what the extent is of her capacity, depend not only on her rationality but crucially on other independent aspects of her psychology. Bayesian maximizers contrast with McClennen's resolute choosers, Gauthier's intenders, and MacIntosh's choosers (MacIntosh 1993, p. 173). The capacity for useful intentions would be complete in agents that were ideal specimens of these purported models of rationality; it would be complete given their rationalities regardless of other independent aspects of their psychologies. In contrast, ideally rational Bayesian maximizers could be quite devoid of the capacity to establish useful intentions to do things that are not antecedently rationally prescribed. They can be devoid of this capacity, but they need not be; indeed, they can be possessed of it in full measure.

Sartre exaggerates, and does not speak for everyone, when he states:

> There exists ... anguish in the face of the past. It is that of the gambler who has freely and sincerely decided not to gamble any more and who when he approaches the gaming table, suddenly sees all his resolutions melt away. ... The earlier resolution of "not playing anymore" is always *there*, and in the majority of cases the gambler when in the presence of the gaming table, turns toward it as if to ask it for help. ... But what he apprehends then in anguish is precisely the total inefficacy of the past resolution. ... What the gambler apprehends at this instant is again the permanent rupture in determinism; it is nothingness which separates him from himself; I should have liked so much not to gamble anymore. ... It seemed to me that I had established a *real barrier* between gambling and myself, and now I suddenly perceive that my former understanding of the situation is no more than a memory of an idea, a memory of a feeling. In order for it to come to my aid once more, I must remake it *ex nihilo* and freely. The not-gambling is only one of my possibilities, as the fact of gambling is another of them, neither more nor less. ... After having patiently built up barriers and walls, after enclosing myself in the magic circle of a resolution, I perceive with anguish that *nothing* prevents me from gambling. (Sartre 1956, pp. 32-3)

Against this I say that while "the native hue of resolution is sicklied o'er with the pale cast of thought" (*Hamlet* III.i) for all persons some of the time and for some persons all of the time, it is not so for all persons all of the time, as it would be if it were a consequence of the metaphysics of consciousness and the will. And it would never be so for an ideally rational person who was in her capacities for action ideally well equipped for life. Such a person would have one or more of the capacities that have been described, or others that provided ways to all possible useful intentions.[4]

6. Postscript

The ways described here in which acts of commitment can work do not have analogs for what one might term "simple acts of faith." And while I believe that acts of commitment are sometimes possible and that I can understand how they can work even for the ideally rational, I am still skeptical about the possibility of simple acts of faith – especially for the ideally rational. Ways have been described in which you might manage, for a nickel, to intend at will and more or less "just like that" to hand me a red pencil; but I can't imagine how you might manage, even for $500, to believe more or less "just like that" that a red pencil was made in Philadelphia. Suppose you have no idea where the pencil was made, so that believing that it was made in Philadelphia (rather than some other place) would be no more contrary to reason than intending to hand it rather than the other one to me.

I understand how persons might be able sometimes "just like that" to make up their minds to *do* future things, even when they see that there are (as things stand) no reasons for doing them. But I cannot understand how persons might manage to make up their minds to *believe* things even when they see that there are no reasons – in the sense of considerations that argue that they are true – for believing them. I cannot understand how persons might simply at will make up their minds in exercises of faith to believe such things. Perhaps this is a conceptual impossibility somehow "tied . . . to the ways in which belief aims at truth" (Bratman 1992, p. 4). I suspect that there is more here than a psychological and technological impossibility. This is a hunch for which Jonathan Bennett has – without, he stresses, complete success – sought an explanatory ground (Bennett 1990, pp. 105ff).

NOTES

1. At a recent National Basketball Association game, a member of the audience (chosen for the color of his shoes) was given one shot from the top of the circle at the opposite end of the court for $1,000,000. He had never tried such a shot before. No one was more surprised than he was when he hit "nothing but net" (though the onlooking professionals were just as surprised). He was lucky, but *did not do it unintentionally.* However, we can also say that he *did not intend to do it* and, I think, that he *did not do it intentionally* (cf. Mele 1992). I conjecture that these locutions behave and are related somewhat differently in contexts of descriptive ascription and denial versus contexts of fault-finding and exculpation.
2. Compare: "if it is rational to form [a] conditional, deterrent intention, then, should deterrence fail and the condition be realized, it is rational to act on it. . . . [This is not to] suppose that, because adoption is utility maximizing, implementation magically becomes *utility maximizing* [as distinct from rational]" (Gauthier 1990, pp. 311–12, emphasis added).
3. Contrary to Hollis and Sugden (1993, pp. 27–8), who say that expected utility maximizers can have only forward-looking reasons. Their argument is based on the character of "acts" in Leonard Savage's foundations as arbitrary functions from "states" to "consequences" (Savage 1972, p. 14). They say that Savage's "is still generally regarded as the most satisfactory statement of the foundations of rational choice theory" (Hollis and Sugden 1993, p. 6). However, foundations such as those provided by Richard Jeffrey with Ethan Bolker, "mono-set" foundations (Fishburn 1981, p. 185) in which acts are propositions "within the agent's power to make true if he pleases" (Jeffrey 1983, p. 84), accommodate all kinds of reasons (cf. Jeffrey p. 211). Along with many philosophers, I favor such foundations, and this not just for their roominess.

4. In conversation in the summer of 1992, Duncan MacIntosh and I agreed that some people, having committed themselves to a course of action that is not reasonable in terms of their initial lights, are disposed to change their preferences and credences in ways that make it reasonable. But such tendencies to rationalize actions (after commitments have been made to them) are foreign to ideally rational persons. Perhaps some ordinary persons can, depending on their tendencies to rationalize, deliberately allow their intentions to get ahead of, and to pull up, their reasons for actions intended. But that is not a way to useful intentions that can be open to ideally rational persons, whose preferences are constrained somewhat (as are their credences) by principles of reflection (Sobel 1987). However – and this is the argument of the present chapter – simultaneous wishful manipulations by acts of commitment both of intentions and preferences, though not of credences, can be open to ideally rational persons (cf. MacIntosh 1993; Sobel 1993b, p. 27, n. 2).

IV

Interacting causal maximizers

13

The need for coercion

Abstract. Very knowledgeable and unerring Bayesians are characterized, and it is shown that problems amenable to coercive solutions are possible for such agents even if they are perfect utilitarians concerned only with what they consider to be the collective good. Even such agents can be bothered by prisoners' dilemmas and frustrated when they would coordinate their actions. Solutions – ways out of or around such problems – by systems of artificial incentives are studied; it is shown that sometimes these need to be directly coercive and that sometimes it is not enough to provide each agent with assurances that all others will do their part in socially useful schemes that all approve. I contend in an appendix that trustworthy agents – agents who endorse as preeminent duties to keep promises – would have noncoercive ways out of, and around, their interaction problems.

0. Introduction

Everyone knows that coercion is necessary for ordinary people. But many philosophers have believed that perfectly good and wise agents could do without it, that perfect agents left alone would inhabit utopias. It is a pleasant thought – though not, I think, true. I believe that even perfect people could need coercion, but in substantiation I offer here demonstration of only a weaker thesis. Hyperrational act-utilitarians are considered, and, without any attempt to defend or to assess their claims to perfection, it is shown that they could need coercion.

Two connected views will be opposed. The first holds that a community of supremely intelligent and well-informed agents, concerned only with the common good and possessed of strong wills, could never need coercion. The second holds that a member of such a community could never need to be coerced to do his part; that knowing of the goodness of a given social arrangement and of the adequacy of its enforcement as regards others, he would do his part willingly; that from his point of view, sanctions – if and when needed by his community – would be needed only

Reprinted with revisions by permission from *Coercion: Nomos XIV* (J. R. Pennock and J. W. Chapman, eds.), Chicago: Aldine & Atherton, 1972, pp. 148–77.

as guarantees that others would follow the rules and do their parts. Both views are mistaken. The need for coercion is more radical than they would suggest. Even hyperrational act-utilitarians could need it, and need it both collectively (a community of such agents could need it in order that the common good be well served) and individually (each such agent could need it in order that he should do his part). The two parts of this thesis are argued in turn, after the agents and communities to be studied are described in more detail.

1. THE HYPERRATIONAL COMMUNITY

We consider communities that satisfy the six conditions described in this section as follows.

1.1. Act teleology

Each member of the community holds that a person ought to do one of the best actions open to him. For more precision, several technical terms are used.

An action x' is a *version* of an action x if and only if x' entails doing x – for example, returning serve backhand is a version of returning serve. An action is *open* to an agent if and only if he can do it by choice or for sure. An action is a *most specific* action open to an agent if and only if it is open to him and no incompatible versions of it are open to him: returning serve is not a most specific action open to a player when he can choose to return backhand or forehand. A *minimal* action open to an agent is an action he can choose to do which, if chosen and initiated by him, could not be stopped short of completion. An agent cannot change his mind about an initiated minimal action.

An agent a's *expected value* for an event e that would take place at t is a probability-weighted average of the values of the possible futures from t were e to obtain. Probabilities are relative to the past up until t and the occurrence of e, and such as a would assign after full consideration; values are those that a would assign after full consideration. Suppose an event e has n possible futures, $F_1, \ldots, F_n$. Let an agent a's probability for future F_i given e and the past be p_i, and let a's value for F_i be v_i. Then a's expected value for e is

$$\Sigma_{i=n}^{i=n} p_i \cdot v_i.$$

A *partition* of an agent's possibilities for action at t is a set of mutually exclusive actions he can do at t which are jointly exhaustive of his possibilities for action at t: An agent must do exactly one action from each partition of his possibilities for action. A partition, each member of which is a most specific action open to an agent, is a *finest analysis* of his possibilities for choice. A finest analysis of an agent's minimal options on an occasion is a finest analysis of his possibilities for choice, each member of which is a minimal action open to him on the occasion.

The practical principle that is endorsed by each member of the community may now be spelled out as follows.

> An action x is a *right* most specific minimal action open to choice by an agent on an occasion if and only if there is a finest analysis K of his minimal options on this occasion such that (i) x is in K and (ii) for each action y in K, his expected value for doing x on this occasion is at least as great as his expected value for doing y on this occasion.[1]

Each member holds that:

> For any agent and occasion, if S is a set of actions containing all and only right actions from a finest analysis of his minimal options on this occasion, then he *ought* to do an action from S on this occasion.

(So each member holds that an agent ought to do an action x if and only if he ought to do an action from the unit set that contains x.)

We cannot assume that there will always exist a unique finest analysis of an agent's minimal options. For example, if an agent can try to light a fire in a certain way and knows that it is raining, then both "trying to light a fire in this way" and "trying to light a fire in this way in the rain" may be for him most specific minimal options; there will be at least two finest analyses of his minimal options — one containing one of these actions, the other containing instead the other action. But though there will not always be a unique finest analysis of an agent's minimal options, the example just stated suggests — and we shall assume — that any two finest analyses of an agent's minimal open actions on an occasion will be *equivalent:* we assume that their members can be paired without remainder, so that the agent performs one member of a pair on the occasion if and only if he performs the other. Furthermore, since actions paired will be open to the agent, it follows that "in the end" he would know that he could not perform one without performing the other. So choices of paired actions will have for him the same expected values. But then, for practical purposes, it cannot matter how an agent's options are described. Applying his maximizing principle to different finest analyses of his minimal options cannot lead to incompatible or even substantially different directives. If applying it to one analysis established that he ought to do x and applying it to another analysis established that he ought to do y, then – given our assumption that finest analyses of an agent's minimal options on an occasion are equivalent – he would do x on this occasion if and only if he did y.

Each agent is a teleologist, a purely forward-looking teleologist. This has implications for his values. When assigning a value to a possible future, he does not regard what would be its past; the value assigned in no way depends upon what would be its past. For illustration, we consider what for us is an important consequence of this stipulation. Suppose that future F_1 would begin with the keeping of a promise that everyone knows was made, and that future F_2 would begin instead with the keeping of a promise that was not made but that everyone thinks was made. For a purely forward-looking agent, nothing has been said that makes F_1 "so far" or ceteris paribus more valuable than F_2. In all things our agents "look solely to the good to follow." Hence they do not value promise-keeping per se. Of course, promise-keeping is only an important example. Neither do they value, or count as obligatory, acts of gratitude, reciprocity, or retribution. A full list would be endless.

The forward-looking restriction is the only assumption we make about values. Thus we do not assume that our agents agree in their values. And we do not insist that they be benevolent. The agents discussed always are universalists and impartially benevolent, but that they are is not essential to any arguments. Altruists or egoists would have served our arguments as well.

1.2. Practical rationality

Roughly, our agents do what they know they ought to do. More precisely, if one of them knows that he ought to do an action from a certain set, then he does an action from this set. And if he does a certain most specific minimal action open to him, then he knows that this action belongs to a set that contains all and only right actions from a finest analysis of his minimal options. *Our agents always do what they know they ought to do, and they never act without such knowledge.*

Suppose that a finest analysis contains several right actions. Given the absence of reasons for doing one of these actions rather than another, we assume indifference (not impotence) and assume that any rational observer will judge each of the right actions equally probable. Other assumptions concerning expectations in cases of indifference suggest themselves: we might assume expectations fitted to the particular case and based, perhaps, on knowledge of causal determinants of the choice. But alternative assumptions need not be explored. When the assumption made plays a role, as in Section 2.2, the argument could be redirected so as to proceed without *any* assumption concerning expectations in cases of indifference. (See note 7.)

1.3. Theoretical rationality

We assume that our agents are reasonable in their assessments of probabilities and, for simplicity, that they agree in these assessments. Further, each agent is to be an accurate and fast reasoner. If he knows certain premises, then he knows of whatever follows from them what he would like to know. Finally, our agents will be orderly in their deliberations, putting first issues first. This condition is difficult to characterize in general and with precision, but what this comes to in at least one case will be apparent in Section 3.2.

1.4. Knowledge of situations

Each agent knows the past and the structure of the situation in which the community finds itself at a given moment: He knows a finest analysis of each agent's minimal options as well as which combinations of these actions are possible. In addition, he knows how each agent views the several possible patterns of action – what expected values the agent assigns to them. Further, he knows the extent (if any) to which these actions are dependent on, and apt to influence, one another.

1.5. Publicity of practical knowledge

In case it is not a consequence of other assumptions, we also assume that all matters relevant to what an agent ought to do are public and knowable by everyone, if knowable by anyone. No one is to have special access to what anyone, including himself, ought to do.

1.6. Community self-knowledge

That the community satisfies conditions 1.1 through 1.5 is to be "common knowledge" in the community: Each member is to know that the community meets these conditions, know that this (i.e., that the community meets these conditions) is known by each member of the community, know that *this* is known by each member of the community, and so on.

This completes our specifications for the community. It is not an objection that *actual* agents are never so well-informed and self-controlled.

Our concern is not with the interaction and coordination problems of actual agents, or with solutions available to actual agents because of their imperfections.[2] Nor is it an objection that *perfect* agents would know more than we have explicitly assumed that ours know, for perfect agents would know at least the things assumed and, as I argue below, agents who know these things can be embroiled in difficulties from which additional information could not extricate them.

2. The community's need for coercion

Reasonable and well-informed as they are, our agents can need sanctions and coercive government. Two cases show this: in the first there is, and in the second there is not, disagreement over values.

2.1. The Farmer's Dilemma

2.1.1. Row and Column live on opposite sides of a chasm: they cannot cross it or move goods across it. Each can either plant wheat that he can eat or flowers that the other can view at a distance and enjoy – each enjoys flowers only when viewed from a distance, so neither can enjoy his own. To simplify, we assume that each must plant something, either only wheat or only flowers. And to complicate, we assume that each is capable not only of planting wheat, or flowers, "for sure," but also of making it probable to any degree p that he plant wheat and to the degree $(1-p)$ that he plant flowers: we assume that a finest analysis of their minimal options contains, in addition to the two "pure" strategies, every "mixed" strategy built upon them. (Strictly speaking, it is unlikely that, for example, planting would ever be a minimal action open to an agent. The case could be made more plausible by taking "planting wheat" as short for "starting to plant wheat," and similarly for "planting flowers.") Such finely tuned agents, such virtuosi, are of course not encountered in real life. But it is good to see that even they can have problems, and it will be clear that the same problems could be more easily established for less well-endowed agents. Finally, regarding their actions, we assume that they are *fully independent*.

Row is a young man, unsubtle and quick to enjoy things of natural beauty. In contrast, Column is an older man, mature and possessed of refined sensibilities. Each would suffer from hunger without wheat, and would take pleasure in the other's flowers. As for their values, in many ways they are in agreement, but there is this difference: Row places a premium on the mature enjoyment that Column would get from looking at flowers, while he counts as relatively insignificant the sort of pleasure flowers afford him. Column values highly the innocent, spontaneous joy of which only Row happens to be capable, while taking a much dimmer view of the sophisticated pleasures that are apt to come his own way.

Row and Column rank the four possible planting patterns as follows: They agree that the pattern (W, W) in which each plants wheat is to be preferred to the pattern (F, F) in which each plants flowers. Neither feels that the other's aesthetic pleasures would be worth their both starving. But they disagree as regards the two mixed patterns. Row assigns *highest* value to the pattern (F, W) in which he plants flowers and Column plants wheat: He ranks (F, W) even higher than (W, W) – he judges Column's aesthetic enjoyment worth one person's discomfort. Row assigns *lowest* value to the pattern (W, F) in which he plants wheat and Column plants flowers: he ranks (W, F) even lower than (F, F) – as has been said, he judges Column's aesthetic enjoyment worth one person's discomfort. Column, given the premium *he* places upon *Row's* sort of aesthetic pleasure, reverses these evaluations for the patterns (F, W) and (W, F). Their situation has the following structure:

	Column W	Column F
Row W	**3** / 3	4 / **1**
Row F	1 / **4**	2 / **2**

The greater the number, the greater the expected value. (The magnitudes have been chosen arbitrarily but without prejudice. Any numbers similarly ordered would serve the argument as well.) Row's expected value for a pattern is entered in bold type in the lower left corner of the pattern's box; Column's in the upper right. Expected values for combinations of mixed strategies are computable. Consider a combination consisting of the mixed strategy $(\frac{1}{3}, \frac{2}{3})$ for Row, under which Row makes his W $\frac{1}{3}$ probable and his F $\frac{2}{3}$ probable, and the mixed strategy $(\frac{3}{4}, \frac{1}{4})$ for Column. Row's expected value for this pattern of strategies is $(3 \cdot \frac{1}{3} \cdot \frac{3}{4}) + (1 \cdot \frac{1}{3} \cdot \frac{1}{4}) + (4 \cdot \frac{2}{3} \cdot \frac{3}{4}) + (2 \cdot \frac{2}{3} \cdot \frac{1}{4})$ or, more simply, $3\frac{1}{6}$.

2.1.2. What will they do? It may seem that what they do should depend upon what they should expect each other to do, but this is not so. It will be clear to each that he should plant flowers *whatever* he should expect the other to do. This result follows from the independence of their actions and the "dominance" for each of the pure strategy in which he plants flowers for sure. (Dominance alone would not support the argument; see

Jeffrey 1965b, p. 8.) Row, for example, could reason: "What I should expect Column to do doesn't depend at all on what I do.[3] As it does not matter what I should expect Column to do, my best bet in any case is to plant flowers for sure. Suppose I should judge it probable to degree p that Column will plant wheat and to degree $(1-p)$ that he will plant flowers. Then my expected value for my pure strategy $(0, 1)$ in which I plant flowers for sure is $4p+2(1-p)=2+2p$, which is greater than my expected value for each of my other strategies $(q, 1-q)$, $1 \geq q > 0$, in which I do not make my flower planting certain: the expected value for such a strategy is $3pq+q(1-p)+4(1-q)p+2(1-q)(1-p)$ or $2+2p-q$."

The situation is one of frustration. It is as if nature had contrived a trap. Agents caught in it might well feel that "there should be a law against such predicaments."[4] They could use a law "with teeth" against planting flowers, and if coercive law were their best remedy then we could say that they *needed* such a law. Of course, even if coercive law were their best remedy, they could prefer to live with their problem unsolved: People can need things that they cannot afford and do not even *want* at the going price.

2.1.3. In the Farmer's Dilemma, actions of reasonable agents operate to frustrate the ends they would have their actions serve. This has an air of paradox. Surely something has gone wrong; some available argument must have been overlooked. It cannot be that, left to their own "natural" devices, these agents are lost. We consider two attempts to evade the conclusions reached in Section 2.1.2.

1. One wants to say, "There must be something wrong with the argument that purports to show that Row, for example, ought to plant flowers. For this argument can be met by one that contradicts its conclusion: 'If I were to plant wheat, then planting wheat would be what I ought to do: for I am a reasonable person – I do only what I know that I ought to do. But if planting wheat were what *I* ought to do, then it would be what *Column* ought to do; the situation is symmetrical. And if planting wheat were what Column ought to do then that is what he would do, for he too is reasonable and knowledgeable. So, if I were to plant wheat, then Column would plant wheat and the pattern (W, W) would obtain; this follows by two applications of hypothetical syllogism. By parity of reasoning, if I were to plant flowers then that is what Column would do, and the pattern (F, F) would obtain. So in actuality my choice is between these two patterns; the other two are not really possible. But then my choice is clear, and so is the conclusion that I ought to plant wheat, not flowers.'" (This argument might be compared to Rapoport 1966, pp. 141-2.)

Several objections suggest themselves. The main one concerns the argument's very first step. Row claims that if he were to plant wheat, then that would be what he ought to do. But in effect it has been argued that this conditional is false. According to the argument in Section 2.1.2, Row ought not to plant wheat; if this is right, then one supposes that if he were to plant wheat (and he is capable of this, even if he should not and would not *do* it) then he would not be doing what he ought to do. We depict Row as seeking to determine what he ought to do, and yet – in the very first step of his argument – we have him take for granted that, since he is a reasonable person *whatever* he did would be right and what he ought to do! The difficulty, I think, is mainly with the notion of a reasonable person, which should apply to one who *does* only reasonable things, not to one who *can* do only reasonable things. Everything that a reasonable person does is reasonable, and Row is a reasonable person. But it does not follow that anything Row did would be reasonable.

This claim – that anything Row did would be reasonable – could be supported only in the totally bizarre circumstance in which, for anything that is not what Row ought to do, doing it would not only be out of character but something of which he was actually incapable. To be brief, perhaps the following suffices against the present objection: Either planting wheat is not what Row ought to do or planting flowers is not what he ought to do, and Row knows this. So he cannot reason *both* that if he were to plant wheat he would be doing what he ought to do, *and* that if he were to plant flowers he would be doing what he ought to do. At the very least, his argument – since it requires both of these conditionals – fails for want of a premise. (Even granting its premises, the argument would be vulnerable. It involves applications of hypothetical syllogism to contrafactual conditionals, and there are in my view good reasons for thinking that this is not a valid form of inference for such conditionals. See Sobel 1970, pp. 436–7.)

2. Even if (F, F) is the inevitable outcome of an isolated Farmer's Dilemma, it can seem that in practice one could anticipate repetition, and that given a series of dilemmas reasonable agents would settle upon (W, W). But, at least under the only knowledge assumption compatible with the spirit of this study, this is not so. The assumption made specific for agents confronted with a series of three Farmer's Dilemmas is:

> Each agent knows that (there are only three Farmer's Dilemmas to come and that when there are only two to come each agent will know that (there are only two to come and that when there is only one to come each agent will know that (there is only one to come))).

Consider agents confronted with a series of three dilemmas. Each can see that the third one, when it comes, might as well be isolated: it will certainly resolve into (F, F). When it comes, each will know that there are no future dilemmas to be considered and perhaps affected by performances in it; and past dilemmas will be of no interest since they could matter only by generating expectations concerning the other's performance in the third dilemma and, as was shown in Section 2.1.2, what an agent ought to do in a Farmer's Dilemma does not depend at all upon what he should expect the other to do.[5] So the last dilemma will resolve into (F, F), and that it will do so will be evident to our agents while in the second dilemma. They will know that for all practical purposes it might as well be last and isolated; and so it too will resolve into (F, F). But then, turning to the first dilemma, our agents will see that *it* might as well be last and isolated, and that each should and will plant flowers for sure in it and indeed in all three dilemmas. (See Luce and Raiffa 1957, pp. 97–102, for a discussion of iterated Prisoners' Dilemmas.)

2.2. The Hunter's Dilemma

2.2.1. Row and Column are in value disagreement in the Farmer's Dilemma, and even if (as I think) this does not mean that either is in error, it is appropriate that we consider a case of value agreement. It can seem especially implausible that men in value agreement and never at cross-purposes should have reason to wish themselves subject to coercion.

We consider a two-man hypothetical community and suppose that its members are wondering, somewhat anxiously, whether they will be able to cope with a situation they see headed their way. Here is the situation, described for simplicity as if it were already upon our agents.

Row and Column have left their fire. They are hunting separately and out of sight of one another, so their actions are independent. Under a finest analysis of minimal options, each has available the pure strategies of continuing hunting and returning to the fire, as well as all mixed strategies built upon these pure strategies. The wind has freshened, making it important that someone return to the fire; but it does not matter who returns. Indeed, under the finest analyses stated, the situation is one of full *practical symmetry:* if a given agent ought to employ a certain strategy or be indifferent as regards several, then the same is true of the other agent. Still, it is important that *someone* return; it would be best if just one did, next best if both did, and worst if neither did. On these, as on all evaluations, the agents are agreed. The situation has the following structure (see matrix at top of next page), where specific magnitudes are again unimportant.

                      Column
                      F   H

                 F | 2 | 3 |
            Row
                 H | 3 | 1 |

2.2.2. What will happen? It will be shown that either Row and Column will change and become less perfect before the dilemma is upon them, or the dilemma in its original form will somehow be headed off. Something will "give," this is certain, for hyperrational agents cannot be in (and act through) a situation exactly like the one described. Hyperrational agents always know what they ought to do (see condition 1.2), and in a Hunter's Dilemma there is nothing such agents *could* know that they ought to do. Every practical knowledge hypothesis can be eliminated as untenable. Row might, one supposes, know that he ought to employ a certain strategy $(p, 1-p)$ in which he makes his returning to the fire probable to degree p and his continuing hunting probable to degree $(1-p)$. Row might also, one supposes, know that he ought to choose indifferently from some nonunit subset of these strategies. We consider these two cases in turn.

1. Suppose Row knows that he ought to employ strategy $(p, 1-p)$, where $1 \geq p \geq 0$. There are three possibilities: $1 \geq p > \frac{2}{3}$; $p = \frac{2}{3}$; $\frac{2}{3} > p \geq 0$. (Given the magnitudes chosen for the structure of the case, $\frac{2}{3}$ is a pivotal value for p.)

Let $1 \geq p > \frac{2}{3}$. That is, assume Row knows that he ought to employ strategy $(p, 1-p)$ where $1 \geq p > \frac{2}{3}$. Then, given the practical symmetry of the situation, this is also the strategy that Column ought to employ, so Column ought to make his own hunting probable to a degree *less* than 1. But given Column's knowledge of Row's practical knowledge and practical rationality (see conditions 1.5, 1.2, and 1.6), Column knows that Row will employ strategy $(p, 1-p)$, thus making Column's expected value (EV) for his own pure strategies F and H, respectively,

$$EV_C(F) = 2p + 3(1-p)$$

and

$$EV_C(H) = 3p + 1(1-p).$$

Column's expected value for $M(q)$, a mixed strategy $(q, 1-q)$ on his part in which $1 > q > 0$, is

$$EV_C[M(q)] = 2pq + 3p(1-q) + 3(1-p)q + (1-p)(1-q)$$

or (regrouping, factoring, and putting equals for equals)

$$EV_C[M(q)] = q[EV_C(F)] + (1-q)[EV_C(H)].$$

It is obvious that unless $EV_C(F) = EV_C(H)$, either $EV_C(F) > EV_C(H)$ and $EV_C(F) > EV_C[M(q)]$ for any q such that $1 > q > 0$, or $EV_C(H) > EV_C(F)$ and $EV_C(H) > EV_C[M(q)]$ for any such q. And so, given that $1 \geq p > \frac{2}{3}$, $EV_C(H) > EV_C(F)$ and $EV_C(H) > EV_C[M(q)]$ for any q such that $1 > q > 0$. But then Column ought to do H for sure, which contradicts the conclusion (reached in the previous paragraph) that he ought to make it probable to a degree less than 1 that he hunts. That $1 \geq p > \frac{2}{3}$ is thus eliminated; that $\frac{2}{3} > p \geq 0$ is eliminated by a similar argument.

It remains only to suppose that Row knows he ought to employ the strategy $(\frac{2}{3}, \frac{1}{3})$. Under this supposition, since Column would know that Row was employing this strategy, Column's expected value for each of his possible strategies – F, H, and $M(q)$ for every q such that $1 > q > 0$ – would be the same: namely, $2\frac{1}{3}$. But then it would not be true that Column ought to employ the strategy $(\frac{2}{3}, \frac{1}{3})$; it would not be true that this would be Column's *only* right action. And so, given the practical symmetry of the situation, it would not be true that Row ought to employ this strategy, which means that – contrary to the supposition being examined – Row could not know that he ought to employ the strategy $(\frac{2}{3}, \frac{1}{3})$.

The conclusion to this point is that there is *not* a certain strategy $(p, 1-p)$ that Row knows he ought to employ.

2. Suppose Row knows that he ought to choose indifferently from some nonunit set of his available strategies. Then Column will judge the strategies in this set equally probable and arrive at probability expectations p and $(1-p)$ for Row's F and Row's H. (Simple averaging determines p if the set if finite; more sophisticated calculations will be required in case the set is infinite.) There are, again, three possibilities: $p > \frac{2}{3}$; $p = \frac{2}{3}$; $\frac{2}{3} > p$. Let $p > \frac{2}{3}$. Then Column ought to do H, Row knows that Column is going to do H, and F is Row's sole right action, which contradicts the assumption that Row knows he ought to choose indifferently from some nonunit set of strategies. That $\frac{2}{3} > p$ is eliminated by a similar argument. It remains only to let $p = \frac{2}{3}$. But then, as was seen previously, each of Column's strategies would have an expected value of $2\frac{1}{3}$ and Column would choose indifferently from them. Knowing this, Row would judge it probable to degree $\frac{1}{2}$ that Column would return to the fire: Column's choosing indifferently from *all* his strategies is equivalent, as far as Row's expectations are concerned, to Column's employing the mixed strategy $(\frac{1}{2}, \frac{1}{2})$.[6] But then F would be Row's sole right action,

which would again contradict the supposition that Row knows he ought to choose indifferently from some nonunit set of his possible actions.

So there is nothing that Row knows he ought to do. There is nothing he *could* know that he ought to do: each possibility has been eliminated. Of course, the same holds of Column. But then, as long as we suppose that Row and Column compose a hyperrational community that satisfies conditions 1.1-1.6, we cannot consistently also suppose that they are in (and acting through) a Hunter's Dilemma. Hyperrational agents always know what they ought to do, and only do what they know they ought to do (see condition 1.2). Our conclusion, put most briefly, is that such agents cannot be in (and act through) such situations, which are for them not possible.

2.2.3. "But," one wants to protest, "if Hunter's Dilemmas are not *possible* for hyperrational agents, surely they cannot be *problems* for them. Could they not dismiss such situations with the simple, true observation, 'this can't happen to us'?" No. For agents who are at a certain time hyperrational, though necessarily not then in a Hunter's Dilemma, cannot by anything that we have assumed be sure that their condition will continue or that they never will be in a Hunter's Dilemma. Indeed, they could realize that they were threatened, that a Hunter's Dilemma lay in their "natural" future, and that one would befall them unless it was somehow headed off. And they could see that were it to descend upon them they would not remain the same as they are but would change, and change for the worse, at least in their own, present eyes. Moral deterioration would be one possibility - they could experience a weakening of resolve or a falling away from pure act teleology; intellectual deterioration would be another - they could come to know less. It is in this way that a Hunter's Dilemma, just because it is impossible for hyperrational agents, can be a problem for them.[7] It is for this reason that they would like Hunter's Dilemmas to be in some way avoided. This could be accomplished by judiciously applied coercion. A law against fires or against hunting would provide a radical solution. A better (perhaps the best) solution would be an enforced system of social roles and special duties.

"But, again, surely laws and threats of sanctions are not needed. Desiring coordination and being reasonable, these agents could talk things out. Since they are out of sight they might need to shout or to settle things in advance; but in any case, with words they could make necessary arrangements. So it is false that Hunter's Dilemmas would pose problems for hyperrational agents or be impossible for them. Such situations would be child's play for them." Not so. They could talk, but there is nothing helpful that they could say to each other – they know each other and their situation too well. Neither could have any news for the other; neither could, even if he wanted to, deceive the other; and they would know that neither values an action just because he has said that he would do it. Thus, if one said, "I'll take care of the fire," a response would be, "How do you *know* you will?" And if one said, "I'll take care of the fire if I can be sure that you will not," a response would be, "True, but not very helpful." If one said "I'll tend the fire no matter what," a proper response would be, "No you wouldn't, not if you were sure that I would!" And finally, if one said, "I'll take care of the fire and that's a promise," a proper response would be, "Empty words, as you well know." Even allowing that they could talk, draft documents, post notices, and in general do the things we do with words, hyperrational agents would have problems with impending Hunter's Dilemmas – problems that coercive laws could solve. In this essential respect, the Hunter's Dilemma is like the Farmer's Dilemma. (Of course there are differences: most notable here is that Farmer's Dilemmas are not impossible for hyperrational agents; such agents can cope with Farmer's Dilemmas, they simply cannot cope well with them.)[8]

3. The individual's need for coercion

Communities composed of good and wise individuals can have use for sanctions and coercive government, as we have seen. But one can still think that if a good and wise agent knows that everyone else is subject to adequate sanctions then he will never himself need to be under threat. It can seem that, if a pattern of threatened sanctions solves a certain social problem, then the sanctions directed against any given good and wise agent will invariably be superfluous, that any such individual would do his part unasked and uncoerced as long as he knew (as he would) of the sanctions to which all *others* were subject. So it can seem, but it is not so. It does not hold for the Farmer's Dilemma, and seems not to hold even for the Hunter's Dilemma. We consider each in turn.

3.1

To simplify, we assume that each pattern of pure strategies has a future that is known to each agent, a future to which each agent assigns a probability of 1. Under this simplification we represent a Farmer's Dilemma by future structure I.

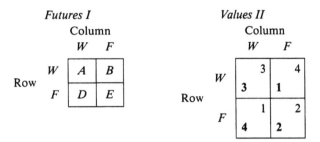

Futures I

	Column W	Column F
Row W	A	B
Row F	D	E

Values II

	Column W	Column F
Row W	3, 3	4, 1
Row F	1, 4	2, 2

A is the future of the pattern of actions (W, W): Each agent judges it probable to degree 1 that if the pattern (W, W) were to obtain then A would be its future. Futures B, D, and E are used in similar ways. The value matrix for the case is as shown in matrix II.

The problem in the Farmer's Dilemma is that, as things stand "in nature," the pattern (F, F) will obtain even though all concerned would prefer pattern (W, W). We shall say that a sanctioned law that secures the jointly preferred pattern (W, W) *solves* the problem of the case. A law against planting flowers, if adequately sanctioned, could accomplish such a solution. Establishing this law and making certain its enforcement would change the futures of three of the four patterns in a manner depicted in matrix III.

Futures III

	Column W	Column F
Row W	A	B+S
Row F	D+S	E+2S

Values IV

	Column W	Column F
Row W	4, 4	3, 2
Row F	2, 3	1, 1

Let $B+S$ be the future of the pattern (W, F) given the sanction against Column's planting flowers; $B+S$ can be thought of as the result of changing the "natural" future B of pattern (W, F) by the introduction of a sanction S at a certain point. Similarly, $D+S$ can be thought of as the result of introducing a similar sanction into D at a certain point, and $E+2S$ can be thought of as resulting from E by the introduction of two similar sanctions at certain points.

Suppose the sanction S would induce the reordering of values that is depicted in matrix IV. Under this assumption, the problem of the case is solved: planting wheat is now a dominant strategy for each agent. The issue is whether or not it is essential to this solution that each agent be subject to sanction if he plants flowers. The answer is that perhaps it is essential. Alternatively, the question is: Would Row, for example, have sufficient reasons - the sanctions against him aside - for participating in the solution pattern (W, W)? The answer here is, Not necessarily.

Suppose that the sanction against Row's planting flowers was literally set aside, while the sanction against Column's planting flowers was maintained. The futures of the four patterns would be as in matrix V.

Futures V

Column
	W	F
Row W	A	B+S
Row F	D	E+S

Values VI

Column
	W	F
Row W	3 / 3	2 / 1
Row Γ	2 / 4	1 / 2

What values might Row and Column assign to these futures? Recalling that Row and Column are benevolent universalists, and supposing that each takes as dim a view of S imposed on the other as he does of S imposed upon himself, several necessary or at least plausible restrictions flow from structures II and IV: It is necessary that the values assigned by each agent to A and $B+S$ be ordered as they are in IV, and that the values assigned by each to A and D be ordered as they are in II. One may expect, though it is not necessary, that the values assigned by each to D and $E+S$ should be ordered as are the values assigned by him in IV to $D+S$ and $E+2S$; and one may expect the values assigned by each to $B+S$ and $E+S$ to be ordered as are the values assigned by him in II to B and E. Further, perhaps S would be a strange sanction if the values assigned by either agent to A and $E+S$ were ordered differently from those assigned by him in II to A and E, or if the values assigned by Row to D and $B+S$ were ordered differently from those assigned by him in II to D and B. But subject even to these restrictions - and certainly no others are necessary - it is possible that the value matrix for future structure V should be VI.

But given the values in matrix VI, Row would not participate in the solution pattern (W, W) even though he knew of the sanctions against Column's F and indeed knew that Column was going to participate in this pattern in the sense that he was going to perform W. Column would plant wheat in this civil state in which only he was coerced. Thanks to the sanction against his planting flowers, planting wheat would be dominant for Column, so he would do his part. And Row could know this. But it would not matter, for even knowing it Row would plant flowers; planting flowers would still be dominant for him.

So when communities that meet our specifications have use for sanctions, it is not necessarily sufficient to an individual's reasons for participating in a well-sanctioned solution pattern that he be sure everyone else is subject to sanction. It *can* be essential to a given sanction solution in which each agent is subject to sanction that each agent actually be subject to sanction.

3.2

Let the future structure and value matrix for a Hunter's Dilemma be as follows.

Futures VII

Column
F H

Row
	F	H
F	A	B
H	D	E

Values VIII

Column
F H

Row
	F	H
F	2	3
H	3	1

The problem with the Hunter's Dilemma is not that our agents would not do well in it, for without corruption of one sort or another our agents could do nothing at all in it. We consider a solution in which would-be Hunter's Dilemmas are converted by an arrangement of sanctions into situations that are not only possible for hyperrational agents but that in their resolutions would leave nothing to be desired. A certain penalty sufficient to secure the optimum pattern (F, H) is imposed against Row's continuing hunting and Column's returning to the fire. The future structure and value matrix for the converted Hunter's Dilemma are, we assume, as follows.

Futures IX

Column
F H

Row
	F	H
F	A+S	B
H	D+2S	E+S

Values X

Column
F H

Row
	F	H
F	3	4
H	1	2

In a Hunter's Dilemma so converted, each agent has a dominant action that brings him into the solution pattern: tending the fire is dominant for Row; hunting is dominant for Column.

Could the sanctions against Row be set aside without destroying the solution? The answer again is, Maybe not. Setting aside the sanction against Row's hunting yields future structure XI.

	Futures XI Column	
	F	H
Row F	A+S	B
Row H	D+S	E

	Values XII Column	
	F	H
Row F	1	3
Row H	2	1

It is necessary that the (agreed) values assigned to $A+S$ and B should be ordered as in matrix X, and that those assigned to B and E should be ordered as in VIII. And it would be strange if $D+S$ were not assigned a lower value than B. Further, one is apt to expect the values assigned to $A+S$ and $D+S$ to be ordered as are those assigned in VIII to A and B. But even subject to these restrictions (and no others are necessary), the value matrix for structure XI could be XII. And Row would not, I think, have sufficient reasons in this partially converted Hunter's Dilemma for doing his part in the solution pattern (F, H); nor, for that matter, would Column. This partially converted dilemma would be, though for different and perhaps less clear and compelling reasons, as problematic (indeed, "as impossible") for hyperrational agents as the original one.

The partially converted dilemma is not marked by practical symmetry. Here is one difference from the original dilemma. And in the partially converted dilemma there is a unique best pattern of actions, which is another difference. Given these differences, especially the second one, it can seem that the partially converted dilemma ought not to be impossible, ought not to be in any way problematic, for our agents. They are concerned to do what is best, and they will know which pattern is best. "Surely they will manage to realize this pattern." But recall that our agents are not "pattern-utilitarians," each concerned to do his part in a best pattern, but rather act-utilitarians, each concerned to do *a* best act. So it is not clear that they would manage to realize the best pattern; on the contrary, it is tolerably clear that they would not so manage.

Whether Row, for example, ought to return to the fire depends on what he should expect Column to do. Row should tend the fire and not hunt if and only if he should judge it probable to a degree greater than $\frac{1}{3}$ that Column will continue to hunt. But it is at least not necessary that

Row should have this expectation. It would be necessary if this expectation could be reached, and if *only* it could be supposed without immediate contradiction; but, in the partially converted dilemma, *several* expectations could be supposed without immediate contradiction. It could be supposed without immediate contradiction that Row should assign to Column's hunting either a probability of 1 or a probability of 0, and if the assumption of equal probabilities in cases of indifference were dropped, a probability of ⅓ could also be supposed without immediate contradiction. So it is not necessary that Row assign a particular probability to Column's hunting, and – what is more important – it seems that no such assignment is possible. Given his knowledge of his community and the structure of the situation, Row would realize that in order to decide what to expect of Column he needs first to discover what Column in the end would expect of him: Row knows that Column never acts without knowledge of his duties, and that what Column ought to do in this situation depends upon what Column should expect him to do. So it seems that in order to decide what to expect of Column, Row must first determine what Column should expect of him; at least this seems right if the decision concerning what to expect is reasonably and responsibly taken, as for our agents it must be. Of course, this could only be the beginning, and there could be no end! Row would know that Column was locked into the same process. Row would be driven to attempt to replicate Column's deliberations, and in particular Column's attempt to replicate Row's deliberations, and in particular Row's attempt to replicate Column's deliberations, *and so on!* Row's impossible task cannot even be put into words. Such labyrinths of mutually dependent expectations are ordinarily terminated by flagging interest, fatigue, lack of time, acts of faith, the assumption of one of these things, the assumption of the assumption of one of these things, and so on. But none of the ordinary terminations are possible for hyperrational agents. It seems that Row could – given the nature of the situation and the character of his community, especially its extensive self-knowledge and the determination of its members to deliberate reasonably and responsibly, putting first questions first – neither avoid nor escape from the labyrinth. (See conditions 1.3 and 1.6.)

It is not necessary that our agents should have any particular expectations in a partially converted Hunter's Dilemma, and given the feedback just traced it seems not *possible* that they should have expectations regarding each other. When individuals are so related in principle and circumstance that their actions depend mutually upon their expectations regarding each other's actions, a regress is generated which, in the limiting case of a hyperrational community, precludes the formation of such expectations. The argument given for this conclusion lacks the rigor one might desire. In particular, it would be nice to have explicit general premises concerning exactly how, under various circumstances, hyperrational agents deliberate and order issues. Still, a case has been made that makes it tolerably clear that hyperrational agents in a partially converted dilemma would not know what to expect of each other, and so would not know what they ought to do. And since this is not possible for our agents (see condition 1.2), it seems that the converted dilemma is as problematic for them as the original unconverted dilemma.[9]

Our thesis is thus reinforced: A good and wise agent cannot always ignore the sanctions he is under and still find sufficient reasons for participating in a desirable and adequately sanctioned arrangement. We cannot say that when coercion is needed, "what reason demands is *voluntary cooperation in a coercive system*" (Hart 1961, p. 193). We cannot say this, at any rate, of our agents. For them at least, the need for coercion is more complete. They can need it not only collectively as a community, as was argued in Section 2, but it seems also individually as participating members, as has been argued in the present section.

Appendix: The Farmer's Dilemma and mutual trust

In the Farmer's Dilemma, what each agent ought to do does not depend at all upon what he should expect the other to do. Even if Row, for example, could trust Column and could be sure that Column would plant wheat, Row would still plant flowers. One is tempted to think that lack of trust is in this case not at all the problem (cf. Tullock 1967, p. 229). But there is a sense in which the problem in the case is one of "want of mutual confidence and security" (Hume 1888, p. 521). Let a *trustworthy* agent be a deontologist who recognizes two ceteris paribus duties, a duty to maximize expected value, and a more stringent duty to keep promises. Trustworthy agents would have a way out of Farmer's Dilemmas. They could, by exchanging promises, convert would-be Farmer's Dilemmas into situations in which planting wheat was reasonable for each. The relatively natural promise-based reasons available at will to trustworthy agents could serve in place of the more artificial sanction reasons we have been considering. (See Gauthier 1967, p. 471.)

But there is a small and soluble puzzle. Suppose that Row and Column were trustworthy and that they could solve their problem by exchanging promises to plant wheat. How could they make the exchange? Though each would prefer that both rather than neither promise to plant wheat, each would most prefer that just the other make this promise. Apparently, just as not planting wheat is dominant in the Farmer's Dilemma, so would not promising to plant wheat be dominant in a preliminary situation in which each could either make this promise or not make it. But then, how could they arrange to exchange promises to plant wheat?

The puzzle has a number of solutions. We consider first a simple but inelegant solution. Suppose there were a document containing the words, "The undersigned agree, agreement effective when two have signed, to plant wheat." Now let it be settled that the document is to be presented first to Row and then to Column. Their problems are solved. By signing, Row can guarantee that Column will sign, that they will both be bound by promises to plant wheat, and that they will both plant wheat. So Row will sign. And then Column will sign. The desired exchange of promises will be accomplished.

Must a document be produced? No. Is it essential that moves be ordered in the contract situation, with Row (Column) "going first" and Column (Row) "going second"? The answer is again, No. Dispensing with the document, each could say in any order or simultaneously, "I promise to plant wheat, my promise effective as soon as both of us have said this." Let S be saying this and $\bar{S}$ not saying this, and suppose each is in a position to S or $\bar{S}$. Their situation has the duty structure displayed in matrix XV. That it has this structure is explained, and then it is shown that in such a structure each would S.

Duties XV

		Column	
		S	$\bar{S}$
Row	S	1, 1	0, 0
	$\bar{S}$	0, 0	0, 0

The numbers represent the presence or absence of duties overall or *sans phrase* – without qualification. (The numbers do not represent expected values, though it can happen that a duty *sans phrase* to do an action can derive from an unopposed expected value-based duty ceteris paribus to do it.) For example, if Column does S then (as we shall see) Row has a duty overall to do S – the **1** in the box for (S, S) expresses this fact. We now turn to the justification of the numbers. Note that if both do S, then each has a promissory duty to plant wheat and each will plant wheat. But if only one does S, or if neither does S, then no promises result and each will plant flowers. Thus, if Column does S, Row has a duty overall to S and no duty overall to $\bar{S}$, for by doing S Row can maximize expected value; the **1** in the (S, S)-box and the **0** in the $(\bar{S}, S)$-box express these facts. If Column does $\bar{S}$ then it does not matter what Row does – he has no duty overall to do S or to do $\bar{S}$; the 0s in the $(S, \bar{S})$- and $(\bar{S}, \bar{S})$-boxes express these facts.

Assuming that Row and Column are hyperrational trustworthy agents, what would they do in a situation of this structure? Each would do S, for each would know that he ought to do S. This is the only practical knowledge that can be supposed without contradiction. Suppose that Row knows that he ought to do S. Then Column knows that Row will do S and – knowing this – knows that he, Column, ought to do S. Knowing this, Row knows that he, Row, ought to do S. The hypothesis that Row knows that he ought to do S "leads back to itself." In contrast, suppose that Row knew that he ought to do $\bar{S}$. Then, as Row would know, Column would know this and be indifferent between his own alternatives. Knowing *this,* Row would judge Column's S and $\bar{S}$ equally probable and so know that he, Row, ought to do S. The hypothesis that Row knows that he ought to do $\bar{S}$ leads to a contradiction, and can thus be eliminated. It could be shown in a similar fashion that each mixed strategy $(p, 1-p)$ regarding S and $\bar{S}$, where $1 > p > 0$, can be eliminated: If Row knew that he ought to employ such a strategy then Column would know that he ought to do S for sure, and so, contrary to the present hypothesis, Row would know that *he* ought to do S for sure. Similarly, Row cannot know that he ought to choose indifferently from some nonunit set of his strategies: From Column's point of view, any such choice is equivalent to Row's employing a mixed strategy; so if Row knew that he was to choose indifferently from some nonunit set of strategies, he would on the contrary (!) know that he ought to do S for sure. So Row would know that he ought to do S. (He knows what he ought to do in every situation in which he finds himself.) Every other practical knowledge hypothesis also leads to contradiction and so can be eliminated. Row would know that he ought to do S and would therefore do S. (But how would he *come* to know that he ought to do S? He could reason in the way I have and by a process of elimination come to know that he ought to do S. He would know that he knew what he ought to do – we suppose that he is in the situation and knows that he knows what to do in every situation in which he finds himself; see conditions 1.2 and 1.6. And he would know that every other practical knowledge hypothesis was eliminable.)[10]

The exchange puzzle can be solved by exchanging mutually dependent unconditional promises to plant wheat. Only formally different solutions would consist in the exchange of independent conditional promises: For example, in signing a certain document, each could promise to plant wheat if the other signs the document. But many superficially similar exchanges would not constitute solutions. We consider two inadequate exchanges.

1. It would not help for each to promise to plant wheat if the other plants wheat. The structure of a would-be Farmer's Dilemma given such promises would be as shown in matrix XVI.

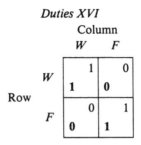

Duties XVI

If either plants wheat then the other has an overriding promissory duty to plant wheat; whereas, if either plants flowers, the other has no promissory duty to plant wheat, can maximize expected value by planting flowers, and so has a duty overall to plant flowers. For reasons rehearsed in Section 3.2, I think that situations of structure XVI are problematic for hyperrational agents. Though trustworthy agents would have a way out of, they could have no reason for getting into, such a situation.

2. It would be no advance for each to promise to plant wheat if the other promises to plant wheat. Given such conditional promises, promise-based duties to plant wheat would exist were either, or both, to make a further unconditional promise to plant wheat. But the situation vis-à-vis these further unconditional promises would have the following structure.

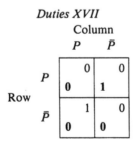

Duties XVII

Both will plant wheat if and only if at least one makes a further unconditional promise to plant wheat. Both will plant flowers if and only if neither makes this further unconditional promise. The crucial point is that nothing is gained by *both* making a further unconditional promise. This is why a further promise is called for from Row, for example, only in case it is not forthcoming from Column. Given the practical symmetry of this situation, every practical knowledge hypothesis can be eliminated. In particular, neither could know that he ought to make a further unconditional promise to plant wheat. For if he knew this, then he would know that the same was required and to be expected of the other. But knowing that the other would make a further promise, he would know – contrary to the hypothesis being tested – that it did not matter whether or not *he* made a further promise. Every other practical knowledge hypothesis is similarly eliminable. A situation of structure XVII would be problematic for hyperrational agents, and trustworthy agents could have no reason for getting into such a situation, though again they would have a way out.

Hyperrational trustworthy agents would find their way out of Farmer's Dilemmas. They could count on each other's promises and they would know what promises to make and how to make them. Probably such agents would find their way out of all interaction problems. It seems they could never need coercion. But they pay a price for their self-sufficiency: Insofar as they are committed to keeping promises regardless of costs and benefits, they are unreasonable. At least this is the judgment of a teleologist.

NOTES

1. I now favor another theory of options for decisions by Bayesian agents. See Chapter 10 in this volume and Section 1.3.1 of Chapter 14.
2. I count conventions (see Lewis 1969) as such solutions. Hyperrational agents, unimpressed as they would be by "salience" and not prone to habits, could not sustain what Lewis calls "convention."
3. I now believe that the argument just started goes wrong after this first sentence. See Chapter 14, Sections 2.2.1 and 2.2.2.
4. See Luce and Raiffa (1957, p. 97). The Farmer's Dilemma is a non-egoistic analog of the well-known Prisoner's Dilemma. For an early consideration of a case of the same structure, see Hume (1888, pp. 520–1): "Your corn is ripe today. . . ."

Though not "egocentric," Row and Column *are* "arrogant": neither in his values takes into consideration the values of the other. It is natural to wonder whether rational nonarrogant agents could find themselves in similar difficulties. Full consideration of such agents and their "value composing" functions lies beyond the scope of this essay, but this much can be said: If there is just one reasonable way of taking into account the values of others when arriving at one's own values, then rational nonarrogant agents cannot be in any value disagreement difficulties. If there are several reasonable ways, then value disagreements are possible for such agents; whether a Farmer's Dilemma is possible depends on the exact characters of these several ways. And if there is no rational value composition function for generating nonarrogant out of arrogant values, then there are no rational nonarrogant agents to consider. For strategic purposes, the important question regarding nonarrogant agents – as for agents of any value persuasion – is what kinds of value disagreements, if any, they can be in.

5. Past performances are not only irrelevant in Farmer's Dilemmas, but I think are irrelevant in general in hyperrational communities. It seems obvious, even if it is not easy to show with rigor, that precedent can play no role in such communities: An agent in a hyperrational community will expect a performance of another to be repeated only if, and to the extent that, it was dictated by reason in the first instance. So it seems that its reasonableness in the first instance cannot depend even in part upon another's being led by it to expect its repetition. For an attempt to articulate arguments against precedents (as well as threats and promises) in hyperrational communities, see Hodgson (1967, chap. II and IV).

6. *A proof:* Consider finite choice sets in which probabilities for F are evenly spaced from 0 through 1. Given size n of such a set, its members can be computed. For $n=3$ they are $(0,1)$, $(½, ½)$, and $(1,0)$; for $n=5$ they are $(0,1)$, $(¼, ¾)$, $(2/4, 2/4)$, $(¾, ¼)$, and $(1,0)$. Suppose that Row will choose indifferently from such a set. Knowing this, Column will make the probability of Row's eventually (after choosing a strategy and exercising it) doing F equal the simple average of the probabilities for F contained in the strategies in the choice set. If the set contains n strategies, then Column will set the probability p of Row's F so that

$$p = \frac{\sum_{i=0}^{i=n-1}\{[(n-1)-i]/(n-1)\}}{n}.$$

The "average" for the infinite case in which Row chooses indifferently from the set of all strategies should be the limit as n goes to infinity of the averages for finite n. Since each average in the series has the value ½, the series limit is of course also ½.

Another route to this result would assign equal infinitesimal probabilities to all mixed strategies. Then ½ would be the simple average in hyperreal probability theory. See Bernstein and Wattenberg (1969).

7. The argument in the text involves an assumption of equal probabilities in cases of indifferent choices. If this assumption were dropped, then perhaps Hunter's Dilemmas could not be shown to be impossible, but they would remain problematic. Let the assumption be dropped. Then perhaps each agent could know that all his actions were right and that it did not matter what he did, and perhaps each agent would judge the other's F probable to degree ⅔. No other practical knowledge and no other expectations concerning F would be possible, but this knowledge and these expectations just might be – at least they could be supposed without immediate and obvious contradiction (even if how they could be "reached" remains a puzzle). However, even allowing this knowledge and these expectations and so allowing Hunter's Dilemmas to be possible for hyperrational agents, Hunter's Dilemmas would remain problematic for such agents. Allowing the stated expectations, each agent would judge it to be ⅔ probable that the other would F, and ⅔ probable that he himself would F. So each agent would view the emergence of one of the best, mixed patterns of action as only 4/9 probable. With coercive organization, on the other hand, a best pattern could be guaranteed.

8. Hyperrational agents could need coercive government. But it is not clear that when they needed it and wanted it they would have it. For one thing, it is possible that hyperrational agents would be incapable of coercive self-government. It has been said that every coercive government needs an inner governing cadre

that is not itself coerced (see Lucas 1966, pp. 76-7), and given the nature of such a cadre's work it seems likely that it would need to negotiate situations that hyperrational agents either could not handle or could not handle well. (Forward-looking teleologists would have other problems with self-government, given their general problems with making good the threats of sanctions and punishment. But these problems might be obviated by the employment of punishment machines rigged in advance.) It is thus entirely possible that hyperrational agents would be incapable of coercive self-government. And though they might sometimes welcome the intervening coercion of a benevolent outsider, there will not always be such an outsider on the scene who is prepared to and capable of intervening. So it is possible that hyperrational agents would sometimes need coercion and have their need for it go unfulfilled.

9. The same conclusion could be reached with no more (and no less) difficulty if the sanctions against Column were set aside. The resultant partially converted dilemma would have future structure XIII and could have value matrix XIV.

Futures XIII

Column

	F	H
F	A	B
H	D+S	E+S

Row

Values XIV

Column

	F	H
F	2	4
H	3	1

Row

10. I now believe that, rather than have Row eliminate each of his options other than S, we should say: Row can come to see that, of his own options, S maximizes by eliminating every other option *for Column* and so settling that Column will exercise option S. See Chapter 14, Section 2.2.5.

14

Hyperrational games

Abstract. Hyperrational games are characterized, and conditions for their resolutions are identified. It is maintained that they can resolve only in kinds of equilibria, and that some do resolve in these equilibria by deliberations that, by processes of elimination, settle a player's expectations concerning the acts of other players. Problems for hyperrational agents are identified: It is held that they would not do well in some situations and that there are some situations that are possible for other agents but with which hyperrational ones simply could not cope.

0. Introduction

This chapter examines normal-form games played by very knowledgeable causal expected value maximizers. Section 1 characterizes these agents and defines hyperrational games. Theorems concerning their resolutions are developed in Section 2. Section 3 discusses games that would be problematic for these agents.

Hyperrational games are highly idealized objects that are approachable but probably never realized. I am interested in the theory of these objects not primarily for the light it promises to cast on actual games, or as a source of prescriptions for actual games, but mainly for contributions I think it can make to explanations and understandings of real agents and to justifications of cultural aspects such as coercive institutions.

1. The concept of a hyperrational normal-form game

1.1

I begin with axioms for all normal-form games, and proceed to ones specifically for hyperrational games.

Sections 1 and 2 are reprinted with revisions from Chapter 5 of *Knowledge, Belief and Strategic Interaction* (Christina Bicchieri and Maria Luisa Dalla Chiara, eds.), Cambridge University Press, 1992, pp. 61–91.

I am particularly indebted to Willa Freeman-Sobel and Włodek Rabinowicz for help with this work.

Axiom 1. *In a pure-strategy game, each of finitely many players has as an option exactly the members of a finite set of strategies.*

Axiom 2. *Each player has an expected value for each possible interaction of strategies.*

World Bayesianism is assumed: Expected values of propositions are probability-weighted averages of values of ways propositions might work out, or worlds in which they might take place:

$$EV(p) = \sum_{\Pr(w \text{ given } p) > 0} [\Pr(w \text{ given } p) \cdot V(w)]$$

(Sobel 1989d [Chapter 1 in this volume]). Sets of strategies and expected values for interactions determine normal-form structures of games. For example, in the following two-person pure-strategy normal-form game,

Game 1

	C1	C2	C3
R1	3, 1	0, 2	2, 0
R2	0, 1	1, 0	4, 3

Row's options are R1 and R2; Column's are C1, C2, and C3. The first number in each box is Row's expected value for the interaction or conjunction of strategies, and the second number is Column's. Thus,

$$EV_R(R1 \& C1) = 3 \quad \text{and} \quad EV_C(R2 \& C3) = 3.$$

Axiom 3. *Strategies of players are causally independent of one another. What a given player does cannot influence what any other player does.*

Axiom 4. *Players are rational in choices of strategies.*

Axiom 5. *Each player knows the game's normal-form structure, and knows that Axioms 3 and 4 are satisfied.*

Axiom 6. *It is "common knowledge"*[1] *among players that Axiom 5 is satisfied: Each player knows that Axiom 5 is satisfied, knows that it is known by each player that Axiom 5 is satisfied, knows that it is known by each player that it is known by each player that Axiom 5 is satisfied, and so on ad infinitum.*[2]

Rationality and knowledge conditions for hyperrational extensive-form games would need to be "subjunctified" and considerably more complicated; see Sobel (1993a [Chapter 16]).

1.2

For mixed-strategy games, or mixed extensions of pure-strategy games, it is assumed that each player has as an option every chance mix of a finite set of pure strategies. To define mixed-strategy games, Axiom 1 is replaced by the following.

Axiom 1m. *In a* mixed-strategy game, *each of finitely many players has as an option every chance mix based on some finite set of pure strategies.*

To have as an option a chance mix is to have the wherewithal somehow to commit irrevocably to chance, with chances fixed as in the mix, which of the strategies mixed is eventually enacted. A *genuine* mixed strategy accords positive chances to at least two distinct pure strategies. A non-genuine or *degenerate* mixed strategy makes some one pure strategy certain; in a mixed-strategy game it is the surrogate for that pure strategy. The pure strategy is not an option in the game, though to avoid circumlocutions I will write as if it were an option identical with the degenerate mixed strategy.

I stipulate as a basic assumption for mixed-strategy games that players' expected values for interactions of mixed strategies can be computed using the formula of utility theory for computing utilities for lotteries given utilities for their prizes. The basic assumption is that, with respect to lotteries that would have as outcomes interactions of pure strategies, players are "risk neutral" in that their preferences for these lotteries are represented by standard utility functions. For example, in the mixed extension of Game 1, the interaction in which Row uses ($\frac{1}{3}$R1, $\frac{2}{3}$R2), the mixed strategy that fixes the chance for R1 at $\frac{1}{3}$ and that for R2 at $\frac{2}{3}$, and in which Column uses ($\frac{1}{6}$C1, $\frac{1}{3}$C2, $\frac{1}{2}$C3), has for Row and Column (respectively) the expected values $\frac{3}{2}$ and $\frac{15}{18}$:

Game 1

	C1	C2	C3	($\frac{1}{6}$C1, $\frac{1}{3}$C2, $\frac{1}{2}$C3)
R1	3, 1	0, 2	2, 0	
R2	0, 1	1, 0	4, 3	
($\frac{1}{3}$R1, $\frac{2}{3}$R2)				$\frac{3}{2}$, $\frac{25}{18}$

Our basic assumption licenses the following calculation for Row's expected value for this interaction of mixed strategies:

$$EV_R[(\tfrac{1}{3}R1, \tfrac{2}{3}R2) \& (\tfrac{1}{6}C1, \tfrac{1}{3}C2, \tfrac{1}{2}C3)]$$
$$= \tfrac{1}{3}\cdot\tfrac{1}{6}\cdot 3 + \tfrac{1}{3}\cdot\tfrac{1}{3}\cdot 0 + \tfrac{1}{3}\cdot\tfrac{1}{2}\cdot 2 + \tfrac{2}{3}\cdot\tfrac{1}{6}\cdot 0 + \tfrac{2}{3}\cdot\tfrac{1}{3}\cdot 1 + \tfrac{2}{3}\cdot\tfrac{1}{2}\cdot 4$$
$$= \tfrac{1}{6} + 0 + \tfrac{1}{3} + 0 + \tfrac{2}{9} + \tfrac{4}{3} = \tfrac{27}{18} = \tfrac{3}{2}.$$

It is a theorem of world Bayesianisms – both Richard Jeffrey's evidential version (in which the term $\Pr(w \text{ given } p)$ in the definition of $EV(p)$ is cast as a conditional probability term $\Pr(w/p)$) and the causal theory assumed for the present study (in which the term $\Pr(w \text{ given } p)$ is cast as the probability term for causal conditionals, $\Pr(p \,\square\!\!\rightarrow w)$) – that, for any lottery L in which the known chance for outcome O_i is c_i:

$$EV(L) = \Sigma_i [c_i \cdot EV(L \& O_i)].$$

However, in contrast with standard utility theory (e.g., that of Luce and Raiffa 1957), it is *not* a theorem that

$$EV(L) = \Sigma_i [c_i \cdot EV(O_i)].$$

World Bayesianisms are roomier than standard utility theories. They make room, for example, for Allais and Ellsberg preferences. So an assumption is needed to license standard calculations for expected values of mixed-strategy interactions (Sobel 1989d [Chapter 1]).

1.3

For hyperrational games, Axioms 1, 1m, and 4 are elaborated; two additional epistemic conditions are inserted between Axioms 4 and 5; and Axiom 5 is readdressed to the expanded and elaborated set of prior Axioms 1 through 4.

1.3.1. Bayesian decision theories that would have agents maximize expected value need to be explicit about "proper" partitions of agent's options, and about what alternatives are relevant for purposes of maximizing comparisons. Such theories, to be complete and exact, need to say precisely which expected values are relevant to the assessments of options.

It might seem sufficient to say that an action maximizes in a manner significant for choice if and only if its expected value is not exceeded by that of any of its alternatives. But this approach encounters problems with relatively indeterminate, disjunctive actions, for expected values of disjunctions (even exclusive disjunctions) on my causal theory are not necessarily bounded by expected values of their disjuncts. Given this fact, although action (A or B) is open whenever A is, one would want to count it as a relevant alternative to A and to B only when it is open in a distinctively disjunctive manner, only when one can choose it without choosing either of them. One reaction to such problems might be to make primary the assessments of "most specific" actions, but this can seem wrong in some cases where the agent can choose to leave open certain specifications, especially specifications pertaining to future stages of (courses of) actions. Another idea might be to make primary the assessments of "most specific minimal" actions that settle nothing for the future that can be left open. But this can seem wrong for cases in which the agent can choose that certain possible future options not be left open. The problem of proper partitions of options, as this problem arises for the causal theory of this study, is addressed in Sobel (1983 [Chapter 10]). In view of difficulties with approaches that would identify agents' options with actions of one kind or another, I propose a theory in which they are identified with certain *choices* for actions.

I assume that hyperrational games are built on proper partitions of options, and for definiteness spell out this assumption in terms of my theory of proper partitions, although the details of that theory – as well as the *correctness* of that theory – are not presently important. Axioms 1h and 1mh are for hyperrational games.

Axiom 1h. *In a pure-strategy hyperrational game, each of finitely many players has a finite set of strategies that are all and only the "precise choices" open to him.*

Axiom 1mh. *In a mixed-strategy game, each of finitely many players has as an option every chance mix based on some finite set of pure strategies, and these are all and only his possible "certain precise choices."*

To spell out this terminology without fully explaining it, I offer these statements:

[A] *precise choice* of an action x [is] a choice of this action that is not accompanied by a choice of any action y such that y entails but is not entailed by x. (Sobel 1983, p. 179 [Chapter 10, Section 4])

[A mixed] choice is . . . a *certain precise choice* if its agent is . . . sure he can make it without making any refined version of it. (Sobel 1983, p. 182 [Chapter 10, Section 4])

An important dividend of including propriety conditions (these or others) on strategy sets is that such conditions should ensure that there can be for a given actual game only one hyperrational idealization, so far as the identities and numbers of agents' options are concerned. This consequence neutralizes certain objections to the use in Axiom 4.1 (to follow) of the "principle of insufficient reason" (Rabinowicz 1986, p. 215) for games in which it is known that a player is confronted with a tie. For example, cloning a maximizing strategy S in a hyperrational game G leads *not* to an equivalent hyperrational game G', but rather to another game where either in place of the option S a player has two versions of S as options, or in addition to S a version of it as an option. The player in G' would have somewhat greater control over his actions than the player in G. Strategy S could be getting one over the plate. Versions of S would then include throwing a curve for a strike and throwing a hard, fast one. Pop!

1.3.2. The definition of a normal-form game leaves open the character of rationality in games. For hyperrational games this issue is settled in favor of causal expected value maximization. Players in these games are causal expected value maximizers, nothing more and nothing less.[3]

Axiom 4h. *Each player in a hyperrational game is rational in the choice of strategies, in that he* (i) *is a self-conscious and deliberate causal maximizer,* (ii) *does only things he knows are rational in causal maximizing terms, and* (iii) *when confronted with several tied maximizing strategies is indifferent between them and makes an indifferent* (unprincipled) *choice of one of them.*

That hyperrational players are causal maximizers raises a problem. For although an option's *evidential* expected value can be analyzed in terms of its evidential expected values in conjunction with the members of any partition of propositions or possible circumstances, not every partition is in this way adequate for the analysis of an option's *causal* expected value. It turns out, however, that – according to the theory of causal expected values assumed in this study – every partition is adequate that satisfies the condition that it is certain that the action is not open in conjunction with both members of any pair of circumstances in it. (This is proved in Sobel 1989a [Chapter 9].) So, in this theory, possible actions of others in a game always make adequate partitions of circumstances, and a strategy's expected value is a weighted average of the expected values of interactions in which it might participate, where weights are its agent's probabilities for remainders of these interactions. For example, in a Prisoners' Dilemma for you and me it will be clear to me that I cannot both confess alone *and* confess along with you; one or the other is possible, depending on what you are going to do, but certainly not both. So your possible actions make an adequate partition of circumstances for mine, and, for example (using a subscripted "m" to stress that *my* expected values and probabilities are at issue),

EV_m(I confess)
$= [Pr_m$(I confess $\square\!\!\rightarrow$ you confess) $\cdot EV_m$(I confess & you confess)$]$
$+ [Pr_m$(I confess $\square\!\!\rightarrow$ you do not confess)
$\cdot EV_m$(I confess & you do not confess)$]$.

Given that I know that your actions are causally independent of mine, this equation reduces to

EV_m(I confess)
$= [Pr_m$(you confess) $\cdot EV_m$(I confess & you confess)$]$
$+ [Pr_m$(you do not confess) $\cdot EV_m$(I confess & you do not confess)$]$.

1.3.3. A special epistemic condition for hyperrational games is about players' expectations concerning the strategies of others whom they know to be confronted with ties.

Axiom 4.1. *When a player i considers the behavior of another player j who i knows is confronted with a tie, player i judges j's pure strategies on which tied strategies are based to be equally probable, and i's probabilities for them sum to 1.*

For pure-strategy games this axiom requires that tied strategies be equally probable, but this requirement is not extended to mixed-strategy games. Rather than require that there be a distribution of equal probabilities to tied strategies themselves in a mixed-strategy game, this axiom requires such a distribution only to all pure strategies on which tied strategies are based. I say that a mixed strategy M is based on pure strategies if and only if each of these, and no others, have positive chances in M. Strictly speaking, pure strategies are not options in mixed-strategy games. Standing in for them, as was noted after Axiom 1m, are degenerate mixed strategies in which single pure strategies are made certain.

It is a consequence of the basic assumption for mixed-strategy games that, if Max is the set of a player's maximizing strategies for a mixed-strategy game, then Max contains all and only the mixed strategies that are based on only pure strategies that correspond to pure strategies in Max. The main lemma for this consequence is that, given the basic assumption for mixed-strategy games, the causal expected value of a mixed strategy M is a weighted average of the expected values of the pure strategies mixed in M. For substantiation, consider Row's mixed strategy $[xR1, (1-x)R2]$ in the following game.

	C1	C2
R1	a, e	b, f
R2	c, g	d, h

By the basic assumption, the normal form for this game includes, for that mixed strategy,

	C1	C2
$[xR1, (1-x)R2]$	$xa+(1-x)c$	$xb+(1-x)d$

Suppose that Row's probabilities include $\Pr_R(C1) = p$ and $\Pr_R(C2) = (1-p)$. Then, by applications of a general partition principle for causal expected values and by elementary algebra,

$$\begin{aligned}EV_R[xR1,(1-x)R2] &= p[xa+(1-x)c]+(1-p)[xb+(1-x)d] \\ &= x[pa+(1-p)b]+(1-x)[pc+(1-p)d] \\ &= x\,EV(R1)+(1-x)\,EV(R2).\end{aligned}$$

Given this lemma, if (for example) $[xR1, (1-x)R2]$ is in Row's Max then: (i) $EV_R[xR1, (1-x)R2] = EV_R(R1) = EV_R(R2)$, for otherwise the expected value of one or the other of R1 and R2 would exceed that of $[xR1,(1-x)R2]$; and so (ii) for every chance y, $[yR1,(1-y)R2]$ is in Max.

As Rabinowicz has noted (1986, p. 228, n. 3), Axiom 4.1 would be "troublesome" if it required for mixed-strategy games that all tied strategies themselves should have equal probabilities. This is because (i) for every real number r no matter how small, the sum of infinitely many r (or, identically, the product $\infty \cdot r$) is not 1 but rather ∞, and (ii) every tie in a mixed-strategy game is between infinitely many mixed strategies. The trouble with a version of Axiom 4.1 that required all tied strategies (and not merely all tied pure strategies) to be equally probable might be dealt with by recourse to equal nonstandard infinitesimal probabilities. I deal with this problem by not raising it. I think that, though interesting, delving into the mathematics of hyperreals would not contribute significantly to the plausibility of the appropriateness of Axiom 4.1 for mixed-strategy hyperrational games. This plausibility is sufficient given the dense and uniform character of sets of tied maximizing strategies in such games (see Chapter 13, note 6).

1.3.4. Here is a second epistemic condition for hyperrational games.

Axiom 4.2. *There is no "private practical knowledge" in a hyperrational game – if any player knows at his time of action that some strategy is uniquely maximizing so that he ought to do it, or knows that some maximizing strategies are tied so that he ought to choose from among them, then these things are known to everyone in the game at their times of action.*

1.4

I believe that Axiom 4.2 is a consequence of other stipulations, for I think that hyperrational players could know nothing of relevance beyond what we (as students of their games) can know, given that we know the normal-form structures of their games and that their games satisfy all conditions for hyperrational games except for Axiom 4.2. And I think that hyperrational players would know everything of relevance that we students of their games can know. Compare:

> In games, players are aware of each other's data and each other's reason, and these data and this reason are their sole means for coming to conclusions. And so it follows from the Transparency of Reason that, in games, if a player knows that he should do something then the other knows that he knows he should do it. (Bacharach 1987, p. 37)
>
> Broadly, the principle of the Transparency of Reason is the claim that if reason suffices for a player who has certain data to come to a given conclusion – say, as to what he should do – then, if a second player believes the first to have these data and to be rational, reason suffices for him to come to the conclusion that the first will come to his. (pp. 36-7)

According to Axiom 4.2, hyperrational players choose with knowledge of and precisely *not* "in ignorance of others' choices" (Bernheim 1984, p. 1009). They do not merely rationalize their choices in terms of internally consistent conjectures concerning others' choices, others' conjectures of others' choices, and so forth. Hyperrational players in the end know not only what maximizes for themselves, but also what maximizes for others.

1.4.1. It is a further, hardly resistible, stipulation for hyperrational games that strategies of players are not only causally independent but also evidentially independent, at least when it is time to act. Each player in a hyperrational game must, when he acts, know what is maximizing for him, and know what was, is, or will be maximizing for others at their times of action. But then, last-minute news to a player – shocking news – that he was not going to do what he knows to be the maximizing (and thus the rational) thing should not be evidence that others also were not going to do what is maximizing for them. Amazing news that a player was going to make a mistake should not be evidence that mistakes were to be rampant – it should not be evidence that mistakes would extend at all beyond his own case.

Consider, for example, the following hyperrational game.

Game 2

		You C1	C2
Me	R1	2, 2	1, 0
	R2	1, 0	0, 1

Consider this game "at the moment of choice or action."[4] Presumably, I know what I should do, for R1 dominates R2. And I should know what you should do, because setting aside R2, as you should be in a position to do, reduces the game to a 1×2 problem in which C1 dominates C2. So for me $\Pr(C1)$ should be either 1 or nearly 1: $\Pr(C1) \simeq 1$. I should not, at the time of action, still have best-response conditional probabilities for your actions, so that not only $\Pr_m(C1/R1) \simeq 1$ but also $\Pr_m(C2/R2) \simeq 1$. For although both of my probabilities $\Pr_m(R2\,\&\,C2)$ and $\Pr_m(R2\,\&\,C1)$ are nearly 0, unless both are exactly 0, the first should be much lower than the second. My probability for doing what I now see to be a very dumb thing *and* your (coincidentally) joining me and also succumbing to the sillies should be much lower than my probability for my "going off" on my own. If $\Pr_m(R2) > 0$ still, at the moment of action, then it should be the case that $\Pr_m(R2\,\&\,C2)/\Pr_m[(R2\,\&\,C1) \vee (R2\,\&\,C2)] \simeq 0$. In any case, $\Pr_m(C2/R2) \simeq 0$ should obtain, notwithstanding that C2 would be your best response to R2 and also what (I am confident) you would do were you to know that I was doing R2. (I assume here nonstandard conditional probabilities defined for all possibilities, and not only for all positively probable ones; cf. Sobel 1987.)

1.4.2. So I think that, in hyperrational games, strategies that are causally independent are also evidentially independent, at least by times for action. Presumably, however, there can be kinds of evidential dependence of actions "early on," supposing that decisions in even hyperrational games can take time to make and that time for decisions is allowed. One interesting hypothesis for early-on conditional probabilities in two-person games would be that each player's initial conditional probabilities are such that news that he was going to do a certain action would mean that the other player was going to employ a best response to it. Some motivation for this hypothesis can be gathered from these words:

> Suppose you assume there is a unique rational choice . . . and that you will end up committing yourself to it, but you haven't yet figured out what [your] choice will be. When you consider the hypothetical news provided by your assumption that you choose strategy A you hypothetically assume that A is the rational act. You keep fixed your assumption that what you will end up committing yourself to will be the rational choice, and assume hypothetically that reasoning legislated choosing A. . . . [Y]ou assume your opponent . . . will have been able to reconstruct the reasoning that leads you to choose A and will have predicted your choice. Thus, you assume that she or he will choose some best response to A when you assume that you will choose A. (Harper 1988, p. 30)

But these grounds for best-response conditional probabilities early on, which are specific to two-person games and not easily extendable even to three-person games, fail to justify these probabilities. The trouble, as Harper himself comes to observe, is with the player's initial assumption of a unique rational choice. There may not be such a choice; even when there is, what seems called for, if early-on best-response conditional probabilities are to be justified, is not its assumption but its discovery. Harper himself comes to quite undercut the motivation just provided for "best-response conditional priors":

> When I hypothetically assumed I would end up choosing strategy A, perhaps all I should have assumed is that A is some strategy in the solution set [i.e., not *the* rational act, but only *a* rational act]. But, this will not allow me to hypothetically assume that an opponent who completely understands the game and the demands of rationality will have predicted my choice. My ground for using a best response prior would seem to be undercut. (p. 38; cf. Rabinowicz 1986, pp. 215-16 and p. 227, n. 1, from whence Harper got this idea)[5]

What suggests itself is not that the idea of best-response priors in two-person games be abandoned, but that it be spelled out somewhat differently. Rather than begin with conditional probabilities that exclude all but best responses to choices of particular strategies, one might more reasonably begin with conditional probabilities that exclude all but best responses to possible indifferent choices from various sets of one's strategies. Given Axiom 4.1, such best responses will be well defined in hyperrational games.

I do not, however, think that initial best-response conditional probabilities, no matter how they are articulated, are especially appropriate for hyperrational games. Indeed, I doubt that anything completely general can be said about appropriate conditional probabilities early on in hyperrational games. Even so, there may be some interest in discussing particular games under one or another hypothesis concerning early-on conditional probabilities: for example, kinds of best-response conditional probabilities and (for some games) kinds of like-response conditional probabilities. I do not here explore these avenues, and in what follows make absolutely no assumptions concerning prior, initial, or early-on conditional probabilities.

Turning from possible early-on conditional probabilities to early-on unconditional ones, it is plausible that when players have no clear idea what they will do – as one might rule should always be the case very early on – they should then judge all pure strategies in each given player's strategy set to be equally probable, or judge all such feasible strategies equally probable. It may be interesting to examine particular hyperrational games under this early-on equal-probabilities hypothesis (see Rabinowicz 1992, where the considerable possible interest of such examinations is demonstrated.) However, I myself favor the idea that hyperrational players are, very early on, quite devoid of determinate expectations or probabilities for one another's strategies, though nothing that follows depends on this idea.

2. Resolutions of hyperrational games

The first theorems about hyperrational games are for *necessary* conditions on resolutions. These theorems say that hyperrational games can resolve only in something like equilibria of sorts. The second group of theorems set out *sufficient* conditions for resolutions. These theorems say that hyperrational games can resolve by various processes of elimination, and would resolve in the choices to which these processes would lead. The two sets of theorems are related somewhat as von Neumann and Morgenstern supposed their "indirect" and "direct" arguments concerning a satisfactory theory for zero-sum two-person games were related. They saw their indirect argument as "[narrowing] down the possibilities to one," but felt that it was "still necessary to show that the one remaining possibility [was] satisfactory" (von Neumann and Morgenstern 1953, p. 148, n. 5). A major difference between my project and theirs is that they sought, and to their satisfaction found, a complete theory of solutions for all two-person zero-sum games. In contrast, I do not seek a complete theory of resolutions for all would-be hyperrational games. Indeed, I think the main interest of the theory of hyperrational games lies precisely in its necessary *in*completeness.

Necessary conditions for, or limitations on, resolutions

2.1

Random strategies are mixed strategies in which certain pure strategies have equal chances. Random strategies, while not strictly identical to indifferent choices from the pure strategies on which they are based, are in a certain way equivalent to such choices in hyperrational games. In a hyperrational game, knowledge that a player will make an indifferent choice, and knowledge that he will employ the random strategy that corresponds to that choice, would give rise to the same expectations and probabilities for pure strategies involved.

The *random extension* of a pure-strategy game is the part of its mixed extension reached by adding to players' options only all random mixed strategies. Random extensions of pure-strategy games are finite. For example, the random extension of a 2×2 game is a 3×3 game, and the random extension of a 3×3 game is a 7×7 one, for there are exactly four distinct random strategies based on a set of three pure strategies.

The *random contraction* of a mixed-strategy game is the random extension of the pure-strategy game of which this mixed-strategy game is the mixed extension. The basic assumption for mixed-strategy games, according to which expected values of interactions of mixed strategies are the usual weighted averages of expected values of interactions of pure strategies, covers random extensions as well as mixed extensions and random contractions. This assumption is always in force, though not all results depend on it.

The *randomization* of a pure-strategy game is its random extension; of a mixed-strategy game, its random contraction. To illustrate, here is the randomization of Game 1:

	C1	C2	C3	[C1, C2]	[C1, C3]	[C2, C3]	[C1, C2, C3]
R1	3, 1	0, 2	2, 0	3/2, 3/2	5/2, 1/2	1, 1	5/3, 1/2
R2	0, 1	1, 0	4, 3	1/2, 1/2	2, 2	5/2, 3/2	5/3, 2
[R1, R2]	3/2, 1	1/2, 1	3, 3/2	1, 1	9/4, 5/4	7/4, 5/4	5/3, 7/6

"[R1, R2]," for example, abbreviates "(½R1, ½R2)"; similarly, "[C1, C2, C3]" abbreviates "[⅓C1, ⅓C2, ⅓C3]."

2.1.1. I proceed to my first theorem concerning hyperrational games. It states that these games can resolve only in equilibria of sorts.

Theorem 1. *A hyperrational game G can resolve only in indifferent choices from (possibly singleton) strategy sets such that the interaction of random strategies that correspond to these choices is an equilibrium in the randomization of G.*

For substantiation, suppose that a *two*-person hyperrational game resolves in indifferent choices from *singleton* sets; suppose that Row does R_i and Column does C_j. Column knows that he ought to do C_j and that this is his maximizing strategy (Axiom 4h). Row also knows this (Axiom 4.2). Row is thus sure that Column will do C_j. Therefore, R_i is Row's unique best response to C_j; if not, he would not be choosing indifferently from the singleton set $\{R_i\}$. Similarly, C_j must be Column's unique best response to R_i. Hence (R_i, C_j) is an equilibrium in the game and thus in its random extension.

This argument can be generalized to hyperrational games of all sizes that resolve in indifferent choices from sets of all sizes. Its generalization to mixed-strategy games depends mainly on results noted concerning the dense and uniform character of sets of maximizing mixed strategies, and the consequent character of a player's expectations for others' pure strategies when he knows that they will make certain indifferent choices. In any hyperrational game (pure or mixed) that resolves in indifferent choices for certain sets of strategies, it is as if each player is, at his time of action, certain that others have used, are using, or will use random strategies corresponding to indifferent choices from their maximizing strategies (see Axiom 4.1 and comments thereon). So the player's own choice needs to correspond to a random strategy that is a best response to those random strategies.

2.1.2. Theorem 1 states that hyperrational games resolve only in equilibria of sorts – roughly, only in equilibria in random extensions or random contractions. It is important that it is possible to be more restrictive in terms of an idea (due to Rabinowicz) of a kind of equilibrium of intermediate strength. Let an interaction $(s_i, ..., s_n)$ in a randomization be an *equilibrium** if and only if it is an equilibrium and "each [player i would lose in utility if he, instead of s_i, played] some *pure* strategy not belonging to . . . [the set of pure strategies on which s_i is based] when other players play their strategies in $(s_i, ..., s_n)$" (Rabinowicz 1992, p. 121, n. 6). Every strong equilibrium in a randomization is an equilibrium*. In contrast, while some weak equilibria of randomizations are equilibria*, others are not. For example, in the randomization (see matrix at top of facing page) of a game discussed in Pearce (1984, p. 1035), neither (weak equilibrium (R1, C1) nor (weak) equilibrium (R2, C2) is an equilibrium*: Against (R1, C1) I note that, though R2 is not in {R1}, Row does not lose by playing R2 when Column plays his strategy in (R1, C1); against (R2, C2) I note that Row does not lose if she defects to R1. In contrast, (weak) equilibrium ([R1, R2], C2) of this randomization is an equilibrium*: There

Game 3

	C1	C2	[C1, C2]
R1	0, 5	−1, 3	−½, 4
R2	0, 0	−1, 3	−½, ½
[R1, R2]	0, ½	−1, 3	−½, 11/4

are no strategies outside {R1, R2} to which Row can defect; and Column loses if he defects to C1, the only strategy to which he *can* defect.

For another example, in the randomization of Game 4,

Game 4

	C1	C2	C3
R1	0, 0	1, 1	1, 1
R2	0, 0	1, 1	1, 1
R3	0, 0	1, 1	1, 1

mixed (weak) equilibrium ([R1, R2], [C2, C3]) is not an equilibrium*. This is so because R3, although not in {R1, R2}, is still a best response to [C2, C3], as the following display of a part of this randomization makes plain.

	C1	C2	C3	[C2, C3]
R1	0, 0	1, 1	1, 1	1, 1
R2	0, 0	1, 1	1, 1	1, 1
R3	0, 0	1, 1	1, 1	1, 1
[R1, R2]	0, 0	1, 1	1, 1	1, 1

Row could defect from ([R1, R2], [C2, C3]) without loss to a strategy outside the base of [R1, R2]. For similar reasons, not one of the six pure-strategy (weak) equilibria is an equilibrium*, and not one of the five displayed pure/mixed-strategy (weak) equilibria is an equilibrium*.

Pure-strategy equilibria are either strong or weak, and a pure-strategy equilibrium in a randomization is strong if and only if it is an equilibrium*. All equilibria that involve mixed strategies are weak, but (as we have seen) some but not all are equilibria*.

Arguments for Theorem 1, when allowed to follow their natural courses, lead as well to the following theorem.

Theorem 1*. *A hyperrational game G can resolve only in indifferent choices from (possibly singleton) strategy sets such that the interaction of random strategies that correspond to these indifferent choices is an equilibrium* in the randomization of G.*

If a hyperrational game resolves, then each player in it makes an indifferent choice from all and only those strategies that are best responses to the indifferent choices he knows that others are making or have made. In a mixed-strategy game, a player's best responses – his maximizing strategies – constitute a set that is closed in a certain manner under mixing: it contains precisely the mixed strategies that are based on pure strategies it contains. In any game that resolves, it will be as if its players see themselves as engaged not in the game itself but rather in its random extension or contraction, and as if each player is, at his time of action, certain that others have used, are using, or will use random strategies corresponding to indifferent choices from their maximizing pure strategies. These random strategies must be best responses each to the others, and must constitute not only an equilibrium, but an equilibrium*, in the game's randomization.[6]

2.1.3. Theorems 1 and 1* may be compared with views of Rabinowicz's regarding two-person mixed-strategy games, views that generalize naturally to n-person games:

A proper demand on any member α of R_a, the set of a's rational strategies, is ... that it should be a best response to R_b. That is, any α in R_a should maximize expected utility on the assumption that b is going to choose one of the strategies in R_b, where all the strategies in R_b are taken to be equiprobable. The analogous demand applies to R_b. (Rabinowicz 1986, pp. 215-16)

If these demands are satisfied, Rabinowicz will say that R_a and R_b make a "probabilistic equilibrium." As a part of any adequate theory of games, Rabinowicz endorses the additional demand that "R_a and R_b should, if possible, constitute a *complete* probabilistic equilibrium," where this condition is met if and only if every strategy α [β] that "is a best response to R_b [R_a] (in the above-explained sense)" is in R_a [R_b] (p. 216). Theorem 1* corresponds to this stronger demand.

Rabinowicz provides a nice proof of a theorem that he says is like a theorem included in class material I distributed in 1981–82. This theorem of Rabinowicz's clarifies the relation between Theorems 1 and 1*, and explains the demand that rational strategies should constitute "complete probabilistic equilibria." Here is the theorem:

If, in a given game, (X_a, X_b) is a complete probabilistic equilibrium, then players a and b have at their disposal subsets P_a and P_b of pure strategies such that
(1) X_a (X_b) coincides with the set of all possible probabilistic mixtures of P_a (P_b); [and]
(2) the "random" strategy on P_a, that is, the mixed strategy assigning equal positive probabilities to all the members of P_a and zero probabilities to all the other pure strategies of a, is in equilibrium with the random strategy on P_b. (pp. 223-4)

Indeed, the random strategy on P_a is in equilibrium* with the random strategy on P_b. Compare: "It is easy to see that $(s_i, ..., s_n)$ is a (strong) equilibrium in the random extension of a pure-strategy game G iff $(X_i, ..., X_n)$ is a (complete) probabilistic equilibrium in G.... [I]nstead of the somewhat misleading term 'strong equilibrium', Sobel prefers a different name for the same concept; he calls it 'equilibrium*'" (Rabinowicz 1992, pp. 120-1, n. 6).

Rabinowicz observes that a

strictly analogous theorem holds in the n-person case [and that knowing] that the Theorem holds greatly facilitates the task of identifying complete probabilistic equilibria. Since each such equilibrium can be associated with an equilibrium pair of random strategies on pure-strategy subsets, the number of possibilities that one has to consider is strictly limited. (Rabinowicz 1986, p. 224)

I proceed in terms of equilibria* and randomizations. However, computing randomizations and scanning for equilibria* is tedious and error-prone. It is therefore useful that, as Rabinowicz has observed in conversation, equilibria* of randomizations of two-person games can be quickly and securely found by exhaustive searches for complete probabilistic equilibria. Using for illustration Game 1,

Game 1

	C1	C2	C3
R1	3, 1	0, 2	2, 0
R2	0, 1	1, 0	4, 3

the following table establishes that this game has a unique complete probabilistic equilibrium, which is ({R2}, {C3}):

Sets of Row's pure strategies	Column's best pure responses to these sets	Row's best pure responses to these responses
{R1}	{C2}	{R2}
{R2}	{C3}	{R2}
{R1, R2}	{C3}	{R2}

It follows that ([R2], [C3]), or, for short, (R2, C3), is the unique equilibrium* in this game's randomization, as inspection confirms.

	C1	C2	C3	[C1, C2]	[C1, C3]	[C2, C3]	[C1, C2, C3]
R1	3, 1	0, 2	2, 0	3/2, 3/2	5/2, 1/2	1, 1	5/3, 1/2
R2	0, 1	1, 0	4, 3	1/2, 1/2	2, 2	5/2, 3/2	5/3, 2
[R1,R2]	3/2, 1	1/2, 1	3, 3/2	1, 1	9/4, 5/4	7/4, 5/4	5/3, 7/6

For another illustration, we revisit Game 3. The following table establishes that this game has the unique complete probabilistic equilibrium ({R1, R2}, {C2}):

Sets of Row's pure strategies	Column's best pure responses to these sets	Row's best pure responses to these responses
{R1}	{C1}	{R1, R2}
{R2}	{C2}	{R1, R2}
{R1, R2}	**{C2}**	**{R1, R2}**

It follows that ([R1, R2], C2) is the unique equilibrium* in this game's randomization, as was recently established by inspection of its randomization.

2.1.4. We proceed to several corollaries of Theorems 1 and 1* that concern cases involving unique equilibria.

Theorem 2. *If the randomization of a hyperrational game has a unique equilibrium, then this game can resolve only in indifferent choices that correspond to the random strategies in that equilibrium.*

Theorem 3. *A hyperrational mixed-strategy game G can resolve only in indifferent choices that correspond to an equilibrium* in its random contraction that is an equilibrium in G itself.*

Theorem 2 follows directly from Theorem 1. Theorem 3 follows from Theorem 1*, given that an equilibrium* in the random contraction of a mixed-strategy game must be an equilibrium – and indeed an equilibrium* – in the game itself.

Theorem 3 is, for the games it addresses, no more restrictive than Theorem 1*. It might seem, however, that it can be strengthened as follows.

¿**Theorem 3'**? *A hyperrational game G, mixed or pure, can resolve only in indifferent choices that correspond to an equilibrium* in its randomization that is an equilibrium in G itself.*

This principle is more restrictive for pure-strategy games than Theorem 1*, since not every equilibrium in the random extension of a pure-strategy game need correspond to an equilibrium in the game itself. But for this reason it is doubtful that the principle is valid. Consider the following pure-strategy game, displayed with its random extension.

Game 5 - Appointment in Samarra

	C1	C2	[C1, C2]
R1	0, 1	1, 0	½, ½
R2	1, 0	0, 1	½, ½
[R1, R2]	½, ½	½, ½	½, ½

It seems that this game might well resolve for hyperrational players in indifferent choices corresponding to the equilibrium* in its random extension. A result to come implies that it *would* resolve for hyperrational players in such choices; see Theorem 9 (cf. Rabinowicz 1986, pp. 218–19, game G6).[7]

Theorem 4. *If a hyperrational mixed-strategy game has a unique equilibrium, then it can resolve only in this equilibrium.*

This theorem follows from Theorem 3.
It might seem that Theorem 4 can be strengthened as follows.

¿**Theorem 4'**? *If a hyperrational game, mixed or pure, has a unique equilibrium, then it can resolve only in this equilibrium.*

Against this conjecture I offer the following pure-strategy game.

Game 6

	C1	C2	C3
R1	2, 2	0, 2	0, 0
R2	2, 0	2, 1	1, 2
R3	0, 0	1, 2	2, 1

This game has the unique equilibrium (R1, C1). Its random extension has two equilibria, (R1, C1) and ([R2, R3], [C2, C3]), but of these only the second is an equilibrium*.

	C1	C2	C3	[C1, C2]	[C1, C3]	[C2, C3]	[C1, C2, C3]
R1	2, 2	0, 2	0, 0	1, 2	1, 1	0, 1	⅔, 4/3
R2	2, 0	2, 1	1, 2	2, ½	3/2, 1	3/2, 3/2	5/3, 1
R3	0, 0	1, 2	2, 1	½, 1	1, ½	3/2, 3/2	1, 1
[R1, R2]	2, 1	1, 3/2	½, 1	3/2, 5/4	5/4, 1	3/4, 5/4	7/6, 7/6
[R1, R3]	1, 1	½, 2	1, ½	3/4, 3/2	1, 3/4	3/4, 5/4	5/6, 7/6
[R2, R3]	1, 0	3/2, 3/2	3/2, 3/2	5/4, 3/4	5/4, 3/4	3/2, 3/2	4/3, 1
[R1, R2, R3]	4/3, ⅔	1, 5/3	1, 1	7/6, 7/6	7/6, 5/6	1, 4/3	10/9, 10/9

By Theorem 1*, Game 6 cannot resolve for hyperrational players in its sole equilibrium, for that (weak) equilibrium does not correspond to an equilibrium* in its random extension. This game can resolve for hyperrational players only in indifferent choices corresponding to the sole equilibrium* in its random extension. Theorem 9 to come implies that it *would* resolve in this manner.

Game 6 poses a problem because its sole equilibrium is weak. Perhaps then the following enhancement of Theorem 4 is valid.

¿**Theorem 4"**? *If a hyperrational mixed-strategy game has a unique equilibrium, then it can resolve only in it; if a hyperrational pure-strategy game has a unique strong equilibrium, then it can resolve only in it.*

Without deciding the status of this principle, I note first that it is not an easy corollary of Theorem 1*. For even if a pure-strategy hyperrational game does have a unique strong equilibrium, its random extension can harbor additional equilibria*. Here is a game illustrative of this point.

Game 7

	C1	C2	C3
R1	5, 5	4, 4	0, 0
R2	4, 4	2, 2	3, 3
R3	0, 0	3, 3	2, 2

The random extension of this game has two equilibria, (R1, C1) and ([R2, R3], [C2, C3]), both of which are equilibria*. (I note that every strong equilibrium is an equilibrium*. The distinction between equilibria and equilibria* is significant only for weak equilibria.)

	C1	C2	C3	[C1, C2]	[C1, C3]	[C2, C3]	[C1, C2, C3]
R1	5, 5	4, 4	0, 0	½, ½	½, ½	2, 2	3, 3
R2	4, 4	2, 2	3, 3	3, 3	½, ½	⁵⁄₂, ½	3, 3
R3	0, 0	3, 3	2, 2	³⁄₂, ³⁄₂	1, 1	⁵⁄₂, ½	⁵⁄₃, ⁵⁄₃
[R1, R2]	9/2, 9/2	3, 3	3/2, 3/2	15/4, 15/4	3, 3	9/4, 9/4	3, 3
[R1, R3]	5/2, 5/2	7/2, 7/2	1, 1	3, 3	½, ½	9/4, 9/4	7/3, 7/3
[R2, R3]	2, 2	5/2, 5/2	5/2, 5/2	9/4, 9/4	9/4, 9/4	5/2, 5/2	7/3, 7/3
[R1, R2, R3]	3, 3	3, 3	5/3, 5/3	3, 3	7/3, 7/3	7/3, 7/3	23/9, 23/9

It is not a corollary of Theorem 1* that Game 7 can resolve only in its unique equilibrium (R1, C1). Theorem 1* leaves open that Game 7 can resolve instead in indifferent choices corresponding to the other equilibrium* in its random extension, ([R2, R3], [C2, C3]).

I leave open the question of whether ¿Theorem 4"?, though not a corollary of Theorem 1*, is a valid limitation on resolutions of hyperrational games in its own right. Considerations presented in Section 3.6 to show that Game 7 would not resolve for hyperrational players, and that it is not a possible hyperrational game, provide reasons for thinking that ¿Theorem 4"? is a valid principle in its own right.

Sufficient conditions for, and ways to, resolutions

2.2

Resolutions of hyperrational games require that players have settled expectations and probabilities for one another's strategies. For only then are expected values for options (and thus maximizing options) *defined*, and to act in hyperrational games the players must know which of their options maximize. In this section ways are considered for hyperrational players to settle their expectations and escape from the labyrinthine courses of deliberation that threaten in ordinary situations where, in order to maximize, a person seeks to determine what others are likely to do while realizing that they may well be engaged in like efforts, and even trying to take into account his awareness of that possibility. Compare:

It is characteristic of social interaction that I cannot maximize my own utility without "taking into account" (in some vague sense which we understand only darkly) what *you* are up to. I must have some theories or intuitions about how you are likely to behave, how you will respond to my actions, and the like. Worse than that, I must be aware that you are likely doing the same thing in regard to me, and I have to take *that* possibility into account as well. But of course you may know that I am aware of this possibility, and you [may] adjust your behaviour accordingly. And so it goes - we are both involved in a tortuous labyrinth of relations, and though we act in this way quite easily and freely, it is better than even money that neither of us could even *begin* to give an explicit account of how we do it. (Moore and Anderson 1962, p. 413)

This section is about ways in which hyperrational players, declining to begin regressions of deliberation or to enter into endless labyrinths of higher- and higher-order reflections, can in some cases - including many unique-equilibrium cases - make up their minds what to expect of one another. I agree that "in order for an agent to make a decision by applying the rule to maximize expected utility, he must assign probabilities to the strategies of his opponents" (Weirich 1991, p. 1). But I argue against the circle-generating contention that "in order to assign probabilities to their strategies rationally, he must [in general] take account of their insight into his reasoning" (p. 1). I contend in what follows that, in the context of hyperrational games, the "strategist's problem of circularity is [*not* in unique-equilibrium cases without dominance or other helping features] insoluble if one insists on the [unadulterated Bayesian] decision principle to maximize expected utility" (p. 4). (Weirich, convinced that a new decision principle is needed to solve the problem of circularity, proposes a principle of ratification that "says to choose an action that maximizes expected utility on the assumption that it is performed"; p. 4.) I contend inter alia that "the vicious circle [only] *seems* inescapable" (Bacharach 1987, p. 19, emphasis added). I think that generally, even if not always and universally, "if a [hyperrational] game has a unique Nash equilibrium, then [it *has* a solution and] that is its solution" (p. 42); Bacharach is more pessimistic (see pp. 42ff).

Theorems to follow depend on the idea that, if a game *can* resolve in a certain way for hyperrational players - that is, if there is a process by which their expectations can be settled sufficiently to resolve the game in some pattern of choices - then the game *would* resolve for them in that way; speedy hyperrational players would find that way, or another to the same end. Let a *potentially resolvable game* be a game whose randomization contains an equilibrium*; let a potentially resolvable game that can, and so would, resolve for hyperrational players be a *possible hyperrational game*. This terminology leaves open whether or not every potentially resolvable game is a possible hyperrational game.

2.2.1. Here is a first simple principle, of very limited coverage.

Theorem 5. *A two-person game resolves for hyperrational players if at least one player has a strongly dominant strategy.*

Consider, for example, the following game with its randomization.

Game 8

	C1	C2	[C1, C2]
R1	3, 2	2, 3	½, ½
R2	2, 2	1, 1	½, ½
[R1, R2]	½, 2	½, 2	2, 2

Row and Column can settle by elimination their probabilities for each other's choices and, having settled these probabilities, can settle the expected values of their own options and so settle or make their own choices. Taking Column first, he can be sure that whatever Row's probabilities at the time of action, Row will not employ R2. For Column can see that R1 is strongly dominant, and knows that Row realizes that Column's actions are causally independent of her own. So Column can set aside R2. When this is done, C2 emerges as uniquely maximizing. Row in turn can by indirect reasoning eliminate the possibility that Column will employ C1. She can reason that if Column did employ C1, he would know that Row, expecting C1, would employ R1. That knowledge would lead to C2's uniquely maximizing. On C1's elimination, R1 emerges as uniquely maximizing. Game 8 can resolve in this way into the interaction (R1, C2), and so it does.

It is not important to the resolution of Game 8 that C2 be uniquely maximizing for Column. Suppose the game changed as follows.

Game 8'

	C1	C2
R1	3, 2	2, 2
R2	2, 2	1, 1

Column can now reach an indifferent choice from his strategies in the way described. When R2 is set aside, there is nothing in the choice between C1 and C2 (and any mixed strategies based on them). And Row can, by indirect reasoning, eliminate the possibility that there will be (at the time of action), something in Column's choice, thus settling her own probabilities at $\Pr(C1) = \Pr(C2) = \frac{1}{2}$.

Here is a second simple principle that is closely related to Theorem 5.

Theorem 6. *A two-person game in which each player has a strongly dominant strategy resolves for hyperrational players in the interaction of these strategies.*

It follows from Theorem 1 that such a game can resolve for hyperrational players *only* in the indicated interaction, for that interaction will be the only equilibrium in the game's randomization. The argument for Theorem 5, especially the part concerned with the settling of Column's probabilities and expected values, indicates how such a game can resolve for hyperrational players in that interaction. Each player can begin with thoughts about the availability to the other player of a strongly dominant option.

2.2.2. I confess to unease with the arguments just given. Column, for example, is allowed to settle by elimination that Row will employ R1. Column is not required to wonder how Row can come to employ R1, nor is he required to explain how Row can settle her expectations for Column (and his expected utilities) so that she can see that R1 uniquely maximizes. Capitalizing on his knowledge of the game's form and hyperrationality, Column finds out by elimination what Row *must* do, and stops there. Similarly for Row.

Each comes first to know that the other will make some choice, to know by elimination that the other player will somehow settle his or her expectations – and thus his or her expected utilities – in certain ways, and make some particular choice. This knowledge of the other's choice settles a player's own expectations and expected utilities, and terminates his or her prechoice cogitations. My players refrain from asking a question that must naturally arise for them, the question whose answer would be, "Why, by eliminating – how else?" Upon determining that their fellows must settle their expectations in certain ways, my players refrain from asking how their fellows will manage to do that, and from asking next how these fellows think *their* fellows will manage to settle their expectations concerning *their* fellows, and so on back and forth and 'round and 'round. My players exercise restraint in their deliberative ratiocinations. It is as if my players have learned that the best answers to some questions are for them: "Don't ask! You don't want to know." Arguments throughout Section 2.2 stand under this cloud. Whether under the circumstances of my players their restraint is admirable is a question to which I return in Section 2.2.6.

2.2.3. Let a strategy *x* be *strictly dominated* if and only if its player has a strategy *x'* such that, for each combination of strategies for all other players, *x'* is better than *x*. Let a game *G* *reduce by a strictly dominated discard* to *G'* if *G'* comes from *G* by deletion of a strictly dominated strategy. For example,

Game 9

	C1	C2	C3
R1	2,4	1,0	0,0
R2	3,1	2,2	3,3
R3	4,0	1,1	2,0

reduces by a strictly dominated discard to

	C1	C2	C3
R2	3,1	2,2	3,3
R3	4,0	1,1	2,0

Let a game *G* reduce by a *sequence* of strictly dominated discards to *G'* if there is a sequence of games, beginning with *G* and ending with *G'*, such that each comes from its immediate predecessor by a strictly dominated discard. For example, Game 9 reduces by such a sequence, first to the 2 × 3 game just displayed and then, by discards in turn of C1, R3, and C2, all the way to

	C3
R2	3,3

Here is a principle built on the idea of dominated discards.

Theorem 7. *If a game G reduces by a sequence of strictly dominated discards to G', and game G' resolves for hyperrational players in a certain way, then game G resolves for hyperrational players in this way.*

If a strategy is strictly dominated in a hyperrational game, then in calculating expected values of strategies all players can assign zero probability to this strategy. Hyperrational players can know that such a strategy must eventually be set aside by its agent and thus by everyone. Hence each player can see that eventually everyone's deliberations will be as they would be from the beginning in the simpler reduced game. Similarly, this holds in turn for that game, if a strategy is strictly dominated in *it;* and so on until no further deletions of this kind are possible.

Applications of Theorem 7 to mixed-strategy games can concentrate on pure strategies, since if a pure strategy is strictly dominated then so is every mixed strategy in which it is assigned a positive chance. It is not difficult to see that an interaction, if it can be reached by a sequence of strictly dominated discards from a game, corresponds to the sole equilibrium* in the game's randomization. Compare: "The outcome is determined by the fact that everybody ignores dominated actions, everybody expects everybody else to ignore dominated actions, and so on" (Lewis 1969, p. 19). There are two differences between David Lewis's idea and mine. First, whereas he seems to have in mind that everybody ignores *his own* dominated options, ..., my idea is that everybody ignores dominated options *of others*, ... I cannot comfortably rationalize a hyperrational player's ignoring his own dominated options until he has settled his expectations for others, and thus his expected values for all his options, so that he can compare and see which of them maximize expected value. He will of course then ignore dominated ones or set them aside, seeing that they are submaximal. But it makes sense that before a hyperrational player has settled his expectations for others he should set aside *their* dominated options as things he is sure they will not do.

The second difference between our ideas is that Lewis's definition of "strictly dominated" is nonstandard: "A *strictly dominated* choice is one such that, no matter how the others choose, you could have made some other choice that would have been better" (p. 17; brackets deleted). Would-be resolutions by discards of Lewis-dominated options are discussed in the appendix to this chapter.

2.2.4. Not every game that can resolve for hyperrational players is covered by Theorem 7. Consider, for example, Game 10.

Game 10

	C1	C2	C3
R1	4, 4	0, 5	0, 3
R2	5, 0	2, 2	0, 3
R3	3, 0	3, 0	1, 1

No strategy is *strictly* dominated, and yet C1 could be discarded by hyperrational players. As both players realize, Row could reason as follows:

If Column is sure that I will use R3 then his best response is C3, and he will not use C1. And if Column is not sure that I will use R3 then – whatever his probabilities for my possible actions R1, R2, and R3 – his expected value for C2 exceeds his expected value for C1, and he will not use C1.

They can thus set C1 aside. The effect of that reduction is a game in which R3 is strictly dominant, and in which C3 is best against R3. Column can be sure that Row will employ R3, and Row can be sure that Column will employ C3. The game can in this way resolve in the interaction (R3, C3).

For a principle that covers such resolutions, let a probability function Pr_i be *admissible* for a player i if and only if, for each other player j, either $Pr_i(s_j) = 1$ for some one strategy s_j of player j, or Pr_i makes several of that player's strategies equally probable with their probabilities summing to 1. Let a strategy be *eligible* if and only if it maximizes under a probability function that is admissible for its agent. Here is the promised principle.

Theorem 8. *If a game G reduces by a sequence of discards of ineligible strategies to G', and game G' resolves for hyperrational players in a certain way, then game G resolves for hyperrational players in this way.*

If a pure strategy is ineligible, then so is every mixed strategy in which it is assigned a positive chance. Thus applications of Theorem 8 can be by inspections of relevant randomizations. For example, inspection of

	C1	C2	C3	[C1, C2]	[C1, C3]	[C2, C3]	[C1, C2, C3]
R1	4, 4	0, 5	0, 3	2, ½	2, ½	0, 4	4/3, 4
R2	5, 0	2, 2	0, 3	½, 1	5/2, ½	1, ½	7/3, 5/3
R3	3, 0	3, 0	1, 1	3, 0	2, ½	2, ½	7/3, 1/3
[R1, R2]	9/2, 2	1, ½	0, 3	11/4, 11/4	9/4, ½	½, 13/4	11/6, 17/6
[R1, R3]	4, 2	3/2, 5/2	½, 2	5/2, 9/4	2, 2	2, 9/4	11/6, 13/6
[R2, R3]	2, 0	5/2, 1	½, 2	13/4, ½	9/4, 1	3/2, 3/2	7/3, 1
[R1, R2, R3]	4, 4/3	5/3, 7/3	1/3, 7/3	17/6, 11/6	13/6, 11/6	1, 7/3	2, 2

confirms that C1 is ineligible: it is not maximum on any row in this randomization of Game 10. Setting C1 aside yields a game in which the resolution (R3, C3) can be reached by the sequence (R1, R2, C2) of dominated discards.

2.2.5. Presumably there are potentially resolvable games not covered by Theorem 8 that would resolve for hyperrational players. One supposes that games with unique equilibria* in their randomizations should resolve for hyperrational players, but there are hyperrational games that have unique equilibria* in their randomizations where every pure strategy is eligible. Here is such a game with its randomization (I owe this modification of Game 7 to Włodzimierz Rabinowicz).

Game 11

	C1	C2	C3	[C1, C2]	[C1, C3]	[C2, C3]	[C1, C2, C3]
R1	5, 5	4, 4	1, 1	$9/2, 9/2$	3, 3	$5/2, 5/2$	$10/3, 10/3$
R2	4, 4	2, 2	3, 3	3, 3	$7/2, 7/2$	$5/2, 5/2$	3, 3
R3	1, 1	3, 3	2, 2	2, 2	$3/2, 3/2$	$5/2, 5/2$	2, 2
[R1, R2]	$9/2, 9/2$	3, 3	2, 2	$15/4, 15/4$	$13/4, 13/4$	$5/2, 5/2$	$19/6, 19/6$
[R1, R3]	3, 3	$7/2, 7/2$	$3/2, 3/2$	$13/4, 13/4$	$9/4, 9/4$	$5/2, 5/2$	$8/3, 8/3$
[R2, R3]	$5/2, 5/2$	$5/2, 5/2$	$5/2, 5/2$	$5/2, 5/2$	$5/2, 5/2$	$5/2, 5/2$	$5/2, 5/2$
[R1, R2, R3]	$10/3, 10/3$	3, 3	2, 2	$19/6, 19/6$	$8/3, 8/3$	$5/2, 5/2$	$25/9, 25/9$

I believe this game would resolve for hyperrational players by eliminations of choices corresponding to nonequilibrium* strategies in this randomization. Row could eliminate by indirect arguments every choice on Column's part that corresponds to a *non*-equilibrium* strategy in this randomization. To take one example, Row could reason that Column will not make a choice that would correspond (as far as Row's expectations are concerned) to the strategy (½C1, ½C2), since if Column did then C1 would tie with C2. Column would realize that Row considered C1 and C2 equally probable and C3 not at all probable, and so would certainly employ R1. Hence C1 would uniquely maximize, and not tie with C2.

Taking another example, Row could reason that Column will not make a choice that would correspond (as far as Row's expectations are concerned) to the strategy (½C2, ½C3), which is an *equilibrium* strategy but not an *equilibrium** strategy. If Column did make such a choice then C2 and C3, but *not* C1, would maximize. Column would realize that Row considered C2 and C3 equally probable and C1 not at all probable, resulting in each of Row's strategies maximizing and her making an indifferent choice from {R1, R2, R3}. But then Column would consider these strategies of Row's equally probable, with the result that C1 *would* maximize; in fact, it would uniquely maximize.

The following principle, one that would have Game 11 resolve in choices corresponding to (R1, C1), is I think valid.

Theorem 9. *If the randomization of a potentially resolvable game G (i.e., a game whose randomization contains at least one equilibrium*) reduces by discards of nonequilibrium* strategies to the randomization of a game G' that resolves in some way for hyperrational players into indifferent choices from (possibly singleton) sets of strategies, then G resolves into those choices.*

Regarding the intent of Theorem 9, I note that discards of nonequilibrium* strategies can be several (indeed all) at one time. Also, included in ways in which game G' may resolve is the way of Theorem 9 itself. It is not difficult to see that Theorem 9 subsumes Theorem 8, for if a strategy of a game is ineligible then it is not an equilibrium* strategy in that game's randomization.

Here is another game (with randomization) that would, I think, resolve for hyperrational players in the way of Theorem 9.

Game 12

	C1	C2	[C1, C2]
R1	1, 1	0, 0	½, ½
R2	0, 0	0, 0	0, 0
[R1, R2]	½, ½	0, 0	¼, ¼

While both (R1, C1) and (R2, C2) are equilibria, only (R1, C1) is an equilibrium*. Row could therefore eliminate, by indirect arguments, every choice on Column's part other than a definite choice of C1. To illustrate, a definite choice of C2 could be eliminated, since if Column made this choice he would know that, expecting it, Row would be indifferent between R1 and R2. Judging these strategies of Row's to be equally probable, Column would choose C1 for sure, not C2 as supposed – C1 would uniquely maximize. An indifferent choice from {C1, C2}, as well as a choice of any available mixed strategy, could be similarly eliminated. (A game of this structure is discussed in Sobel 1972, pp. 173–4 [Chapter 13, Appendix].)

2.2.6. There remains much unfinished business. This includes not only consideration of other ways in which games might resolve for hyperrational players (e.g., further eliminations may be possible for some symmetrical games – see Section 3.4), but also ministration to the unease to which I have confessed, following Game 8, with all the ways of resolution defended in this Section 2.2.

I maintain that hyperrational players, in order to settle their own expectations and expected utilities, use their knowledge that their fellows will, in the end, have somehow settled *their* expectations, and thus *their* expected utilities. Hyperrational players in two-person games are in effect allowed to settle their expectations for their fellows by elimination of what would be for their fellows unratifiable choices – they are allowed to eliminate possible choices of their fellows that they can see that their fellows could not know they were making. But I do not propose this as a way in which hyperrational players might suppose that their fellows make up their minds what to do. Similarly, speaking now of agents in general and not just of hyperrational ones, I hold that to be rational an action must not only maximize relative to all relevant alternatives, but also be the subject of a ratifiable choice. Yet I do not confine relevant alternatives to subjects of possible ratifiable choices, or propose as a method of deliberation that one first eliminate unratifiable options and then find what maximizes among options that remain (see note 5). Nor do I think that hyperrational players might reasonably do something similar, and find out what they ought to do by eliminating options of their own that they see could not be known by them to maximize.

There is a cloud over the ways of resolution I endorse, where hyperrational players in two-person games in effect eliminate unratifiable options of their fellows. There is an air, if not of circularity and begging the question, at least of incompleteness or lack of candor about these ways of resolution. It is clear, as maintained in Section 2.1, that hyperrational games can resolve only in indifferent choices corresponding to equilibria* in their randomizations, but it is not as clear (and one is not as easy about) just *how* some hyperrational games can resolve in such indifferent choices, or by what processes of reasoning hyperrational players can sometimes settle their expectations for one another's actions as well as their expected utilities for their own actions.

What is to be made of this problem? – perhaps a virtue. That is what I want to do. Perhaps one should not expect it to be entirely clear how hyperrational players would manage their affairs. Ordinary interactions are facilitated by all manner of natural and cultural props that I believe would be irrelevant to hyperrational interactions; habits, kinds of suggestibility, conventions, shared senses of "salience," precedence, inductive inference (cf. Bacharach 1987, p. 43), and so forth either have no place or lack force (not already reflected in expected values) in hyperrational communities. There is so much that is of importance (independently of expected values) to ordinary interactions that would be irrelevant in the hyperrational case that the wonder should be that hyperrational players can interact at all.

Under the sway of these thoughts, I welcome the idea that hyperrational players would manage to interact by narrowing down possibilities for their fellows' actions. Their play, which might be cast as *Labyrinths Declined,* in fact strikes me as just right for them, and if not the whole of the story for them then at least its first and main part.[8] I believe that the kind of eliminative thinking attributed here to hyperrational players is evident in our own procedures, though so covered over by many more direct and effective deductions as to be hardly noticeable. My view is that this kind of thinking is the main part (if not all) of what, at the hyperrational limit, would be left of ordinary interactive thinking. What is remarkable is not that what would be left at this rarefied limit would be, at least in its main part, so roundabout and restrained. What is marvelous, I think, is that anything at all would be left, and that hyperrational players would have at least some ways to interact.[9]

3. Problems for members of hyperrational communities

We are interested in problems that are possible for members of hyperrational communities. As will be seen, in some games that are possible for them *they do not do well.* And there are games that can arise for ordinary players that are not possible for the hyperrational, and in which, in their extreme hyperrationality, *they could do nothing at all.* Confronted with what would be such games, something about these agents would have to give. A hyperrational community would never be in situations of certain kinds, and a hyperrational community that always did well would never be in ones of certain other kinds.

Games in which they would not do well

3.1

I begin with two well-known games, displayed here with their randomizations.

Game 13 - Prisoners' Dilemma

	C1	C2	[C1, C2]
R1	2, 2	4, 1	3, ½
R2	1, 4	3, 3	2, ½
[R1, R2]	½, 3	½, 2	½, ½

Game 14 – Gasoline war

	C1: $1	C2: 90¢	C3: 80¢	[C1,C2]	[C1,C3]	[C2,C3]	[C1,C2,C3]
R1	5, 5	1, 7	1, 6	3, 6	3, 13/2	1, 13/2	7/3, 6
R2	7, 1	4, 4	2, 6	11/2, 5/2	9/2, 7/2	3, 5	13/3, 11/3
R3	6, 1	6, 2	3, 3	6, 3/2	9/2, 2	9/2, 5/2	5, 2
[R1, R2]	6, 3	5/2, 11/2	3/2, 6	17/4, 17/4	15/4, 9/2	2, 23/4	29/6, 29/6
[R1, R3]	11/2, 3	7/2, 9/2	2, 9/2	9/2, 15/4	15/4, 15/4	11/4, 9/2	11/3, 4
[R2, R3]	13/2, 1	5, 3	5/2, 9/2	23/4, 2	9/2, 11/4	15/4, 15/4	14/3, 17/6
[R1, R2, R3]	6, 7/3	11/3, 13/3	2, 5	29/6, 29/6	4, 11/3	17/6, 14/3	35/9, 35/9

Game 13 can resolve only in its sole equilibrium (Theorem 1). And, given that R1 and C1 are strictly dominant, there is a way to this resolution for hyperrational players (Theorem 6). Game 14 can resolve only in *its* sole equilibrium. And, by the sequence of strictly dominated discards (C1, R1, R2, C2), there is a way to this resolution for hyperrational players (Theorem 7).

For a third game we have, with its randomization,

Game 3

	C1	C2	[C1, C2]
R1	0, 5	−1, 3	−½, 4
R2	0, 0	−1, 3	−½, 3/2
[R1, R2]	0, 5/2	−1, 3	−½, 11/4

As observed in Section 2.1.2, and established in a different way in Section 2.1.3, ([R1, R2], C2) is the sole equilibrium* of the randomization of this game. Hence it can resolve only in choices corresponding to this equilibrium* (Theorem 1*), and it would resolve in this way (Theorem 9) even though this resolution is suboptimal and dispreferred by both players to the optimal (weak) equilibrium (R1, C1).[10]

Here for good measure and to complete the present group is another game with its randomization, employed previously against ¿Theorem 4'? (see matrix at top of facing page). Game 6 can resolve only in indifferent choices corresponding to the sole equilibrium* in its randomization, ([R2, R3], [C2, C3]). And it would resolve in this way notwithstanding

Game 6

	C1	C2	C3	[C1, C2]	[C1, C3]	[C2, C3]	[C1, C2, C3]
R1	2, 2	0, 2	0, 0	1, 2	1, 1	0, 1	2/3, 4/3
R2	2, 0	2, 1	1, 2	2, 1/2	3/2, 1	3/2, 3/2	5/3, 1
R3	0, 0	1, 2	2, 1	1/2, 1	1, 1/2	3/2, 3/2	1, 1
[R1, R2]	2, 1	1, 3/2	1/2, 1	3/2, 5/4	5/4, 1	3/4, 5/4	7/6, 7/6
[R1, R3]	1, 1	1/2, 2	1, 1/2	3/4, 3/2	1, 3/4	3/4, 5/4	5/6, 7/6
[R2, R3]	1, 0	3/2, 3/2	3/2, 3/2	5/4, 3/4	5/4, 3/4	3/2, 3/2	4/3, 1
[R1, R2, R3]	4/3, 2/3	1, 5/3	1, 1	7/6, 7/6	7/6, 5/6	1, 4/3	10/9, 10/9

the presence of an equilibrium, the weak equilibrium (R1, C1) that both players would prefer.

The familiar trouble with these games is that their resolutions for hyperrational players would be suboptimal. Assuming that members of a hyperrational community could see one of these games coming and that "heading it off" was in some way possible, they would have reasons, though not necessarily sufficient ones, for heading it off. They would also have reasons, though again not necessarily sufficient ones, *not* to head off these games but instead, if this is possible, to make themselves (in some particular chosen manner) other than hyperrational, at least for these games.

Games in which they could not do anything at all

3.2. A quasi-prisoners' dilemma: Variable-sum game with multiple equilibria and practical symmetry

Here is the game, together with its randomization:

Game 15

	C1	C2	[C1, C2]
R1	1, 1	3, 1	2, 1
R2	1, 3	2, 2	3/2, 5/2
[R1, R2]	1, 2	5/2, 3/2	7/4, 7/4

This game is formally symmetrical: interchanging R1 and C1, and R2 and C2, yields another representation of the same game, in which (despite her name) Ms. Row's strategies are assigned columns, and Mr. Column's are assigned rows. Let the game under consideration also be *practically* symmetrical in the sense that a strategy (the first or second one) of a player maximizes if and only if the corresponding strategy of the other player maximizes.

Games of this sort can arise for ordinary communities. Even very well informed and organized and rational communities can know they are in such games. But though hyperrational players could be headed for such a game, they could never be in one, for they could not act in one. In other words, games of this sort cannot be hyperrational games. Here are two arguments for this conclusion.

Suppose that a hyperrational community were in such a game. The game could resolve only in choices corresponding to an equilibrium in its randomization (Theorem 1). But of the five equilibria, four are excluded by the stipulation of practical symmetry. For example, this condition rules out that though both R1 and R2 maximize for Row, only C1 maximizes for Column: it rules out choices corresponding to the equilibrium ([R1, R2], C1). So the game could resolve for hyperrational players only in definite choices of R1 and C1 based on these strategies' being uniquely maximizing for their agents. But it could not resolve in this way, for if it did then Row and Column would be certain of C1 and R1, and so neither R1 nor C1 would be uniquely maximizing for its agent.

For a simpler argument that shows that Game 15 with practical symmetry cannot be hyperrational, observe that the displayed randomization contains exactly two equilibria*, (R1, [C1, C2]) and ([R1, R2], C1). Practical symmetry rules out choices corresponding to these.

For somewhat similar comments about a zero-sum mixed-strategy game displayed here with its random contraction,

Game 16

	C1	C2	[C1, C2]
R1	1, −1	2, −2	½, −½
R2	1, −1	0, 0	½, −½
[R1, R2]	1, −1	1, −1	1, −1

see Rabinowicz (1986, p. 216).

3.3. The Hunter's Dilemma: Double-sum game with multiple equilibria and practical symmetry

We consider the game displayed here with its randomization:

Game 17

	C1	C2	[C1, C2]
R1	2, 2	3, 3	½, ½
R2	3, 3	1, 1	2, 2
[R1, R2]	½, ½	2, 2	9/4, 9/4

and again assume practical symmetry. This game would not resolve for hyperrational players because it could do so only in choices corresponding to an equilibria in its randomization (there are two), and these resolutions are ruled out by the game's practical symmetry. This game is discussed in Sobel (1972, pp. 159-63 [Chapter 13, Section 2.2]).

3.4

Games 15 and 17, complete with practical symmetry, are not possible for hyperrational players. Indeed, I believe that given merely their *formal* symmetries these games are not possible for hyperrational players, for I believe that formal symmetries of hyperrational games imply practical symmetries. Ordinary maximizers, whose probabilities for each others' strategies can be partly hunches or based partly on things other than features of their games, can differ in their probabilities in formally symmetrical games. But for reasons implicit in the comments on Game 6 (see Section 3.6), I think that differences in probabilities for others' strategies are not possible in hyperrational formally symmetrical games, and if this is right then differences in sets of maximizing options are not possible in such games either.

More work is needed on symmetry restrictions on resolutions. One point made by Hans Herzberger (1978, pp. 185-6) is that these restrictions, precisely framed, should extend to games that by reorderings and renumberings of strategies turn into formally symmetrical games. Here is an example.

Game 17'

	C1	C2
R1	3, 3	2, 2
R2	1, 1	3, 3

This game turns into Game 17 by permuting C1 and C2. A second point is that symmetry restrictions should extend to games reached by admissible transformations of individuals' expected value functions (Rabinowicz 1992, p. 122, n. 10); a third point is that these restrictions should not be specific to two-person games (*ibid.*). And a fourth point is that "a fuller development" of the argument from formal to practical symmetries for hyperrational players is needed (Herzberger 1978, p. 186).

3.5. *A two-person zero-sum mixed-strategy game with a unique equilibrium*

We consider Game 18, the mixed extension of the following pure-strategy game.

	C1	C2
R1	3, −3	1, −1
R2	2, −2	4, −4

Game 18 has the random contraction

	C1	C2	[C1, C2]
R1	3, −3	1, −1	2, −2
R2	2, −2	4, −4	3, −3
[R1, R2]	½, −½	½, −½	½, −½

Since "maximin" strategies in mixed-strategy Game 18 are (½R1, ½R2) and (¾C1, ¼C2), the part of this mixed-strategy game that is relevant for present purposes is

	C1	C2	(¾C1, ¼C2)
R1	3, −3	1, −1	½, −½
R2	2, −2	4, −4	½, −½
(½R1, ½R2)	½, ½	½, −½	½, −½

Mixed-strategy Game 18 is not a possible hyperrational game. This follows from Theorem 1, since there is not an equilibrium in this game's random contraction. For a second argument, this impossibility follows from Theorem 3, since the sole equilibrium of this mixed-strategy game is not an equilibrium* in its random contraction. (For related comments on zero-sum games, see Sobel 1975, p. 682, n. 4; McClennen 1978, pp. 350-1; and Rabinowicz 1986, pp. 218-19, 220.)

3.6. A double-sum pure-strategy formally symmetrical game with a unique strong equilibrium

Game 18 shows that unique equilibria do not ensure resolutions for hyperrational communities. But perhaps the trouble with Game 18 stems from its being a zero-sum game, from the weakness of its unique equilibrium, or from its equilibrium involving mixed strategies. Perhaps absent these features, resolutions for hyperrational players would be assured by unique equilibria. Perhaps

> if there is no considerable conflict of interest, the task of reaching a unique coordination equilibrium is more or less trivial. It will be reached if the nature of the situation is clear enough so that everybody makes the best choice given his expectations, everybody expects everybody else to make the best choice given his expectations, and so on. (Lewis 1969, p. 16)

One might guess that when there is

> a unique coordination equilibrium and predominantly coincident interests [then] common knowledge of rationality is all it takes for an agent to have reason to do his part of the one coordination equilibrium. He has no need to appeal to precedents or any other source of further mutual expectations. (p. 70)

(Lewis does not say *how* "common knowledge is all it takes." Possibly he supposed that it would allow everybody to ignore nonequilibrium strategies, to expect everybody else to ignore them, and so on; cf. p. 19.)

To challenge Lewis's claim, I offer for consideration the following full-agreement, pure-strategy game in which there is a unique strong equilibrium. The structure of the game with its random extension is displayed at the top of the next page. Assume that players in Game 7 are restricted in their information regarding one another, and that they know of each other only what we can deduce from the normal form of their situation, its status as a game, and their maximizing characters. Ordinary communities – and even very well informed, organized, and rational communities – can play Game 7; the question is whether it is possible for hyperrational communities. It is not possible if, as it seems, there is no way in which hyperrational players could settle their expectations concerning one another, or the expected values of their options, and so no way in which they could maximize, which is the only thing that hyperrational agents can do.

Game 7

	C1	C2	C3	[C1, C2]	[C1, C3]	[C2, C3]	[C1, C2, C3]
R1	5, 5	4, 4	0, 0	9/2, 9/2	5/2, 5/2	2, 2	3, 3
R2	4, 4	2, 2	3, 3	3, 3	7/2, 7/2	5/2, 5/2	3, 3
R3	0, 0	3, 3	2, 2	3/2, 3/2	1, 1	5/2, 5/2	5/3, 5/3
[R1, R2]	9/2, 9/2	3, 3	3/2, 3/2	15/4, 15/4	3, 3	9/4, 9/4	3, 3
[R1, R3]	5/2, 5/2	7/2, 7/2	1, 1	3, 3	7/2, 7/2	9/4, 9/4	7/3, 7/3
[R2, R3]	2, 2	5/2, 5/2	5/2, 5/2	9/4, 9/4	9/4, 9/4	5/2, 5/2	7/3, 7/3
[R1, R2, R3]	3, 3	3, 3	5/3, 5/3	3, 3	7/3, 7/3	7/3, 7/3	23/9, 23/9

Players who knew no more than we can know could have, I think, only conditional knowledge of values of their options. Row, for example, could know that if Column chooses C1 then R1 is best, and that if Column chooses indifferently from {C2, C3} then R2 and R3 are best. But there seems no way in which Row could know *unconditionally* what is best. The interaction (R1, C1) stands out and is salient for us, and we can assume that it would stand out for hyperrational players. It is the sole equilibrium in the game and the best one in its random extension. However, for single-minded players who want only to maximize, none of this is of any obvious or immediate relevance. Hyperrational players are maximizers, and are impressed by and moved by nothing other than maximum expected values. They do not even "tend to pick the salient as a last resort" (Lewis 1969, p. 35): They are *maximizers,* that is all; they do not have last-resort routines when maximization is stymied; they do not even have nonmaximizing tie-breaker routines (cf. Gilbert 1989). Peculiarities of the interaction (R1, C1), of which there are plenty, are relevant to Row's problem – for example, *if* they are somehow relevant to her expectation concerning Column's choice. But *that* observation, tempting first tentative steps down endless paths, counsels despair.

In Game 7, players know only what we (as students of their game) know; this game is, I think, not a possible hyperrational game. Indeed, a bolder conclusion – addressed to all games of the normal form of Game 7 – seems warranted, since it seems that hyperrational players should never know anything of relevance about their games beyond what we can gather from those games' structures and hyperrationality. I believe that they could not, for example, find *evidence* for performances to come (from others of their kind) in the *past* performances of these others.

[T]hey would know each other and their situations too well to learn from experience in past situations what actions to expect in a current situation. . . . They [would] know each other and the structures of their situations so well that the only inductive projections they *might* make [would be ones] that could for them *serve no purpose*. . . . Hyperrational maximizers . . . would . . . know what to expect of each other without regard to the past. (Sobel 1976, in Campbell and Sowden 1985, pp. 307, 312 [Chapter 15, Sections 0 and 1])

Their expected values can be affected by the past, for hyperrational players can have values that are *not* exclusively forward-looking (Sobel 1976, in Campbell and Sowden 1985, pp. 312-13 and p. 316, n. 1 [Chapter 15, Sections 1 and 2]). They can, for example, be traditionalists. But their choices are determined entirely by their present expected values (through which all things that matter practically to them are "filtered"), and can be anticipated by their hyperrational fellows – given their knowledge of these present expected values – and anticipated quite *without regard to their grounds and the past*. Late entrants in series of *hyperrational* games would not be disadvantaged by their tardiness, even if it were common knowledge that their late entry made them ignorant of past plays of which other players had perfect memories.

3.7. Multiple-equilibria, double-sum coordination problems

These games are problematic for reasons similar to those adduced for Game 7. Hyperrational players could not, I think, coordinate in them.

Game 19 – Meeting at the station

	C1	C2	[C1, C2]
R1	3, 3	1, 1	2, 2
R2	1, 1	2, 2	½, ½
[R1, R2]	2, 2	½, ½	½, ½

Game 19' – Meeting at another station

	C1	C2	[C1, C2]
R1	2, 2	1, 1	½, ½
R2	1, 1	2, 2	½, ½
[R1, R2]	½, ½	½, ½	½, ½

Note that even if Game 19' would for hyperrational players resolve (which I doubt) in indifferent choices corresponding to its mixed equilibrium*, it would remain problematic in the way of the first games considered: its resolution would be suboptimal. This holds similarly for the following game, displayed with its randomization:

Game 20

	C1	C2	[C1, C2]
R1	1, 1	2, 0	½, ½
R2	0, 2	3, 3	½, ½
[R1, R2]	½, ½	½, ½	½, ½

Even if this game would resolve for hyperrational players in indifferent choices corresponding to the random equilibrium* in its randomization, it would remain problematic in that its resolution would be suboptimal. In this case, unlike Game 19', this compromise resolution would correspond to an interaction that was Pareto inferior to a *unique* Pareto optimal interaction.

3.8

Game 20 is game 4 in Rabinowicz (1992), wherein interesting iterative processes for resolutions are examined (without endorsement). Game 11' and Game 20 would resolve in their mixed suboptimal equilibria* by every method Rabinowicz studies.

The processes studied have hyperrational players beginning deliberations in a game with, for each other player, tentative equal probabilities for that player's strategies or for some subset of his strategies – for example, strategies on which his equilibrium* strategies in the game's randomization are based. Different starting points determine different iterative processes. The processes depend on its being common knowledge that initial tentative equal probabilities give rise *first* to tentative dispositions of all players to indifferent choices from strategies that maximize given their initial probabilities, *second* (if these strategies differ from those that were initially considered to be equally probable) to new tentative equal probabilities for now tentatively maximizing strategies of other players, *third* to tentative dispositions to indifferent choices from strategies that maximize given *these* probabilities, *fourth* (if these strategies differ from those that were secondly considered to be equally probable) to new tentative equal probabilities for *these* tentatively maximizing strategies, and so on.

The processes are "discrete" and have players "seeking the good" (Skyrms 1990, pp. 30-1). They reach "fixed points" in some games at which strategies that maximize given current tentative equal probabilities are the ones that are tentatively judged to be equally probable. Dispositions and probabilities at such points become fixed and no longer tentative. In other games, these processes cycle interminably. A theoretical proposal of interest would be that a certain one of these iterative processes – one with a peculiarly appropriate starting point – leads for an acceptable elaboration of hyperrationality to resolutions.

I have difficulties with the psychology of these processes. At the first stage, for example, dispositions though tentative are to be exclusive. Players are to have absolutely no dispositions, tentative or not, to choose strategies other than those that maximize given those initial tentative probabilities. That is odd if, as it seems, players will at this stage still view those probabilities as only tentative. Similar difficulties relate to the probabilities envisioned at the second stage. It is odd that no probability is allowed to strategies other than initially maximizing ones when, as will sometimes be the case, they are viewed as only tentatively maximizing. And so on for further stages.

Perhaps these difficulties can be dealt with by supposing only "progressively more focussed," but never absolutely exclusive, interim dispositions to indifferent choices, as well as interim equal probabilities for certain strategies. (Such changes would render Rabinowicz's hyperrational deliberation dynamics "more Bayesian"; Skyrms 1990, p. 36.) Perhaps some such response to those difficulties can be made that is both mathematically and philosophically satisfactory. If so, iterative processes might be held to supplement ways of resolution such as I have described. Possibly, maximum eliminations could provide for reductions of games prior to the initiation of an iterative process that could then be addressed to the resultant maximally reduced game.

Rabinowicz has confessed to doubts concerning the identification of uniquely appropriate starting points for unreduced games: "It is not clear whether there is any 'non-question-begging reasoning' that allows the players to settle on a particular starting-point in their deliberation processes" (Rabinowicz 1992, p. 119). My idea supposes results of all possible eliminations to be always unique, and contemplates them as a way to an appropriately reduced game to which iterative maximizing deliberation is to be addressed, rather than to appropriate starting points (in Rabinowicz's sense) for iterative maximizing deliberations within given games. Hyperrational deliberation on this model would be a two-stage procedure (cf. Skyrms 1990, p. 52) in which all possible eliminations of strategies for players took temporal precedence over iterative "metatickle enhanced" causal maximizing.

It may be that a satisfactory ramified theory of hyperrational resolutions can be worked out along some such lines. But even if one can be, I am confident that hyperrational players, though left with fewer problems overall, will be left with some problems of both of the main kinds I have canvassed: suboptimal-resolution problems (e.g., Games 13, 14, 3, and 6 of Section 3.1, and perhaps also Games 11' and 20 of Section 3.7) and no-resolution problems (e.g., Games 15, 17, and 18 of Sections 3.2, 3.3, and 3.5 respectively, even if neither Game 7, 11, 11', nor 20 of Sections 3.6 and 3.7).

The significance of these problems

What interest can there be in the fact that hyperrational players would not always do well in certain games, and that there are even games with which they could not cope, games in which they could do nothing at all? Here are brief indications of two kinds of answers to this question.

1. If one thinks that people tend naturally to be maximizers, then one may feel that the problems hyperrational maximizers would have with certain games are relevant to *explanations* of certain social institutions and structures of conventions and rules, and as well of natural dispositions to certain patterns of preferences and resolutions. (See Ullmann-Margalit 1978, reviewed in Sobel 1980.)

2. If one thinks that maximizing is rational, then one may feel that the problems hyperrational maximizers would have with certain games are relevant to *justifications* of various conventions and rules (and social practices and institutions, including practices and institutions for popular and political sanctions) and for the coercive enforcement of rules as very useful (if not strictly needed) parts of best mixed-system solutions to similar problems in real communities. (See Sobel 1972 [Chapter 13].)

3. Opposed to justificatory ideas, but complementing explanatory ones, is this other view of problems for hyperrational maximizers: Observing that even hyperrational maximizers would have problems with certain games – that there are even games with which they simply could not cope – one may feel that this shows that there is something *wrong* with such agents and their single-minded determinations to maximize. One may feel that these problems show that maximizing is not always rational, that in some social contexts reason requires that one perform strategies that do not maximize. (Consider Braybrooke 1985; Gauthier 1974, 1975, 1985a; and McClennen 1972, 1978.)

I believe in explanatory and justificatory roles for the theory of hyperrational games, especially the part to do with problems for hyperrational players, and oppose the idea that from the fact that hyperrational players would have problems it can be gathered that there is something wrong with them, and in particular with their austere principle of practical rationality. What follows from the fact that hyperrational players would have problems is, I think, simply that perfect practical rationality and very extensive knowledge are not sufficient to fruitful interaction, and that some problems of interaction must be solved by other means.

APPENDIX: LEWIS DOMINATION

A.1

Consider the following pure-strategy game.

Game 21

	C1	C2	C3
R1	1, 5	5, 1	1, 4
R2	5, 1	1, 5	5, 4
R3	4, 1	4, 5	4, 4

There is no equilibrium in this pure-strategy game.

Can this game be solved by a process of discarding strategies that, while not dominated, suffer from a related disability? Take, for example, C3. It is not dominated by either C1 or C2 singly, but it is, one might say, dominated by C1 and C2 together: C1 is better than C3 against R1, and C2 is better than C3 against both R2 and R3. Whichever of Row's strategies Column expected her to use, Column would have a better answer to it than C3. It is tempting to think that Row could, on these grounds alone, be confident that Column would not use C3, and that on similar grounds Column could be sure that Row would not use R3, so that their problem would effectively reduce to the (constant-sum pure-conflict) structure shown here:

	C1	C2
R1	1, 5	5, 1
R2	5, 1	1, 5

This seems to have been David Lewis's view.

A *[strictly] dominated* choice is one such that, no matter how the others choose, you could have made some other choice that would have been better.... *[W]hich* other choice would have been better for you may depend on how the others chose.... The outcome [of a game may be] determined by the fact that everybody ignores dominated actions, everybody expects everybody else to ignore dominated actions, and so on. (Lewis 1969, pp. 17–19)

Probably, however, the temptation should be resisted. Take C3. It is dominated by C1 and C2 together, but even so its expected value is not necessarily inferior, regardless of the probabilities Column assigns to Row's actions. If he assigns the probabilities $Pr(R1) = \frac{1}{2}$ and $Pr(R2) = \frac{1}{2}$ (so that $Pr(R3) = 0$), Column's expected values for C1, C2, and C3 are, respectively, $\frac{6}{2}$, $\frac{6}{2}$, and $\frac{8}{2}$. Similarly, if Column assigns the probabilities $Pr(R1) = \frac{1}{3}$, $Pr(R2) = \frac{1}{3}$, and $Pr(R3) = \frac{1}{3}$, then Column's expected values for his three actions are, respectively, $\frac{7}{3}$, $\frac{11}{3}$, and $\frac{12}{3}$. (Though C3 is Lewis dominated in pure-strategy Game 21, it is "rationalizable" in the sense of Bernheim 1984 and Pearce 1984.)

Lewis dominated strategies can sometimes be ignored. Consider again Game 10:

	C1	C2	C3
R1	4, 4	0, 5	0, 3
R2	5, 0	2, 2	0, 3
R3	3, 0	3, 0	1, 1

C1 is not only weakly dominated by C2 in this game; C1 is also Lewis dominated by C2 and C3 together. This can suggest that perhaps a strategy can be eliminated, by the kind of argument used in Section 2.2.4 to eliminate C1 from Game 10, when a strategy is both weakly dominated *and* Lewis-dominated. The following game shows that this is not so.

Game 22

	C1	C2	C3	C4
R1	1, 5	5, 1	1, 4	1, 4
R2	5, 1	1, 5	5, 4	5, 4
R3	4, 1	4, 5	4, 4	4, 4
R4	4, 1	4, 5	4, 4	4, 6

If Column is sure that Row will use R4, then C3 is not maximizing. So an argument of the kind used for Game 10 can be initiated. But if column is *not* sure that Row will use R4, then C3 may be maximizing along with C4 (for then possibly Column's probabilities include $\Pr(R1) = \Pr(R2) = \frac{1}{2}$).

A.2

Returning to Game 21, it is noteworthy that C3 is not only not dominated by either pure strategy C1 or C2 in it, but that it is also not dominated in the mixed extension of Game 21 by any mixture of C1 and C2. (*Proof:* Consider the mixture $[pC1, (1-p)C2]$. If this mixture dominated C3, then $5p + (1-p) > 4$ by Row 1 and $p + 5(1-p) > 4$ by Row 2. From the first inequality it would follow that $p > \frac{3}{4}$, and from the second that $p < \frac{1}{4}$.)

In contrast, C3 is dominated in the mixed extension of

Game 23

	C1	C2	C3
R1	1,4	4,1	1,2
R2	4,1	1,4	5,2
R3	2,1	2,5	2,2

by the mixed strategy ($\frac{1}{2}$C1, $\frac{1}{2}$C2):

	C1	C2	C3	($\frac{1}{2}$C1, $\frac{1}{2}$C2)
R1	1,4	4,1	1,2	$\frac{5}{2}, \frac{1}{2}$
R2	4,1	1,4	4,2	$\frac{5}{2}, \frac{1}{2}$
R3	2,1	2,4	2,2	$2, \frac{5}{2}$

Whatever probabilities Column assigned to Row's strategies, the expected value of mixed strategy ($\frac{1}{2}$C1, $\frac{1}{2}$C2) would be $\frac{5}{2}$, greater than the expected value of 2 that C3 would have. This fact could provide Row with a sufficient reason for ignoring C3 in the mixed extension of Game 23 wherein the mixed strategy ($\frac{1}{2}$C1, $\frac{1}{2}$C2) is open to Column; in that game, C3 would be strictly dominated in the usual sense.[11] But it seems not to be a sufficient reason for Row's ignoring C3 in pure-strategy Game 23, where Column is (by hypothesis) restricted to the pure strategies C1, C2, and C3, and lacks the wherewithal to execute mixtures of these.

NOTES

1. The term is David Lewis's, but the idea is different. Common knowledge, as explained by Lewis, might better be termed "common reasoned belief." Furthermore, common knowledge as explained by Lewis will come to sequences of higher-order expectations (beliefs) that have *a certain kind of basis and mode of generation*. "[C]ommon knowledge is [not] the only source of higher-order expectations" (Lewis 1969, p. 59). Also, common knowledge, in Lewis's sense, will in general *not* come to higher-order beliefs ad infinitum (cf. Bacharach 1987, p. 52, n. 12). Common knowledge with a particular "basis" (Lewis 1969, p. 56) that "manifests a modicum of rationality" (p. 57) will come to higher-order expectations that are cut off by "the limited amount of rationality indicated" (p. 57).
2. I think that the requirement of common knowledge does no work that could not be done by a requirement of "sufficient mutual knowledge," where mutual knowledge (cf. Bacharach 1989) is like common knowledge but is ad finitum to some integer n, rather than strictly ad infinitum. An n sufficient to a particular game would be a function of the number of players in this game and the numbers of their strategies.
3. Cf. "[O]rthodox bayesianism [features] a commitment to the view that expected utility considerations, reflecting as they do the contributions that opinion and valuation make to rational preferability, *exhaust* the considerations relevant to rational decisionmaking" (Kaplan 1989, p. 67, n. 16).
4. Take "at" as short for "just before" and the whole as short for "at any moment prior to the action during a period bounded above by the time of the action, during which period the agent's credences and preferences do not change." It is a precondition for hyperrational games that for each player there is such a period.
5. Harper explores the consequences for game theory, not of a simple causal maximizing theory, but of a somewhat qualified theory that makes *non*maximizing actions rational in some situations. He observes that, in some situations, patterns of conditional probabilities can be such that an action that would maximize causal expected value would not still do so after one had revised one's probabilities for circumstances by conditionalizing on news that this action was to take place. In that case the action, though maximizing, would not be *ratifiable* (Harper 1988, p. 28). The version of causal decision theory that Harper favors prescribes actions that, amongst ratifiable ones, are maximizing. In Harper's theory, ratifiability is a lexically prior condition to maximizing, and sometimes the rational act is not one that maximizes among all options relative to time-of-choice preferences and probabilities.

    It is relevant to my interest in hypotheses concerning early-on conditional probability, and relevant in particular to the fact that I am less interested in them than is Harper, that the somewhat adulterated causal decision theory that I favor makes ratifiability not a prior condition to but, rather, a coordinate condition with maximizing. The theory I favor rules that an action is rational if and only if its expected value is maximum and it is ratifiable in the sense that a decision for it would be ideally stable (Sobel 1983 [Chapter 10], 1990a [Chapter 11]).
6. By Theorem 1* a hyperrational game resolves only in choices corresponding to equilibrium* strategies of its randomization. These strategies make a proper

subset of equilibrium strategies of randomizations that are "rationalizable" in the sense of Pearce (1984) and Bernheim (1984). Consider the randomization of Game 11 in Section 2.2.5: ($½C2, ½C3$), though an equilibrium strategy and rationalizable (every equilibrium strategy is rationalizable; Pearce 1984, p. 1034), is not an equilibrium* strategy. One explaining difference between our approaches is that while I insist upon indifferent choices from (and equal expectations for) tied strategies, Pearce and Bernheim allow but do not mandate equal-expectation conjectures in such cases (Pearce 1984, p. 1035). Another difference is that each of my players has knowledge of the others' indifferent choices, whereas one of their players has only a "(possibly incorrect) *conjecture* about others' strategies" (p. 1030).

7. As stated in Section 1.3.2, hyperrational players are causal maximizers, nothing more and nothing less. Thus they do not conform completely to the adulterated causal decision theory that I favor, according to which an action is rational if and only if it not only maximizes but is ratifiable; see note 5.

   "Superrational" players – players who were ideally rational according to that adulterated theory – could not cope with a version of Game 5 in which even indifferent choices of particular strategies were good predictors early on of like particular strategies on the other player's part. Only a restricted form of Theorem 9 would be valid for "superrational" games.

8. See Rabinowicz (1992), and comments thereon in Section 3.8 (this chapter) for ways in which hyperrational players might be supposed to do more than merely eliminate possibilities for their fellows' actions.

9. In connection with what must have been an extraordinary interaction, "Dizzy Gillespie... [on the occasion of] a funny, rambunctious duet with the drummer Max Roach [said] . . . , 'I understand where he's going before he does, and he understands the same thing about me'" (*The New Yorker,* September 17, 1990, p. 48). They "go together" not by habit or projection of traditions. With common knowledge that they seek together to sing, they do, as if by magic.

10. Cf.: "(R1, C1) is a Nash equilibrium that Pareto dominates all other Nash equilibria of [Game 3]. . . . Some game theorists, then, would single out (R1, C1) as the solution of [Game 3]. Opposition is bound to come from others who would insist that in the face of Row's indifference between R1 and R2 (regardless of Column's strategic choice), Column should consider it equally likely (according to the principle of insufficient reason) that R1 and R2 will be played. Column would then choose C2, which is *not* his strategy in the Pareto dominant equilibrium" (Pearce 1984, p. 1035).

11. It has been observed and proved that "in [mixed strategy] 2-person games, a strategy is strongly dominated if and only if there is no conjecture to which the strategy is a best response (see Appendix B, Lemma 3). Hence for 2-person [mixed strategy] games, rationalizable strategies are those remaining after the iterative deletion of strongly dominated strategies. This does *not* hold for" larger games (Pearce 1984, p. 1035; see also Bernheim 1984, p. 1012, n. 6).

15

Utility maximizers in iterated prisoners' dilemmas

Abstract. Barring question-begging assumptions of such "untraditional" motives and values as respect for precedent or a disposition to reciprocate cooperative overtures, hyperrational maximizers would be defeated in sequences of prisoners' dilemmas just as they are in isolated ones. They would fail to optimize not only in sequences of finite known lengths but also in sequences of constantly indefinite lengths. Such agents would know without experience what to expect of one another and could not by their actions in a round affect expectations in future rounds or expected values in future rounds. Traditionally valued hyperrational maximizers have problems aplenty with prisoners' dilemma situations. But this does not show that such agents are at fault in their maximizing ways of thinking. Nor does it show that there is something wrong with purely forward-looking "traditional" values and motives, and that agents possessed only of such are "missing something."

0. INTRODUCTION

Maximizers in isolated prisoners' dilemmas are doomed to frustration. But in Braybrooke's view, maximizers might do better in a series, securing Pareto optimal arrangements if not from the very beginning, at least eventually. Given certain favorable special conditions, it can be shown according to Braybrooke, even without question-begging motivational or value assumptions, that in a series of dilemmas maximizers could manage to communicate a readiness to reciprocate, generate thereby expectations of reciprocation, and so give rise to optimizing reciprocations which, in due course, would reinforce expectations, the net result of all this being an increasingly stable practice to mutual benefit. In this way neighbors could learn to be good neighbors: They could learn to respect each other's

Reprinted with revisions by permission from *Dialogue* 15 (1976), pp. 38–53.
  This chapter is a revised and expanded version of comments made on a paper presented by David Braybrooke to the Institute on Moral and Social Philosophy sponsored by the Canadian Philosophical Association in Toronto, June 3–5, 1974. "The Insoluble Problem of the Social Contract" (Braybrooke 1985) is a revised version of that paper.

property, to engage in reciprocal assistance, and perhaps even to make and keep promises covering a range of activities. So maximizers are, Braybrooke holds, capable of a society of sorts. Out of the ashes they might build something. But even under favorable conditions they could not build much and most conditions would defeat them almost entirely, for many-person dilemmas whether isolated or in series would be quite beyond them, question-begging motivational assumptions aside. In settings at all like those in which we live, utility maximization assuming only traditional motivation is self-defeating without remedy, and most certainly not an adequate basis for social life.[1] The probable inference from this, Braybrooke has suggested, is that maximization is inadequate as a conception of individual rationality:[2] for surely truly rational agents would be, if not perfectly well designed, at least better designed for communal living than utility maximizers would be assuming only traditional, non–question-begging values or motivation. Or, if this inference is denied and maximization is not challenged as a conception of rationality, we can conclude instead that the advice, "Be rational," is incomplete, and when addressed to groups should be supplemented by, "And be somewhat 'untraditional' in your values; respect at least some rules that would make coordination and cooperation possible even in many-person interaction problems" (Braybrooke 1985, p. 299).

In my view the case against utility maximization given only traditional values is at once both stronger and weaker than Braybrooke maintains. It is stronger in that barring question-begging untraditional motivational assumptions, maximizers would not optimize in iterated two-person prisoners' dilemmas any more than in isolated ones. Barring untraditional values, that a situation will or may come up again cannot matter to hyperrational maximizers in ways they could invariably exploit: they would know each other and their situations too well to learn from experience in past situations what actions to expect in a current situation. And the case against utility maximization given only traditional values is weaker in that the inference from its admitted inadequacy as a basis for society to its inadequacy as a principle of individual rationality is a non sequitur. Nor is the recommendation, recently put in place of this inference, of untraditional coordination- and cooperation-enabling values well founded. These points are developed in turn in what follows.

1. The defeat of utility maximizers in iterated prisoners' dilemmas

We consider two series: first, one of definite and known length; then, one of indefinite and unknown length. Regarding each, the issue will be the performance that would be forthcoming from *hyperrational* maximizers – unerring maximizers who know they are unerring, know the utility structures of their situations, know they know these things, and so on – maximizers about whom it is an important fact that they expect of each other nothing more nor less than maximization.[3]

Let the matrix for an isolated dilemma be as follows.

	B	
	b_1	b_2
a_1	**2** / 2	**4** / 1
a_2	**1** / 4	**3** / 3

Row's utility for an outcome (for example, for the outcome associated with his employment of a_1 while B employs b_1) is given in boldface type in the lower left corner of the cell for that outcome (in the example, Row's utility is 2). Column's utilities are given in lightface type in the upper right corners. Assume, following Braybrooke, that each dilemma in the series to be considered involves ordered, nonsimultaneous moves. For simplicity, let A always make the first move. (In following Braybrooke we depart, of course, from the tradition in which Prisoners' Dilemma is a game in normal form, with moves being taken simultaneously.) The *moves* open in each round of the series for A are a_1 and a_2 and for B are b_1 and b_2. However, we suppose that each player has in each round a choice between many *strategies*, each of which determines, perhaps conditionally on the history of the series, a move in the round at hand. Included among strategies open to A are four categorical strategies on the model of

α_1: a_1 regardless of moves or strategies in previous rounds (if there have been previous rounds)

In addition to these categorical strategies, we assume that our hyperrational maximizers can somehow employ various conditional strategies in which a move in a round is made conditional on moves or strategies made or chosen in earlier rounds (if any). Here are two examples:

α_5: a_2 if either b_2 in each previous round or if this is the first round; otherwise, a_1

α_6: a_2 if *either* both b_2 in each previous round and not a_2 in the ten just previous rounds *or* this is the first round; otherwise, a_1

Each dilemma in the series considered is thus an $n \times n$ game, where n, the number of strategies open to each player, is vastly greater than 2. But how are we to understand the choice in a round of a strategy, as distinct from the execution in a round of a move? We assume the players do both things; we eschew the artificiality of slips of paper upon which strategies are submitted to an umpire who then makes the move. How then is the choice of a strategy, as distinct from a move, to be understood? Perhaps as the execution of a move *with a certain intention or plan in mind,* or *conceived under a certain description or rule.* However, it is well to emphasize that the choice of a strategy is only for the round at hand: A player remains free in subsequent rounds to choose the same strategy again or a different one; choice of a strategy in a round entails no commitment to, and is not tantamount to, choosing it in all subsequent rounds or even in the next round (barring certain untraditional values). (Strategy choices in rounds of a series are quite different from strategy choices in the "supergame" for the series in which each player chooses *once* for *all* rounds, making moves in rounds conditional, perhaps, on moves in earlier rounds. See Luce and Raiffa 1957, pp. 99-102. The relevance of the supergame – and what hyperrational maximizers would do in it – to the *series* of games and what such maximizers would do in *it* is not clear.)

We assume that in each round in a series the history of the series before this round, not only the moves made but the strategies used, is common knowledge. We assume this even though it is, I think, hard to see how what strategies could be common knowledge among hyperrational maximizers with traditional values. *Other* agents could simply tell each other what their strategies, their intentions and plans, were, but it will be plain from what follows that, with Hodgson, I doubt that hyperrational traditional maximizers would be capable of that kind of communication. We of course suppose throughout that no rules for truth-telling, promise-keeping, or the like are actually enforced. We suppose throughout that our maximizers are in *this* sense in a state of nature. So we assume, despite difficulties, complete knowledge of the history of the series. Furthermore, we allow expectations for a round to be affected by such historical knowledge; more precisely, we do not assume that they are *not* so affected, though it is an open secret that I will argue that they are not. But we do assume (though this assumption will be suspended once) that *utilities* in the $n \times n$ matrix for a round are not affected by strategies employed or moves made in earlier rounds – except insofar as these strategies and moves affect probabilities, not values, of consequences of strategy pairs in the round. We assume, for example, that our players do not in the present place any value on *consistency* with their past performances (though this assumption too is suspended once); and that they do not value *reciprocation* – doing unto others what others have done unto one. Indeed, though this detail is not exploited in the arguments to come, for definiteness we assume that utilities in the $n \times n$ matrix for a round in the series to be considered here are a function of the utilities in the 2×2 matrix for the isolated dilemma, according to the following rule: The entry for A in cell (α_i, β_j) of the matrix for round r is

$$(2p^1 + \cdots + 2p^k + \cdots) + (4q^1 + \cdots + 4q^k + \cdots)$$
$$+ (1r^1 + \cdots + 1r^k + \cdots) + (3s^1 + \cdots + 3s^k + \cdots),$$

where p^k is *how likely* the pair (a_1, b_1)'s occurring in exactly k rounds, including and subsequent to round r, *would be made by* the use of strategies α_i and β_j in round r (given the history of the series before round r);[4] q^k, s^k, and t^k are similarly related to the pairs $(a_1, b_2), (a_2, b_1)$, and (a_2, b_2), respectively. Of course, whether or not strategy choices in round r can influence moves in subsequent rounds in the hyperrational community, without affecting values of possible consequences (and in this way utility matrices) for future rounds, is the issue that concerns me. If there is no such influence then p^k, for example, is simply the (unconditional) probability of the pair (a_1, b_1)'s occurring in exactly k rounds including and subsequent to round r. The entry of B in cell (α_i, β_j) is determined similarly; more precisely, interchange the constants "1" and "4" in the displayed sum.

For series of *definite* length, we consider a series of length 3 and assume that this length is common knowledge, and that it is common knowledge from the beginning and throughout the series that in each round it is common knowledge which round is then current and thus how many rounds remain. It is clear that this series resolves into the moves a_1 and b_1 in each round, and that any series of definite known length resolves similarly for hyperrational maximizers. Consider first the last round. Strategically, it might as well be isolated and will resolve into a_1 and b_1. In the last round the series has no future that might somehow be affected by strategy choices or moves made. And the past of the series is then irrelevant. The history of the series to this round will not, we assume, affect the utility matrix for this round. And even if the history of the series could generate expectations regarding strategies of the other party in this round, this history would not thereby be relevant, since expectations are in the last round plainly irrelevant. Neither would then need to know what the other had done, or was going to do, in this the last round, since each would have dominant strategies all pairs of which would yield a_1 and b_1. The last round might as well be isolated and will resolve into a_1 and b_1. And this will be obvious in the second, or next-to-last, round. It will be known then that there are no future rounds for which possible effects need to be taken into account; it will be known that, in the relevant sense, the next-to-last round has no future, that strategically it might as well be last. So this round too will resolve as if isolated. And so will the first round.

So much for series of definite lengths. In them each dilemma resolves for hyperrational maximizers as if isolated. (What "intelligent players" would do depends on what such players are like. Identifying several we might experiment, though results of such experiments would of course be irrelevant to our problems, unless we had reason to think that at least some intelligent players were hyperrational maximizers. However, it seems likely that this ideal is only approached, never realized, in actual communities. See Luce and Raiffa 1957, p. 101.) *Indefinite* series, in which Braybrooke is interested, may work differently. At any rate, the argument used for the definite case does not apply. It was crucial to that argument that ideal maximizers would know when they were in the last round, the next to the last round, and so on.

To fix the discussion, we consider a series which is indefinite under the following rule: There is a probability p such that at each round the probability of another round is p. Of course, there are other, more realistic rules. One could allow p to decrease as the series progresses. But the simpler specification serves present purposes, and if anything makes reaching the conclusion I seek more difficult than more realistic rules would.

What will hyperrational maximizers do in such series? The answer is that they will do just what they do in isolated dilemmas, because the *chance* that there are dilemmas to come can, for hyperrational maximizers, make no more difference than can the known fact that there are a certain number more to come. (It would be strange if this were not so – if, in particular, hyperrational maximizers who knew only that there was a chance of more dilemmas would behave differently and do *better* than hyperrational maximizers who knew the facts.)

Dilemmas to come could make a difference for hyperrational maximizers only if, when involved in a series, such maximizers could through their strategy choices in early rounds teach each other what to expect in later rounds. If this were possible, then choices would not only resolve the round at hand but might generate expectations of similar choices in future rounds – expectations which could, in due course, affect choices in these future rounds. Players, through present choices, might be able to control the conduct of future rounds, affecting each other's expectations, each other's actions, and even their own future actions. They might in effect set precedents for themselves, making reasonable (by doing similar actions now) future actions which would otherwise have been unreasonable. This is how the fact or chance of future dilemmas might be thought to make a difference. And so it might for many agents, even for many maximizers; but not for hyperrational maximizers. Given the thoroughness of their knowledge of one another and their situations, and the way in which they would form expectations regarding one another's actions (assuming only traditional values), the processes described cannot work for them. As long as they are and remain hyperrational and traditional, they know each other too well to teach each other what to expect or to set for themselves effective precedents.

Members of hyperrational communities of maximizers expect from each other at a given time choices that are reasonable, that is, maximizing. But then, assuming traditional values, hyperrational maximizers in situations that are possible for them[5] can invariably form expectations concerning each other's conduct, and do so without regard to past performances in like situations.[6] Consider the first or earliest situation of some kind. What is done – the strategy that is used, the random choice made among strategies – is reasonable in maximizing terms and so for our maximizers is clearly expectable without regard to this situation's past, for in the relevant sense the first situation of some strategic kind has no past. But then, expectations of repetitions of the performance in this situation in analogous subsequent situations must be similarly formulable without regard to *their* pasts: What made the response reasonable and expectable in the first case necessarily attaches to it in every subsequent analogous case with regard to which hyperrational maximizers might form expectations (partly) on the basis of responses in earlier cases. So reasonable past performances can make no difference to present expectations; they can be ignored. And of course unreasonable past performances either have no place in our scheme or can be contemplated only by assuming that our presently ideal hyperrational maximizers may have once been other than ideal hyperrational maximizers. But even if unreasonable (i.e. not maximizing) past performances are entertained by us in this way, it is plain that hyperrational maximizers would ignore them as quite irrelevant to present expectations. Therefore, hyperrational traditional maximizers would know what to expect of each other without regard to the past.

This conclusion can be put somewhat differently: Hyperrational maximizers would not learn what to expect of each other from experience or by induction. They would, we can assume, reason inductively about natural events, but they would form expectations regarding each other's *actions* in a different way. They know each other and the structures of their situations so well that the only inductive projections they *might* make are projections that for them could serve no purpose. Hyperrational maximizers might project each other's responses in one type of case to future cases quite like the base one in all respects that bear on what utility maximizers ought to do. Hyperrational maximizers would not project to other cases, cases not on all fours strategically with the base case; they would realize that what was done in the base case by one of their own was no evidence at all for what to expect in relevantly (i.e. strategically) different cases. But they might project to cases strategically quite like the base case, though such projections would be idle since what made certain responses reasonable in the base case would make like responses in these future cases reasonable and to be expected *without* projection. Hyperrational maximizers, to return to an earlier form of words, would (supposing still only traditional values) know what to expect of each other without regard to the past. But then – and this is the really pertinent result – through present choices they could not generate or make a difference to future expectations. They could not teach or set precedents.[7] At least this is so barring certain motivational assumptions that are quite bizarre on a traditional view but quite human nonetheless. Consideration of one such assumption may serve to clarify and reinforce my position.

Suppose that A places a premium on "consistency," on doing again whatever he has done before whether or not it was uniquely reasonable, or reasonable at all, in the first place. (Doing again under what circumstances? Let us say: doing again in situations that would have, if isolated, the same matrix as the initial situation would have if it were isolated.) A is not a creature of habit. Rather, he is an agent with a sense for what might be termed "personal traditions." Given a premium on constancy, A can, by what he does in early rounds, affect his utilities for possible outcomes of interaction in later rounds. He does with his actions what certain other maximizers (for example, *truthful* maximizers[8] who value saying and having said truths, and *trustworthy* maximizers who value doing what they have promised to do) can do with words. Indeed, if the premium on consistency is great enough, A can by employing a strategy even commit himself to it: by employing it he can, the premium being great enough, render it dominant evermore. So he might, for example, commit himself to α_3. Suppose he did. Agent B, knowing that A was committed to α_3, would (if the fixed chance p of a next trial is great enough) have no choice but to respond with a strategy that yielded b_2. Or A might commit himself to α_4, in which case (again, if the numbers are right) B would have no choice but to respond with a strategy that yielded b_2. (I have supposed that only A has the passion for consistency. Were it shared and were the dilemmas constituted of simultaneous choices, I suspect that neither could know what to do and that the series, for reasons of a sort explained elsewhere, would not resolve; see note 5. I suspect that the same is true of some of the supergames considered by Luce and Raiffa 1957, p. 102.)

So a hyperrational maximizer, if possessed of a certain quite human (albeit untraditional) motivation, could through his actions "teach" other hyperrational maximizers what to expect of him: His actions could themselves give them good reasons for expecting certain actions from him. But hyperrational maximizers who lacked his passion would also lack his teaching and precedent-setting capacities. It is true that they too would be consistent, choosing always as they have chosen, but this would be due not to concern for constancy but to the obvious fact that what is reasonable and maximizing in a case is so in every repetition of it.

> Hyperrational maximizers can, through choices in a round in a series of interaction problems, affect expectations in future rounds only by affecting the utility matrices for these future rounds by affecting values of consequences of possible strategy n-tuples in these rounds.

For example, choices in early rounds in a series of dilemmas might affect expectations in later rounds if the entry in a cell of the $n \times n$ matrix for this later round were determined by a mathematical formula like the line displayed previously, and if players could, through their early choices, affect the constants in this formula. It is only in such ways that hyperrational maximizers can teach by doing, or set effective precedents for themselves. But then, traditional maximizers cannot, in general, teach or set precedents. (Utilities for later rounds can, even supposing only traditional values, be functions of outcomes of – and so choices in – earlier rounds, but not in ways that make possible as certain untraditional values do, the controlled manipulation, through one's choices, of future expectations. In the absence of such values, utilities for outcomes of interaction in later rounds will generally depend not on an *individual's* choices in previous rounds but, rather, on outcomes of, and payoffs in, previous rounds – that is, in two-person interactions, on *pairs* of choices in previous rounds.) So the presence or absence, or chance great or small, of future rounds can make no difference for hyperrational traditional maximizers. Dilemmas in a series, whether of definite or indefinite length, will defeat such maximizers in just the way in which all should agree that isolated dilemmas will defeat them.

2. The significance of this defeat

Hyperrational utility maximizers who possess only traditional values will not optimize in prisoners' dilemmas whether or not they (the maximizers) are repeat offenders. More generally, to their disappointment and frustration, they will often fail to sustain what would be mutually beneficial social arrangements. On this general point Braybrooke and I agree, though I give maximization even lower marks here than he does. But what *follows* from this? Here we disagree. Braybrooke has suggested that what follows is that maximization is an inadequate conception of individual rationality, that individual rationality is either something more or something else. But clearly there is a gulf between premise and conclusion. There is suppressed, I think, the assumption that *truly* rational and well-informed agents would do as well by their several and collective interests as any agents could, given those interests in their circumstances. Even though it would be nice if this principle of harmony were true, it is at least not yet available as a premise for criticizing conceptions of rationality. Given the complexity of interaction problems and the current state of theories about them, we must, I think, allow that reason may *not* conquer all. Utility maximizers are sure that reason alone has its limits, and even their opponents should allow that - so far as anyone now knows - this remains a possibility.

Utility maximization may be inadequate as a conception of individual rationality. Perhaps there is more to individual rationality than preferences for outcomes that are representable by a utility function and actions in accordance with these preferences, where actions are (as if) viewed as gambles on possible outcomes. Rationality may, for example, impose additional constraints on an agent's ends taken either singly or as a system. And perhaps, moving in the other direction, utility maximization is not even always *part* of practical rationality. Perhaps, for example, a rational agent would sometimes, when possessed of only weakly based assessments of probabilities, refuse to choose amongst his actions as if they were gambles on possible outcomes. Utility maximization may be, in any number of ways, inadequate as a conception (or total theory) of individual rationality; I do not dispute this. All I claim is that the issue - whether or not (and if so, how) utility maximization is inadequate as a conception of individual rationality - cannot presently be settled by showing how it is inadequate in what is on the face of it a quite different and possibly unrelated way: as a basis for communal life.

In "The Insoluble Problem of the Social Contract," Braybrooke no longer presses the inference here criticized, but instead - without questioning the adequacy of maximization as a conception of rationality - suggests that the inadequacy for social living of maximization given only traditional values shows the need for and (in a sense) the wisdom of untraditional values. *Complete* advice, we are told, to a *group* of people - advice addressed to them as a group that they are to take, if at all, only collectively - would include not only the admonition to be rational but also the direction to respect rules, respect for which would enable members of the group to coordinate and cooperate in pursuit of their various ends. Thus complete collective advice, we are told, would include the recommendation of certain untraditional motives or values.

What about this most recent conclusion of Braybrooke's? I think that it is still objectionable, though in ways different from the old conclusion. Complete advice to persons in groups includes, he tell us, the admonition that they respect certain rules and respect them – not (evidently) for what some would perceive to be their intrinsic merits, but instead for certain ulterior reasons; no intrinsic merits are suggested by Braybrooke. A group of persons who do not already respect such rules is to harken to some set of them for the good, it seems, that can flow from the group's respecting them. Now I suppose that this sort of thing is possible and that, as persons can make themselves *believe* things for ulterior (non-evidential) reasons, so persons for ulterior reasons can make themselves *respect* things. Such self-manipulation, though not easy, is in various measures for some persons sometimes possible. But I would always hesitate to recommend it, and would in every case hope for better means to the ends that it would serve perhaps only precariously, if it were general through the community. It would always be more decent, and often it would be safer, to do without such double-dealing. But what alternatives are there, assuming only traditional values? One simple-minded alternative to manufacturing respect for certain rules that is sometimes available would be arranging for the *enforcement* of these rules. Of course, the project presupposes at least some already established ways of coordination and cooperation, but then this does not after all distinguish it from the project of collective self-manipulation. How is that to be accomplished in a social vacuum? Each self-manipulator-to-be might require assurance that others were "with him." ("But what remedies *would* be available to hyperrational traditional maximizers in a social vacuum?" Possibly none; from which might follow the collective advice, "By all means stay *out* of that condition!" That advice would at least be addressed only to groups that might have the means to take it.)

My criticisms of the collective advice Braybrooke advocates, and of the unqualified character of his advocacy, are that this advice would not always work, that even when it worked it would not always be the best advice possible, and finally that even when it was the best advice it could entail substantial costs and risks which Braybrooke does not mention (cf. Sobel 1975, p. 685). I resist the temptation to discuss further, and in a general way, advice addressed to groups or (as I have termed it) "collective advice," and to discuss what is involved in a group's taking such advice.

I also resist a greater temptation to elaborate upon the intrinsic merits of rules that would enable persons to coordinate and cooperate. Most persons on reflection would in fact discover considerable intrinsic merit in some such rules, and perceive as grotesque and disfiguring a restriction to traditional, purely forward-looking values – that is, an *indifference,* for purposes of present decisions, to the past, to even one's own personal past. Elaborating on this theme, however, would be a different kind of story from the present one – important and (if well told) capable of honestly and openly changing people's minds, but having nothing to do with the social limitations of traditional, purely forward-looking values assuming maximization (to reverse the order this once) with which this chapter has been concerned.

NOTES

1. The tradition in question is that of Hobbesian state-of-nature analysis, in which traditional values are always, though generally without explicit notice, assumed to be exclusively forward-looking: in this tradition it is axiomatic that proper or *rational* values are for states of affairs without regard to their pasts. The motto in this tradition for proper values could be: What is done is done. Thus the seventh Law of Nature is, *"That in Revenges* (that is, retribution of Evil for Evil.) *Men look not at the greatnesse of the evill past, but the greatnesse of the good to follow"* (Hobbes 1950, p. 126). This law could hardly appeal to men who place a value on retribution of evils done, and a value that varies with the greatness of that evil. But then such values are contrary to reason and, equivalently, the seventh law of nature is (though Hobbes does not in *Leviathan* insist on this point) already unconditionally binding in nature. Far from being a help, however, this is a part of the problem in nature. More precisely, the general exclusion of backward-looking values is an important part of the problem, or at any rate my reconstruction of the problem, in nature. Regarding the supposed irrationality of backward-looking revenge, and the special status (pointed out to me many years ago by W. D. Falk) of the seventh law, consider that this law is not only somehow "consequent to the next before it, that commandeth Pardon, upon security of the Future time," but *"Besides,* Revenge without respect to the Example, and profit to come, is triumph, or glorying in the hurt of another, tending to no end; (for the End is alwayes somewhat to come;) and glorying to no end, is vainglory, and contrary to reason; and to hurt without reason, tendeth to the introduction of Warre; which is against the law of Nature, and is commonly stiled by the name *Cruelty"* (pp. 126-7). Cruelty is contrary to reason whether or not one has sufficient security that others will not be cruel: that is, cruelty is contrary to reason even in nature. Consider Hobbes, *De Cive or the Citizen* (New York, 1949, p. 56: the note to Part I, Chapter 3, paragraph 27.) And, by implication, *all* rational values are exclusively forward-looking, "for the End is always somewhat to come."

   And so lines from *Leviathan* are in a footnote to this chapter, which of course is itself but a footnote to that great work. I owe thanks to Willa Freeman-Sobel

for many helpful suggestions and discussions concerning in particular, but not only, the idea of forward-looking values.
2. The suggestion, not repeated in Braybrooke (1985), is made in the following words in predecessors of that paper presented to the Canadian Philosophical Association (Toronto, 1974) and the American Political Science Association (Chicago, 1974): "How much regard and respect should we have for rationality when it is not associated with suitable motivations? I think, strictly limited respect. Perhaps we could take the line that rationality by itself has no indefeasible claim to respect. Perhaps, however, we would do better to hold that since rationality, conceived minimally as operating without motivations which eliminate the problem, leads to self-defeat in Prisoner's Dilemmas and the contract problem for the 'multitude' (also in Sobel's variants), rationality so conceived is an inadequate conception of rationality. Ironically, the minimal conception has attracted many thinkers because it seemed both hard-headed and ethically neutral. . . . Yet it will not work, even in its own terms: it is, as the chief merit of studying the contract problems may be to tell us, collectively self-defeating in its own terms, which cannot be charged against conceptions more ancient, and wiser."
3. I discuss the hyperrational community in greater detail in Sobel (1972, pp. 149-53 [Chapter 13 in this volume, Section 1]).
4. The awkwardness of the formula "how likely ψ would be made by ϕ" is deliberate. This formula is not supposed to express the conditional probability of ψ given ϕ, but is intended to express the result of adjusting the (unconditional) probability of ψ by the likely influence (if any) of ϕ on ψ. The lesson, I hold, of Newcomb's Problem is that some such measure is to take the place of conditional probabilities in expected utility calculations. But more of this in another place.
5. The qualification is necessary. Not every situation possible for agents is possible for hyperrational maximizers. Arguments in support of this proposition, and examples of situations that are not possible for hyperrational maximizers, can be found in Sobel (1972 [Chapter 13]; 1975).
6. Note that the argument being constructed for series of two-person prisoners' dilemmas of indefinite, unknown lengths is actually completely general. Its conclusion covers *all* series, whether of definite known or indefinite under some rule and unknown length, of *all* kinds of interaction structures. The conclusion that hyperrational maximizers cannot teach or set precedents – barring untraditional motivational assumptions – covers, for example, not only series of structures in which dominant strategies are not jointly Pareto optimal, but also series of coordination problems in which no strategies are dominant and in which there might therefore at first appear to be more scope for teaching and precedents.
7. For an excellent discussion (with which I am in substantial disagreement) of teaching effects in a hyperrational communty of maximizers, see Gibbard (1971, pp. 156-75). Gibbard writes, "Perhaps Hodgson is saying it is irrational to rely on induction from past acts of promise-keeping, but I see no reason why it should be irrational" (p. 170). My claim is of course *not* that, among hyperrational maximizers, induction from past acts is irrational, but that it is always unnecessary. The heart of the matter, if I am right, is that the only projections hyperrational maximizers could make from one another's past acts would be unneeded projections to cases quite like the base cases.

8. See Lewis (1972, pp. 17–19). Assume that "you" and "I" are hyperrational maximizers. (So "you" and "I" are not you and I.) Let them be in a coordination problem such that it is in the interest of each to push red if and only if the other pushes red: let (R, R) be for each one of the two best equilibria. "You" has said, "I pushed red." Can "I" believe him? An affirmative answer, Lewis assures us, is consistent with what has so far been assumed, since it is consistent with their hyperrationality that "you" and "I" should be, and know each other to be, truthful in the extensional sense of saying only true things (when it is best to instill true beliefs). Lewis does not say that "you" and "I" might, consistent with their hyperrationality, know each other to be truthful in my sense (truthful in the sense of valuing saying truths), but of course they might be – and know each other to be – truthful in this sense. And supposing that they are truthful in my sense would provide one explanation of why they are extensionally truthful, and one explanation of how they *know*, presumably even without experience and the very beginning of their relationship, that they are extensionally truthful. One way of explaining this would be to say that they are by their *values* truthful. Is there another way, given that they, "you" and "I," are hyperrational maximizers?

"But as a matter of fact we know we are extensionally truthful. You and I know this. How else could we talk? This is 'knowledge that [we] do in fact possess' (p. 19). So why not inquire *realistically* into how we know? Perhaps that would reveal 'another way.'" Perhaps. But since we are not "you" and "I," since we are not hyperrational maximizers, it is just as likely that an empirical investigation would only confuse the issue.

Consider Mackie's (1973) discussion of coordination, and consider a symmetrical situation with two best equilibria:

	2	1
2	2	1
1	1	2
	1	2

Mackie asks, "How do we solve this apparently insoluble problem?" In practice Mackie observes, "one person happens to move . . . before the other [who] then adapts his own movement to fit in with that of the first" (p. 293). That seems right; often, at any rate, we negotiate such situations in the way Mackie describes. But hyperrational maximizers never just *happen* to move, they need reasons. And there are situations possible for hyperrational maximizers in which each must move without prior observational knowledge of the other's move. (That is true even of us.) Mackie holds, incidentally, that coordination problems possessed of unique best equilibria, for example,

	2	1
2	2	1
1	1	3
	1	3

343

resolve without the participants' monitoring each other's moves, but he does not say *how* such situations resolve supposing "each is aiming simply at maximizing utility" (p. 291). For reasons not spelled out, he holds that it is obvious that hyperrational maximizers would do well in these situations, so he does not say how they would manage or even how *we* manage in such situations. I discuss, in Sobel (1975, pp. 681-2, n. 3) a way that does *not* work for hyperrational maximizers, and maintain further that no way does work for such maximizers.

16

Backward induction arguments: A paradox regained

Abstract. According to a familiar argument, iterated prisoners' dilemmas of known finite lengths resolve for ideally rational and well-informed players: They would defect in the last round, anticipate this in the next-to-last round and so defect in it, and so on. But would they anticipate defections even if they had been cooperating? Not necessarily, say recent critics. These critics "lose" the backward induction paradox by imposing indicative interpretations on rationality and information conditions; to regain it, I propose subjunctive interpretations. To solve it, I stress that implications for ordinary imperfect players are limited.

0. INTRODUCTION

I respond in this chapter to recent criticisms of backward induction reasoning, and especially to criticism by Philip Pettit and Robert Sugden. The introduction to their paper, which I present here in full, provides points of reference for this chapter.

Suppose that you and I face and know that we face a sequence of prisoner's dilemmas of known finite length: say n dilemmas. There is a well-known argument – the backward induction argument – to the effect that, in such a sequence, *agents who are rational* and *who share the belief that they are rational* will defect in every round.[1] This argument holds however large n may be. And yet, if n is a large number, it appears that I might do better to follow a strategy such as tit-for-tat, which signals to you that I am willing to cooperate provided you reciprocate. This is the backward induction paradox.

Although game theorists have been convinced that permanent defection is the rational strategy in such a situation, they have recognized its intuitive implausibility and have often been reluctant to recommend it as a practical course of action. We believe that their hesitation is well-founded, for we hold that the argument for permanent defection is unsound and that the backward induction paradox is soluble. (Pettit and Sugden 1989, p. 169, emphasis added)

Reprinted with revisions by permission from *Philosophy of Science* (forthcoming).
 I am grateful to Allan Gibbard, Philip Pettit, Włodek Rabinowicz, Willa Freeman-Sobel, and Robert Sugden for comments, to Merrilee Salmon for editorial advice, and to an anonymous referee for insightful criticism and suggestions.

I spell out ungenerous indicative interpretations of emphasized premises in Section 1, and present in Section 2 an argument for permanent defection that is certainly not good when these premises are so interpreted. Section 3 contains subjunctive interpretations that are adequate for such arguments and, I contend, descriptive of perfectly rational players. Section 5 recycles the standard solution to the paradox thus regained. It affirms that none of us are perfect, and that even if ideally rational and well-informed players would be doomed to permanent defection, very rational and well-informed players can sometimes do better and cooperate at least in long sequences, and until late rounds. Section 4 comprises an interlude in which I maintain that in place of common beliefs, game-theoretic models can feature common knowledge. "Of course they can," one might say, "for they almost always *do*." One might say this and no more but for recent challenges, such as Pettit and Sugden's, to common-knowledge models.

Frank Jackson (1987) discusses surprise-examination paradoxes. The difference I work between indicative and subjunctive interpretations of rationality and knowledge assumptions is somewhat like a difference of importance to Jackson between belief assumptions BEC and BIC for an "easy" paradox that he claims to solve, and certainty assumptions CEC and CIC for a "hard" one. What is certain in the strong sense *would still be certain* whatever positively probable condition were to obtain as the examination week unfolded. The difficulty of the "hard" paradox (which he does not solve) lies, Jackson implies, at the heart of the surprise-examination conundrum (p. 122).

1. Indicative rationality and belief premises

Players in sequences are to be rational. I take this to mean that each *maximizes causal expected utility* in each round. (Though I identify rational behavior with behavior that is maximizing *and* for which decisions would be stable on ideal reflection – Sobel 1990a [Chapter 11 in this volume] – the second condition coincides with the first in prisoners' dilemmas, and can be presently ignored.)

Players are also to "share the belief [subjective certainty] that they are rational" (Pettit and Sugden 1989). This, we are told, is to be a common belief, something that each believes, that each believes that each believes, and so on without limit. It is stressed that this premise says only that these beliefs obtain "at the start of the game" (p. 172), but to give backward induction reasoning a chance, and without prejudice to Pettit and Sugden's main objection, we want a stronger premise according to which common beliefs in players' rationality are "compounded forward to the end of the sequence" thus: For a sequence to come of length n, in each round k ($1 \leq k \leq n$) it will be a common belief that:

($n-1$) it will be a common belief in round $n-1$ that the players will be rational in round n,

($n-2$) it will be a common belief in round $n-2$ that the players will be rational in rounds $n-1$ through n, and that ($n-1$),

($n-3$) it will be a common belief in round $n-3$ that the players will be rational in rounds $n-2$ through n, and that ($n-2$) and ($n-1$),

and so on to

($n-(n-k)+1$) = ($k+1$) it will be a common belief in round $k+1$ that the players will be rational in rounds $k+2$ through n, and that ($k+2$), ..., ($n-2$) and ($n-1$),

where for exposition k is taken to be less than $n-4$. This premise compounds through the sequence common beliefs that go one way - forward - and the compounding stops short of the last round. I cannot motivate, as required to give backward induction reasoning a chance, common beliefs that go both ways or that are compounded through the last round, but without prejudice to Pettit and Sugden's objections such beliefs can be assumed. The belief premise (to be impugned as inadequate) may be stated as follows.

Common beliefs in rationality compounded all ways through and beyond the sequence

(1) It is always - before, during, and after the sequence - a common belief that players are rational in every round of the sequence.

(2) It is always - before, during, and after the sequence - a common belief that (1).

(3) It is always - before, during, and after the sequence - a common belief that (2).

And so on without end.

If backward inductions are to have a chance, beliefs need to extend to things other than rationality. For example, premises need to ensure that throughout sequences players have beliefs concerning:

(a) the causal independence of actions in rounds;
(b) where they are in the sequence, and how many rounds are to come;

(c) possible payoffs in rounds; and
(d) how expected utilities for plays in sequences of rounds are related to payoffs for plays in rounds sequenced.

For simplicity I suppose that these further matters are subjects not merely of common belief but of common knowledge, compounded all ways through and beyond the sequence, and suppose further that this knowledge not only obtains but would obtain throughout any possible play of a sequence. For concreteness, players can be in isolation booths with buttons they can press (defect) or not (cooperate), payoffs can be dollars, and utilities for sequences can be sums of payoffs. Assume the payoff structure displayed in the following matrix:

	Press	~Press
Press	$1, $1	$3, $0
~Press	$0, $3	$2, $2

2. A BACKWARD ARGUMENT FOR DEFECTION

2.1

For rational players who believe they are rational it can *seem* that, in each round k ($1 \leq k \leq n$) of a sequence of length n, each player could reason in the following manner to the conclusion that his defection maximizes in that round. (For exposition, k is taken to be less than $n-2$.)

(*i*) We will defect in the last round, *and we would defect in that round whatever we had done in previous rounds.*

In the last round, all that will matter to determinations of expected utilities of actions will be possible payoffs in it, and that is all that would matter no matter what we had done in previous rounds. As we will (and would in any case) realize that in the last round, strategicially the dilemma in it could as well be isolated. In it, as in every round, we will (and would in any case) see that defection dominates in terms of payoffs in it and that our actions are causally independent of one another, so that defection will (and would in any case) maximize expected utility. Finally, since we will (and would in any case) be rational in this round, we will (and would in any case) defect in it.

(*ii*) In the next-to-last round we will, *and would whatever we had done in previous rounds,* believe that (*i*).

Why? Because we will (and would in any case) believe that defection will (and would in any case) maximize in the last round, and that we will (and would in any case) be rational in the last round.

(*iii*) We will, and would whatever we had done in previous rounds, defect in the next-to-last round.

Why? Because, by (*ii*), in the next-to-last round we will (and would in any case) believe that our actions in the only future round will be (were going to be) causally independent not only of one another but of actions in the then *preceding* round – this current next-to-last round. That is, we will (and would in any case) see in the next-to-last round that strategically it could as well be last and itself isolated. (Cf.: "Things are clear on the last trial . . . ; hence the penultimate trial . . . is now in strategic reality the last"; Luce and Raiffa 1957, p. 98. "The last round might as well be isolated. . . . And this will be obvious in the . . . next-to-last round"; Sobel 1976, in Campbell and Sowden 1985, p. 310 [Chapter 15, Section 1].)

(*iv*) In the next-to-next-to-last round we will, and would whatever we had done in previous rounds, believe that (*i*) and (*iii*).

Why? For the kinds of reasons detailed under (*ii*), but ramified so that they showed that we will (and would in any case) believe that we will (and would in any case) in each coming round (a) have beliefs and values that will make (would make) defection rational in it; and (b) be rational in it.

(*v*) We will, and would whatever we had done in previous rounds, defect in the next-to-next-to-last round.

We will (and would in any case) see that actions in subsequent rounds are (were) going to be causally independent of actions in this next-to-next-to-last round, so that strategically it could as well be the next-to-last round or, for that matter, the very last round. We will (and would in any case) see that the dilemma in *it* could as well be isolated.

And so on to:

($2(n-k)$) In round k we will, and would whatever we had done in previous rounds (if any), believe that (*i*), (*iii*), ..., ($2(n-k)-1$). That is, in each round to come we will (and would whatever we had done in previous rounds) defect.

The argument may be recast in strong mathematical induction form (cf. Sorensen 1986, p. 342) as follows.

Basis. Ideally rational and well-informed players will, and would whatever they had done in previous rounds, defect in round n of a sequence of n prisoners' dilemmas. (This is (*i*) of the previous argument.)

Inductive step. For every k such that $1 \leq k < n$, if ideally rational and well-informed players will (and would whatever they had done in previous rounds) defect in rounds $k+1$ through n, then they will (and would whatever they had done in previous rounds) defect in round k.

For this we have the main lemma: For every k such that $1 \leq k < n$, if ideally rational and well-informed players will (and would whatever they had done in previous rounds) defect in rounds $k+1$ through n, then in round k they will (and would whatever they had done in previous rounds) *believe* that they will, and so would (whatever they had done in previous rounds) defect in rounds $k+1$ through n. (Grounds for this are illustrated under (*ii*) and ramified under (*iv*).)

Therefore: Ideally rational and well-informed players will defect in every round.

2.2

It has been said that "the [backward induction] argument for permanent defection is unsound" (Pettit and Sugden 1989, p. 169). I say that the argument just given certainly does overreach the premises stressed, when these are strictly and narrowly interpreted. According to the first stressed premise, the players are rational. For a sequence of dilemmas that lies entirely in the future, this premise when narrowly interpreted says no more than that in each round each player *will* perform an action that, given his then-current beliefs concerning possible consequences of his actions, maximizes expected utility. According to the second premise narrowly interpreted, the first premise *is* itself a matter of common belief compounded through the sequence. The problem with the argument from these *indicative* premises is that they do not even begin to ground the several *subjunctive* moves it involves.[2] Pettit and Sugden find in this problem "the solution to [the backward induction] paradox," which they say parallels "the solution offered by Frank Jackson to a version of the surprise-examination paradox [his 'easy' paradox]" (p. 171).

To illustrate, the part emphasized in (*ii*) is not supported. The premises do entail that in rounds previous to the next-to-last round each player will believe that in the next-to-last round they will believe that we will defect in the last round. But these premises leave open the possibility that players do *not* believe that – no matter what they had done in rounds previous to the next-to-last one – in that next-to-last round they *would* believe they were going to defect in the last round, let alone that they would believe that they would defect in the last round *no matter what* they were to do in the next-to-last round. The premises leave open that players believe they will believe in the next-to-last round that they will defect in the last round largely *because* they believe that they *will* believe in the next-to-last round that they will have defected in all rounds leading up to it, and that they were going to defect in it. Consistent with that opinion, however, they might believe that they are disposed to form expectations regarding each other by very simple induction, so that previous acts of cooperation would lead them in the next-to-last round to expect actions in the last round to be further acts of cooperation.

Also left open is the more important possibility that they believe that previous acts of cooperation could lead them to suspect that actions in the last round were not destined to be causally independent of actions in the next-to-last round. For example, they could believe that some previous acts of cooperation would lead a player to suspect in the next-to-last round that the other player had been, and would continue to be, disposed to reciprocate cooperation – that he had been, and would continue to be, following a tit-for-tat strategy. Given such suspicions, the penultimate trial would in his view *not* be, in strategic reality, the last trial.

The stressed premises, interpreted strictly and narrowly (i.e. indicatively), fail to support (*ii*). Indeed, interpreted strictly and narrowly they do not support even (*i*). The first premise, indicatively construed, leaves open whether the players would be rational in a round no matter what actions they had taken in previous rounds. Left open, for example, is that some actions that were irrational given then-current beliefs concerning consequences might, if performed, establish habits and dispositions not to maximize, or that they might affect others and make them cooperative and nonmaximizing in future rounds. Even for a length-2 sequence, the premise that players *will* be rational and maximize in each round does not entail that a player *would* be rational and defect in the last round no matter what players had done in the first round, even if they had behaved irrationally and cooperated in the first round.

3. Subjunctive rationality and belief premises

> "There can be no doubt that the conclusion [defection in every round] follows from the stated premises, but the premises deserve scrutiny" (Skyrms 1990, p. 130).

3.1

The premises of Section 1 are woefully inadequate to the argument of Section 2. Matters could have been worse: These premises could have been not merely inadequate to the conclusion that players will permanently defect but actual obstacles to that conclusion – barriers to reasoning to permanent defection. They could even have implied that resourceful players would *not* permanently defect. But these indicative premises do not make such troubles, and to repair the argument it is possible simply to strengthen their construction. It is possible to interpret Pettit and Sugden's words generously in ways that make them adequate to the subjunctive moves in our reasoning (set out in Section 2) and thus better candidates for what intelligent proponents of resolutions by backward induction reasoning can be supposed to have had more or less clearly in mind.

Pettit and Sugden disagree. They claim that their premises are barriers that preclude reasoning to permanent defection: "the players are [not only] not necessarily in a position [but] indeed are necessarily not in position, to run the backward induction" (1989, p. 174). They suggest that for such reasoning players would need (a) "the belief that the common belief in rationality would survive even if cooperative moves were played." And they maintain that the premises they state imply that (b) "neither of the players *can* believe that the common belief in rationality will survive whatever moves the players make" (p. 174, emphasis added).

There is a sense in which, given certain assumptions that we should be willing to make, neither player can believe that a common belief in their rationality would survive no matter what; but this is not a sense in which it is necessary, for backward induction reasoning, that players *should* believe that a common belief in rationality would survive no matter what. That is, while there is a sense of common belief in rationality in which (a) is needed and a sense in which, given assumptions I am willing to make, (b) is true, these senses are different; the sense in which (b) is true leaves open that players can believe that, come what may, they would have the common beliefs needed for backward induction reasoning.

I concede that if (as we should) we take for granted that players would in every round remember what plays they had made in previous rounds, then a common belief in everyone's rationality in all rounds – including rounds already completed – could not survive the *last* round, whatever moves the players had made in rounds including it. And if we take for granted also that the players would, no matter what, remember in every round what they had *believed* in all previous rounds, then a common belief in everyone's rationality in all rounds, including rounds already completed, could not survive even the *first* round, whatever moves (holding beliefs constant) players had made in it. But this does not mean that our players *cannot* believe that a common belief in rationality sufficient for backward induction reasoning would survive every round, no matter what moves had been made in previous rounds and no matter what moves were made in the then-current round. For such a conclusion would require only that common beliefs (compounded in a certain manner) in players' rationality would survive in all the rounds (if any) *still to come,* no matter what moves the players had made in rounds completed.

3.2

Pettit and Sugden lose the backward induction paradox by casting its premises as indicatives, and seek to bury it by insisting that for paradox common beliefs that go both ways must survive no matter what. To regain the paradox, I frame – and briefly defend as appropriate – considerably stronger, largely *subjunctive* rationality and belief premises that for the most part go only one way: forward.

According to our indicative rationality premise, each player is rational in the sense that in each round he will perform an action that, given his then-current beliefs, would maximize expected utility. For a stronger premise, I let a player be *resiliently rational* in a round if and only if he is rational in it and, for every subsequent round (if any), he is rational in it and would be rational in it no matter what the players – himself included – had done in previous rounds. A resiliently rational player would, in any subsequent round, even if he (even if everyone) were irrational in the current round and every intervening round, finally come to his senses. Using this definition, I propose the following as a stronger rationality premise for a sequence of known finite length of prisoners' dilemmas.

Resilient rationality through the sequence

Each player is not merely rational but resiliently rational throughout the sequence in the sense that, for each round, he will – and would no matter what the players, himself included, had done in previous rounds – be resiliently rational in it.

This premise is in the idealizing spirit of classical game theory. Whereas rationality in the thin sense of ever maximizing can be superficial, and rationality in the less thin sense of a persistent underlying disposition to such actions can be eradicable, resilient rationality is deep-seated and includes not only a display of rational actions, but a deeply entrenched and ineradicable disposition to such actions that would assert itself no matter what insults it had suffered. To say that players are not merely (and perhaps only coincidentally) rational, but *resiliently* rational, is to say that they are very rational indeed. (A strong-willed and nonaddictive nonsmoker not only does not smoke, but, even if he were to smoke once or twice, would not smoke anymore. If perfectly and completely resilient, then, for every n, he would not smoke anymore even if he were to smoke n times.)

3.3

According to the indicative belief premise – common beliefs in rationality compounded all ways through and beyond the sequence – the original thin rationality premise is a common belief compounded through and beyond the sequence in a certain manner. For a strengthened common-belief premise I conjoin with that indicative premise the proposition that the new resilient rationality premise is a common belief compounded in a certain subjunctive manner forward to the end of the sequence.

Common beliefs in resilient rationality, compounded robustly forward to the end of the sequence

For a sequence to come of length n, in each round k ($1 \leq k < n$) it will be, and would be "no matter what" (this from now on is short for "no matter what they had done in previous rounds, if any"), a common belief that:

$(n-1)$ it will be, and would no matter what be, a common belief in round $n-1$ that the players are then resiliently rational,

$(n-2)$ it will be, and would no matter what be, a common belief in round $n-2$ that the players are then resiliently rational, and that $(n-1)$,

$(n-3)$ it will be, and would no matter what be, a common belief in round $n-3$ that the players are then resiliently rational, and that $(n-2)$ through $(n-1)$,

and so on finally to

$(n-(n-k)+1) = (k+1)$ it will be, and would no matter what be, a common belief in round $k+1$ that the players are then resiliently rational, and that $(k+2), \ldots, (n-2)$ and $(n-1)$, where for exposition k is taken to be less than $n-4$.

This strengthened belief premise is suited to resiliently rational players. Such players would be rational in a round, and indeed resiliently rational in it, no matter what they had done in previous rounds. The new belief premise attributes confidence in appropriately robust immediate non-inferential *appreciations* of themselves and their fellows in each round that would obtain regardless of what they had done in previous rounds. It elaborates on the idea that in any round they would not only be resiliently rational but would believe in their resilient rationality, and this notwithstanding contrary evidence provided by patterned or even by constant past failings. The thought that underlies this premise is that ideal players would, no matter what, believe in a round that they would, no matter what, in every subsequent round see through any prima facie evidence provided by their behavior in previous rounds for their not being resiliently rational in it, and find it possible always in one way or another to discount such evidence. Ideal players would always, no matter what, believe that they would always, no matter what, know themselves and each other that well.

3.4

It has been said of extensive-form games that "if the same player has to move at different points in the game, we want that player's knowledge to be the same at all of his information sets" (Bicchieri 1989, p. 336). A propos this I note that, because it is not needed, it is not part of my common-belief premise that a player's beliefs concerning the rationality of players and concerning their beliefs are to be the same at all information sets, from the first one on, that he actually reaches. More importantly, because it is not needed and because of threats of inconsistency were it included, it is most certainly not part of my common-belief premise that beliefs of a player should be the same at all information sets, including in particular ones that will not be reached. I want to leave open the possibility that information sets that will not be reached might be reached by irrational play in earlier rounds, as well as by rational play based on beliefs disproved in later rounds and hence jettisoned. If I were to require that at each possible information set my agents should commonly believe certain things about past rounds, then I would have them commonly believe only what will be (or would be) *true* things about their beliefs and play in these rounds. It is not part of that premise that players, no matter what they had done or believed in previous rounds, would have common beliefs concerning their rationality and beliefs that went "in both directions" (Bicchieri 1989, p. 336), common beliefs which – depending on what had been done and believed in previous rounds – might be false. Connectedly, I allow that acts of cooperation would provide evidence that *parts* "of the announced conditions of the game do not really hold" (Sorensen 1986, p. 345), and that "each of the players has [repeated] opportunit[ies] to undermine the other's knowledge of [parts] of the game situation" (p. 345). But I take care that no acts provide evidence that would undermine parts that are needed for backward reasoning to defection in every round.

I do not want or need that the beliefs of a player should be the same not only at all of the information sets he actually enters but also at his merely possible information sets. Nor do I understand why anyone would want that, or think that it was somehow unavoidable or traditional for the topic, or partly definitive of an extensive-form model worth investigating. This feature of Bicchieri's model seems to be important to her conclusion that assuming more levels of knowledge than is needed "leads to an inconsistency" (Bicchieri 1989, p. 336), that it leads "to a discrepancy between what is observed and what is known about the other player" (Bicchieri 1990, pp. 77-8). That conclusion might be compared with one of Pettit and Sugden's: "For any act of cooperation by one player in round $n-j$, where $0 \leq j \leq n-1$, the partner, if rational, must respond with a belief that causes the common belief in rationality to break down at level $j+1$ or at some lower level" (Pettit and Sugden 1989, p. 174). They write of common beliefs that would go both ways, without noticing that backward reasoning to permanent defection does not *need* such extensive common beliefs.

On a similar note, Philip Reny writes that "there is ... a large class of extensive form games, in which it is not possible for rationality to be common knowledge throughout the game" (Reny 1989, p. 362). It is consistent with my rationality and common-belief premises that the rationality of all players throughout the game should be a common *true* belief throughout the game *if* "throughout the game" is taken in the sense of "at every information set that is actually reached." But this is not consistent with those premises if "throughout the game" is taken in the sense of "at absolutely every information set."

3.5

My resuscitation of backward induction arguments does not assume that out-of-equilibrium acts of cooperation would invariably be viewed as "trembling hands" – random and uncorrelated accidental deviations that are for this reason not projected or taken as arguments against current rationality. It assumes that not even histories of constant past deviations and past irrationality would be projected by perfectly rational players; not even unbroken histories of deviations – viewed not as random and accidental but as persistently confused, unreflectively imitative, corruptly principled, or grounded sometimes in one of these ways and sometimes in another – would be projected. It assumes that all such projections would be contravened because, even after exceptionless runs of irrational actions had taken place and regardless of what might have been their grounds, our agents would not only *in fact* come to their senses and take proper charge of their actions at last and evermore but would also immediately *see* that they had all come to their senses and, regardless of grounds for past deviations, would be taking proper charge of their actions at last and evermore. (What *might have been* the explanatory grounds for several acts of cooperation, whether scattered amongst acts of defection or presented in exceptionless runs, would be what *are* the grounds for these acts in "nearest" worlds in which they take place; the identities of such worlds depend on details of particular stories for sequences of prisoners' dilemmas.) My players cannot build reputations and thereby "influence each other" (Sorensen 1986, pp. 345-6). However, this is not because they "play in total ignorance of each other's moves" but rather because, whatever they knew that they had done, they would still at sight know each other for the resiliently rational players that – even if only at last, and for the first time – they in fact would be.

For theorists friendly to a game model that includes resilient rationality – rationality that would always, no matter how long it had been dormant, assert itself and be real in this relentless way – such robust and immediate recognition and confidence should seem a natural complement and a necessary part of a balanced and complete model of perfection. It would be a strange model that supposed that no matter how long and how near to completely it had been dormant, rationality would always assert itself firmly and forevermore, but also allowed that that it had asserted itself would not always be appreciated or believed, at least not right away. The natural complementing assumption is that, on the contrary, not only would each perfectly rational player realize that he had emerged from his night of irrational abandonment, but each would see also that all of his fellows too were resiliently rational still, again, or at last.

Ken Binmore has written, "in similar games after similar histories, players should normally be expected to choose similar . . . actions" (1987, p. 200). He writes, I assume, of ordinary players who come to know one another, and to learn what to expect of one another, largely from experience. In contrast, agents who are perfect maximizers would, according to the present account, always know one another immediately on sight, and do so either without any experience of one another in past encounters (if they have only just met) or independently of any experience they may have had of one another (if they have a history of encounters).

Is this a wildly unrealistic hypothesis whose consequences must be completely irrelevant to ordinary players and real situations? Is there no continuity between the ideal condition here proposed and natural ones? Consider animals in the wild, in particular young animals (I am thinking of the cub in the movie *The Bear*). Do they at first not know what to make of the growls, barks, and licks, of the grimaces, smiles, and caresses of other animals, especially other animals of their own kinds? Do they come to know what to make of these only eventually and from experience? And are their readings of one another sensitive to every aberrant past performance, or can they be resistant to change even in the face of counterevidence? I think that natural animals – babies in their mothers' arms, for example – are sometimes in their readings of other animals highly resistant to such changes, and disposed not to project behavior that is "out of natural character." The robust and immediate confidence of my ideal agents in one another's resilient rationality can be viewed as a doxastic disposition whose natural resistance to change, which could be of survival value, has been rendered extreme. That is appropriate as a feature of an idealization.

These new premises – the strengthened rationality premise that calls for resilient rationality, and the strengthened common-belief premise that calls for robust and immediate forward-looking self- and fellow recognition compounded to the end of the sequence – are, I suggest, descriptive of perfectly rational and well-informed agents. And they evidently provide sufficient rationality and common beliefs in rationality for arguments of the kind explained in Section 2.1 concerning permanent defection in prisoners' dilemmas of known finite lengths.

For the record, I have elsewhere argued (without dependence on backward reasoning) for a more general conclusion. I maintain that "hyperrational maximizers" who have only forward-looking values would always "know each other too well to teach each other what to expect or to set for themselves effective precedents" (Sobel 1976, in Campbell and Sowden 1985, p. 311 [Chapter 15, Section 1]), and that, as a consequence, dilemmas "in series, whether of definite [known finite] or *indefinite* length, will defeat such maximizers" (p. 314, emphasis added).

4. Interlude: Defense of common-knowledge premises

4.1

It is evident that all references to beliefs in my common-belief premise are, given this premise and the new resilient rationality premise, references to what are (or would be under the conditions in which they would take place) *true* common beliefs. And so one supposes that these references could be to agents' common *knowledge* of the things of which my premise says they have common beliefs.

Pettit and Sugden think that there are special properties of and problems for common-knowledge idealizations. They suggest for one thing that given common knowledge – as distinct from mere belief – among players that they are rational, "each [*could*] run the backward induction and each will . . . defect" (1989, p. 181). My difficulty with this is that an indicative common-knowledge premise analogous to the initial indicative common-belief premise would not put players in a position to run backward inductions. Perhaps Pettit and Sugden are influenced by the mistaken idea that common *knowledge* of rationality in all rounds would necessarily survive no matter what, as if knowledge were always "strong" in a manner analogous to Jackson's "strong certainty" (Jackson 1987, p. 122).

Pettit and Sugden (1989, pp. 181-2) also suggest that since stipulations of a common-knowledge idealization would entail that the players will permanently defect, these stipulations cannot, on pain of incoherence, include stipulations concerning what would happen were the players to cooperate, or stipulations concerning what players are to think would happen were they to cooperate. The suggestion is that antecedents of conditionals that would say what would happen if players were to cooperate would be inconsistent, given stipulations that entailed that they would not cooperate. My difficulty here is that even if stipulations of common knowledge were inconsistent with players cooperating, the antecedents that had them cooperating, though counterfactual, would not be inconsistent and so pointless or uninteresting. Regarding the pointlessness of inconsistent suppositions, I note that it is a valid principle of Lewis-Stalnaker conditional logic that supposing an inconsistency makes *everything* the case: For any inconsistent or impossible p and *any* proposition q, the conditional ($p \square\!\!\rightarrow q$) is true. "A more adequate [but more complicated] theory would allow truth-value gaps" and render these conditionals neither true nor false (Sobel 1970, p. 432).

The general point is that an ideal case (indeed, any case or model) can be defined by subjunctive conditions along with indicative ones, and antecedents of subjunctive conditions can be incompatible with implications of conditions for the case. The only restriction on useful and interesting subjunctive conditions is that their antecedents be possible or entertainable for purposes of counterfactual supposition. Consider that we can of course think coherently about a case in which everyone knows that everyone is going to vote, and in which everyone also knows that even if anyone were not to vote still everyone else would. Similarly, we can think about a case in which everyone knows that everyone is going to vote, but that if anyone were not to vote then no one would.

To set conditions that logically imply that players *will* defect - for example, conditions that ensure that they will as a consequence of backward induction reasoning defect, or conditions that entail that they know (never mind how) that they will defect (never mind why) - is not to make it "a matter of logical necessity [that they] *must* defect" (Pettit and Sugden 1989, p. 181). For if it were then, given such conditions, to ask what would happen were a player not to defect *would* be equivalent to asking "what would happen under [the] inconsistent hypothesis" that he does not defect and does defect (p. 181). It is a valid principle of subjunctive conditional Lewis–Stalnaker logics that for any propositions *p*, *q*, and *r*,

$$\Box p \to \Box[(q \,\Box\!\!\to r) \leftrightarrow ((p \,\&\, q) \,\Box\!\!\to r)].$$

4.2

Perhaps Pettit and Sugden can be read as taking an "imported view" of common-belief conditions, and proposing that it is uniquely appropriate to take an imported view of common-knowledge conditions in game models. This would, without making defection logically necessary, have some of the effects of doing that: It would have the effect of making suppositions of cooperation inconsistent.

Hubin and Ross contrast imported views of conditions with exported views. To take an exported view of conditions of a practical puzzle or model is to hold "that [they] simply [do], but need not, obtain" in the puzzle or model (Hubin and Ross 1985, p. 441). "On the imported view ... we consider as the outcome of an action what would happen were you to perform the action ... when the puzzle conditions obtain" (p. 443). To take an imported view of a condition of a puzzle is to hold that this condition is necessary, not absolutely and logically, but conditionally on all actions in the puzzle. It is to suppose the condition holds "no matter what act is performed" (p. 445). For any action *a*, condition *c*, and state *s*, an imported view of *c* has the effect of importing *c* into the antecedents of conditionals that say what would be if *a*,

$$(a \,\Box\!\!\to s) \leftrightarrow [(a \,\&\, c) \,\Box\!\!\to s];$$

an imported view of *c* makes *c* independent of *a*,

$$c \,\&\, [(a \,\Box\!\!\to c) \,\&\, (\sim a \,\Box\!\!\to c)].$$

It can be seen that when the distinction matters – when it makes a difference which view one takes of a condition – taking an imported view of a condition of a puzzle or model of someone else's making can, with due respect to Hubin and Ross, be a plain mistake. (They hold that the imported view is right without exception for all puzzle conditions: "In solving a hypothetical practical problem, the stipulated puzzle conditions must be assumed to hold no matter what act is performed"; Hubin and Ross 1985, p. 445.) For example, while it is right to take an imported view of the independence conditions of standard Newcomb Problems (these conditions say that predictions made are causally independent of actions, and that contents of the boxes are causally independent of actions), taking an imported view of correctness conditions for predictions made would be a mistake. It would mistake the intents of makers of standard Newcomb Problems to suppose that the prediction made of my action is not only correct but would be so no matter which action (of the things I can do) I were to do. For it is part of all standard Newcomb Problems that, whatever has been predicted, both actions are still possible, and that if I were to act in an unpredicted way then the prediction made would of course not be correct. (It is part of "limit" problems that though the prediction be correct and the predictor infallible in the sense of never-erring, the prediction is not necessarily correct and the predictor is not infallible in the mysterious sense of being incapable of error. Cf. Sobel 1988b [Chapter 5].) Similarly, while imported views can be taken of common-belief conditions for standard game models, taking imported views of common-*true* belief conditions and common-*knowledge* conditions, as perhaps Pettit and Sugden can be read as doing, would mistake intents of makers of standard game models. When it is supposed, for example, that it is common knowledge that players will behave rationally, it is taken for granted that they do not *have* to behave rationally, and that if they were not to behave rationally it would not *per impossibile* be known that they had behaved rationally.

4.3

Binmore strikes what seems a discordant note similar to Pettit and Sugden's: "Conventional arguments . . . require counterfactuals of the form, 'If a rational player made the following sequence of *irrational* moves, then . . .'" (Binmore 1987, p. 198). (Cf. p. 183: It "seems inevitable to me" that "out-of-equilibrium behavior is to be treated in terms of mistakes.") Binmore also suggests (without saying this straight out) that if players are perfectly rational in the ways required by these conventional arguments for equilibrium behavior (e.g., permanent defection in sequences of prisoners' dilemmas), then any suppositions made in these conventional arguments that would have players carrying out sequences of out-of-equilibrium irrational acts are not merely counterfactual but instances of "absurd speech" (p. 179). In fact, however, the supposition that a player who is rational makes an irrational move, while counterfactual, is not absurd or contradictory; this easy point is one that Binmore himself seems at times to acknowledge. Rational players *do not* make irrational moves, and that is necessary; but this is not to say that rational players *cannot* make irrational moves, or that irrational moves are impossible for rational players.

4.4

There are well-known ambiguities concerning scopes of modal operators that may be relevant here. Thus,

"A rational player cannot do an irrational thing"

can have the true sense of "it is logically necessary that if a player p is rational, then he does no act a that is irrational"; roughly symbolized,

$$\Box[(Rp\,\&\,Ia) \to {\sim}Dpa].$$

And it can, perversely, have the false sense – indeed, the preposterous (and problematic, in that it would feature quantification into a modal context) sense – of "if a player is rational then it is logically necessary that he does nothing that is irrational"; roughly symbolized,

$$[(Rp\,\&\,Ia) \to \Box{\sim}Dpa].$$

If a player is rational then it would be merely out of character for him to do an irrational thing and "impossible" only in the very weak sense that is consistent with his doing six impossible things before breakfast. Compare the sense in which it is impossible for a really honest man to tell a lie.

Further to the possible relevance to our text of well-known modal ambiguities, consider: "[If] rationality is a matter of common knowledge [for players, then] it is not something on which they could be mistaken" (Pettit and Sugden 1989, p. 180). This could have the false and ridiculous sense of "if players p know that they are rational, r, then it is not logically possible for them to be mistaken about r"; roughly symbolized, this is

$$[Kpr \to {\sim}\Diamond Mpr]$$

or, equivalently,

$$[Kpr \to \Box{\sim}Mpr].$$

And it could also have the true (albeit completely unremarkable) sense of "it is logically necessary that if players p know that they are rational, r, then they are not mistaken about their rationality r"; roughly symbolized, this is

$$\Box[Kpr \to {\sim}Mpr].$$

Possible confusions aside, we should say (though it will rarely need saying) that if a rational player were to make an irrational move, he would in making it not be rational. What else that we should say would obtain if a rational player made certain irrational moves will depend on our "background theory" (Binmore 1987, p. 189). In the idealization I have framed, the background theory features resilient rationality and spells out several other things we should say.

5. THE PARADOX AND ITS SOLUTION

> "The conclusion that rational players must defect seems logically inescapable, but at the same time it is intuitively implausible. That is the backward induction paradox" (Pettit and Sugden 1989, p. 171).

The standard (and, I think, correct) solution to the "paradox" argues that while this predicament is logically inescapable for perfectly rational players, it is of only limited predictive and prescriptive relevance for real players, including very well-informed and reasonable ones. Perfect players would be trapped in permanent defections to their mutual disadvantage. The conclusion that they defect in every round is logically inescapable. But imperfect players, even very intelligent and well-informed maximizing players, can be well-advised on maximizing grounds to cooperate at least in early rounds. After all, real and thus imperfect maximizers can sometimes – even when they have well based probabilities informed by appreciations of the subtleties of game-theoretic analyses – interact without error in non-equilibria even in "static" one-off games. (Cf. Binmore 1987, p. 211; Bernheim 1984.)

What a player's maximizing play would be in a round of a given sequence of dilemmas depends on his probabilities and preferences; it depends especially on his probabilities for possible effects of his play on his opponent's plays in future rounds. And these probabilities of real players will often have little to do with what they would expect, and realize they would expect, were they possessed of robust common beliefs compounded through the sequences in their resilient rationality. Real players, including very thoughtful ones, may with good reason think that they and their opponents are not resiliently rational, and that robust common beliefs that say they *are* are certainly not compounded through the sequence – it will often be obvious to real players that they exist far from the perfect rationality and common-belief conditions of game-theoretic idealizations.

This familiar solution to the backward induction paradox has advantages over those that would pick sides and say either that permanent defection is the rational strategy for all rational and well-informed players, or that not even ideally rational and well-informed players would necessarily defect in every round. The first would-be solution is certainly wrong: it ignores the dependence of a player's rational strategies on his subjective probabilities. And the second would-be solution at least runs into difficulty, given how widely and implausibly it would cast logical aspersions. To be interesting it would need to say that there is "nothing in" the continuing tradition of idealizations that purport to support backward induction resolutions, not even when these idealizations are interpreted generously.

Rather than take sides, I have tried to give each its due by explaining conditions of rationality and common beliefs that would put players satisfying them into positions from which they could reason backward to defection, and by stressing that this result does not have direct and unmoderated implications (predictive or prescriptive) for real players. However, though the main result of this chapter – namely, that perfect players would defect in sequences of dilemmas – lacks direct and unmoderated implications for real players, it is not, I think, completely irrelevant to predictions and prescriptions for them. Coupled with the plausible idea that real players in *short* sequences are apt to be nearly ideal in the sense of my conditions, this result predicts that even when cooperation is established in a sequence, it is apt to be unstable as the sequence winds down (cf. Luce and Raiffa 1957, p. 98). If (as I think) real players are sometimes nearly ideal maximizers in short sequences, then the main result of this chapter, together with details of its grounds, is something real players might usefully take into account in their deliberations – when beginning short sequences, and when in sequences that have run to points at which only short segments remain.[3]

Finally, and on a very different note, details of my recovery project (in particular, of my subjunctively strengthened rationality and belief premises) may be suggestive of exact points of departure for studies of species of bounded, less than perfect, rationality in finitely repeated prisoners' dilemmas.[4] I think that in the case of the backward induction paradox, even if not with the "surprise examination" paradox (see Montague and Kaplan 1974, p. 271), interests of theory are best served not by finding ways to deny it but by articulating conditions that make it genuine.

NOTES

1. Brian Skyrms (1990, p. 130) presents such an argument. Similar arguments have been offered for sequences of other games: "The paradoxical result[s]," have been said to be "due to [assumptions of] complete and perfect information" (Kreps and Wilson 1982, p. 276; cf. Milgrom and Roberts 1982, p. 283).
2. The argument discussed in Hollis and Sugden (1993, pp. 22ff) is similarly defective. This argument would have a certain game that could continue through three decision nodes terminate for rational and informed players at its first node. And yet this argument's implicit rationality and belief premises "tell us nothing . . . about what would happen, were the second and third nodes

reached" (p. 22 – Hollis and Sugden proceed to discuss the argument without saying that it is vitiated by these words).
3. Challenging the backward induction argument, which they observe "holds [for what it is worth] however large n may be," Pettit and Sugden observe that "if n is a large number, it appears that I might do better to follow a strategy such as tit-for-tat" (1989, p. 169). These words suggest that they may think that defection can be reasonable when n is small. How, if they believe in it, could they explain that size sensitivity? They could not say that players in short sequences often approach conditions of perfect rationality and information that support backward inductions to permanent defection. For they think that conditions of perfect rationality and information do not support backward inductions, and their structures have nothing to do with the size of n.
4. "The issue ... is whether this puzzle [of defection round after round in finitely repeated prisoner's dilemmas] can be resolved in the context of rational, self-interested behavior. The approach we adopt is to admit a 'small amount' of the 'right kind' of incomplete information" (Kreps et al. 1982, p. 246). Two illustrations are given. In one, players begin with small probabilities for their opponents' not simply maximizing (which is what they in fact will do) but playing tit-for-tat (p. 247). In the other they have small probabilities for their opponents' having cooperative preferences (which they in fact do not have), for their enjoying "cooperation when it is met by cooperation" (p. 251). These small initial probabilities are shown to be "right kinds" of "incomplete information" (read, "misinformation"). Details of my hyperrational model could suggest other hypotheses.

References

Adams, Ernest W. (1975), *The Logic of Conditionals: An Application of Probability to Deductive Logic.* Dordrecht: Reidel.
Armendt, Brad (1986), "A Foundation for Causal Decision Theory," *Topoi* 5: 3-19.
Bach, Kent (1987), "Newcomb's Problem: The $1,000,000 Solution," *Canadian Journal of Philosophy* 17: 409-25.
Bacharach, Michael (1987), "A Theory of Rational Decision in Games," *Erkenntnis* 27: 127-55.
  (1989), "Mutual Knowledge and Human Reason," paper presented at the Workshop on Knowledge, Belief, and Strategic Interaction (Castiglioncello, Italy).
  (1992), "Backward Induction and Beliefs about Oneself," *Synthese* 91: 247-84.
Bennett, Jonathan (1990), "Why Is Belief Involuntary?" *Analysis* 50: 87-107.
Bernheim, B. Douglas (1984), "Rationalizable Strategic Behavior," *Econometrica* 52: 1007-28.
Bernstein, Allen R., and Frank Wattenburg (1969), "Nonstandard Measure Theory," in W. A. J. Luxemburg (ed.), *Applications of Model Theory to Algebra, Analysis, and Probability.* New York: Holt, Rinehart & Winston, pp. 186-94.
Bicchieri, Cristina (1989), "Backward Induction without Common Knowledge," in *PSA 1988*, vol. 2. East Lansing, MI: Philosophy of Science Association, pp. 329-43.
  (1990), "Paradoxes of Rationality," in P. A. French et al. (eds.), *Midwest Studies in Philosophy*, vol. 15: Philosophy of the Human Sciences. Notre Dame, IN: University of Notre Dame Press, pp. 65-79.
Binmore, Ken (1987), "Modeling Rational Players: Part I," *Economics and Philosophy* 3: 179-213.
Bolker, Ethan (1967), "A Simultaneous Axiomatization of Utility and Subjective Probability," *Philosophy of Science* 34: 330-40.
Bouyssou, Denis, and Jean-Claude Vansnick (1988), "A Note on the Relationship between Utility and Value Functions," in B. Munier (ed.), *Risk, Decision and Rationality.* Dordrecht: Reidel.
Brand, Myles (1984), *Intending and Acting.* Cambridge, MA: MIT Press.
Bratman, Michael E. (1992), "Practical Reasoning and Acceptance in Context," *Mind* 101: 1-15.
Braybrooke, David (1985), "The Insoluble Problem of the Social Contract," in Campbell and Sowden (1985), pp. 277-306. (First published in *Dialogue* 15, 1976, pp. 3-37.)
Campbell, Richmond, and Lanning Sowden, eds. (1985), *Rationality, Cooperation, and Paradox: The Prisoner's Dilemma and Newcomb's Problem.* Vancouver: University of British Columbia Press.
Castañeda, Hector-Neri (1983), "Reply to Bratman," In J. Tomberlin (ed.), *Agent, Language, and the Structure of the World: Essays in Honor of Hector-Neri Castañeda.* Indianapolis: Hackett, pp. 395-409.
Christman, John (1993), "Defending Historical Autonomy: A Reply to Professor Mele," *Canadian Journal of Philosophy* 23: 281-9.

Eells, Ellery (1981), "Causality, Utility, and Decision," *Synthese* 48: 285-329.
  (1984a), "Newcomb's Many Solutions," *Theory and Decision* 16: 59-105.
  (1984b), "Metatickles and the Dynamics of Deliberation," *Theory and Decision* 17: 71-95.
  (1984c), "Causal Decision Theory," in P. D. Asquith and P. Kitcher (eds.), *PSA 1984*, vol. 2. East Lansing, MI: Philosophy of Science Association, pp. 177-200.
  (1985a), "Causality, Decision, and Newcomb's Paradox," in Campbell and Sowden (1985), pp. 183-213.
  (1985b), "Reply to Jackson and Pargetter," in Campbell and Sowden (1985), pp. 219-23.
  (1985c), "Levi's 'The Wrong Box'," *Journal of Philosophy* 82: 91-104.
  (1987), "Learning with Detachment: Reply to Maher," *Theory and Decision* 22: 173-80.
  (1989a), "The Popcorn Problem: Sobel on Evidential Decision Theory and Deliberation-Probability Dynamics," *Synthese* 81: 9-20.
  (1989b), "Comments on Jordan Howard Sobel's 'Non-Dominance and Third-Person Newcomblike Problems'," unpublished paper presented at the APA meetings (Chicago), April 28, 1989.
Eells, Ellery, and Elliott Sober (1986), "Common Causes and Decision Theory," *Philosophy of Science* 53: 223-45.
Ellsberg, Daniel (1954), "Classic and Current Notions of 'Mathematical Utility'," *Economic Journal* 64: 528-56.
Falk, Arthur E. (1985), "Ifs and Newcombs," *Canadian Journal of Philosophy* 15: 449-81.
Farrell, Daniel M. (1989), "Intention, Reason, and Action," *American Philosophical Quarterly* 26: 283-95.
  (1992), "Immoral Intentions," *Ethics* 102: 268-86.
Festinger, L. (1957), *A Theory of Cognitive Dissonance*. Evanston, IL: Row, Peterson (reissued 1962; Stanford University Press).
Fishburn, Peter (1970), *Utility Theory for Decision Making*. New York: Wiley.
  (1973), "A Mixture-set Axiomatization of Conditional Subjective Expected Utility," *Econometrica* 41: 1-25.
  (1974), "On the Foundations of Decision Making under Uncertainty," in M. Balch et al. (eds.), *Essays on Economic Behavior under Uncertainty*. Amsterdam: North-Holland.
  (1981), "Subjective Expected Utility: A Review of Normative Theories," *Theory and Decision* 13: 139-99.
  (1988), "Expected Utility: An Anniversary and a New Era," *Journal of Risk and Uncertainty* 1: 267-83.
Gardner, Martin (1973), "Mathematical Games: Free Will Revisited, with a Mind-bending Paradox by William Newcomb," *Scientific American* (July). (For a report by Robert Nozick on letters to *Scientific American* about this column, see "Mathematical Games," March 1974.)
Gauthier, David (1967), "Morality and Advantage," *Philosophical Review* 76: 460-75.
  (1974), "The Impossibility of Egoism," *Journal of Philosophy* 71: 439-56.
  (1975), "Reason and Maximization," *Canadian Journal of Philosophy* 4: 411-33. (Reprinted in *Moral Dealing*.)
  (1984), "Deterrence, Maximization, and Rationality," *Ethics* 94: 474-95. (Reprinted in *Moral Dealing*.)
  (1985a), "Maximization Constrained: The Rationality of Cooperation," in Campbell and Sowden (1985), pp. 75-93.
  (1985b), "The Unity of Reason: A Subversive Reinterpretation of Kant," *Ethics* 96: 74-88. (Reprinted in *Moral Dealing*.)
  (1986), *Morals by Agreement*. Oxford: Clarendon Press.
  (1990), *Moral Dealing: Contract, Ethics, and Reason*. Ithaca, NY: Cornell University Press.

Gibbard, Allan (1971), "Utilitarianism and Coordination," doctoral dissertation, Harvard University.
  (1981), "Two Recent Theories of Conditionals," in W. L. Harper et al. (eds.), *Ifs: Conditionals, Belief, Decision, Chance, and Time*. Dordrecht: Reidel, pp. 211-47.
  (1990), *Wise Choices, Apt Feelings*. Cambridge, MA: Harvard University Press.
Gibbard, Allan, and William L. Harper (1978), "Two Kinds of Expected Utility," in Hooker et al. (1978), pp. 125-62. (Reprinted with abridgement in Campbell and Sowden 1985, pp. 133-58.
Gilbert, Margaret (1989), "Rationality and Salience," *Philosophical Studies* 57: 61-78.
Ginet, Carl (1962), "Can Will Be Caused?" *Philosophical Review* 71: 49-55.
Hammond, Peter J. (1988), "Consequentialist Foundations for Expected Utility," *Theory and Decision* 25: 25-78.
Harper, William (1985), "Ratifiability and Causal Decision Theory: Comments on Eells and Seidenfeld," in P. D. Asquith and P. Kitcher (eds.), *PSA 1984*, vol. 2. East Lansing, MI: Philosophy of Science Association, pp. 213-28.
  (1986), "Mixed Strategies and Ratifiability in Causal Decision Theory," *Erkenntnis* 24: 26-36.
  (1988), "Causal Decision Theory and Game Theory: A Classic Argument for Equilibria Solutions, A Defense of Weak Equilibria, and a New Problem for the Normal Form Representation," in W. L. Harper and B. Skyrms (eds.), *Causation in Decision, Belief Change, and Statistics*. Dordrecht: Kluwer, pp. 25-48.
Hart, H. L. A. (1961), *The Concept of Law*. Oxford: Clarendon Press.
Herzberger, Hans (1978), "Coordination Theory," in Hooker et al. (1978), pp. 125-62.
Hobbes, Thomas (1950), *Leviathan*. New York: E. P. Dutton.
Hodgson, D. H. (1967), *Consequences of Utilitarianism*. Oxford: Clarendon Press.
Hollis, Martin, and Robert Sugden (1993), "Rationality in Action," *Mind* 102: 1-35.
Hooker, C. A., J. J. Leach, and E. F. McClennen, eds. (1978), *Foundations and Applications of Decision Theory*. Dordrecht: Reidel.
Horgan, Terence (1985a), "Counterfactuals and Newcomb's Problem," in Campbell and Sowden (1985), pp. 159-82. (Reprinted from *Journal of Philosophy* 78, 1981, pp. 331-56.)
  (1985b), "Newcomb's Problem: A Stalemate," in Campbell and Sowden (1985), pp. 223-34.
Horwich, Paul (1985), "Decision Theory in Light of Newcomb's Problem," *Philosophy of Science* 52: 431-50.
Hubin, Don, and Glenn Ross (1985), "Newcomb's Perfect Predictor," *Noûs* 19: 439-46.
Hume, David (1888), *A Treatise of Human Nature*. Oxford: Clarendon Press.
Hurley, Susan L. (1989), *Natural Reasons: Personality and Polity*. New York: Oxford University Press.
  (1991), "Newcomb's Problem, Prisoners' Dilemma, and Collective Action," *Synthese* 86: 173-96.
Jackson, Frank (1987), *Conditionals*. Oxford: Basil Blackwell.
Jackson, Frank, and Robert Pargetter (1983), "Where the Tickle Defense Goes Wrong," *Australasian Journal of Philosophy* 61: 295-99. (Reprinted with a postscript on metatickles in Campbell and Sowden 1985, pp. 214-19.)
Jacobi, N. (1993), "Newcomb's Paradox: A Realist Resolution," *Theory and Decision* 35: 1-17.
Jarrow, Robert (1987), "An Integrated Axiomatic Approach to the Existence of Ordinal and Cardinal Utility Functions," *Theory and Decision* 22: 99-110.
Jeffrey, Richard C. (1965a), "New Foundations for Bayesian Decision Theory," in Y. Bar-Hillel (ed.), *Logic, Methodology and Philosophy of Science: Proceedings of the 1964 International Congress*. Amsterdam: North-Holland, pp. 289-300.
  (1965b), *The Logic of Decision*. New York: McGraw-Hill.

(1974), "Frameworks for Preference," in M. Balch et al. (eds.), *Essays on Economic Behavior under Uncertainty.* New York: Elsevier.

(1981a), "Choice, Chance, and Credence," in G. H. von Wright and G. Floistad (eds.), *Contemporary Philosophy: A New Survey,* vol. 1: Philosophy of Language and Philosophical Logic. The Hague: Nijhoff, pp. 367-86.

(1981b), "The Logic of Decision Defended," *Synthese* 48: 473-92.

(1982), "The Sure Thing Principle," in *PSA 1982,* vol. 2. East Lansing, MI: Philosophy of Science Association.

(1983), *The Logic of Decision,* 2nd ed. Chicago: University of Chicago Press.

(1986), "Judgmental Probability and Objective Chance," *Erkenntnis* 24: 5-16.

(1987), "Risk and Human Rationality," *The Monist* 70: 223-36.

(1988a), "Biting the Bayesian Bullet: Zeckhauser's Problem," *Theory and Decision* 25: 117-22.

(1988b), "How to Probabilize a Newcomb Problem," in James H. Fetzer (ed.), *Probability and Causality.* Dordrecht: Reidel.

(1992), "Newcomb Kinematics," unpublished manuscript, November 2.

Kahneman, David, and Amos Tversky (1979), "Prospect Theory: An Analysis of Decision under Risk," *Econometrica* 47: 263-91.

Kaplan, Mark (1989), "Bayesianism without the Black Box," *Philosophy of Science* 56: 48-69.

Krantz, David, R. Duncan Luce, Patrick Suppes, and Amos Tversky (1971), *Foundations of Measurement,* vol. 1: Additive and Polynomial Representations. New York: Academic Press.

Kreps, D. M., P. Milgrom, J. Roberts, and R. Wilson (1982), "Rational Cooperation in the Finitely Repeated Prisoners' Dilemma," *Journal of Economic Theory* 27: 245-52.

Kreps, D. M., and R. Wilson (1982), "Reputation and Imperfect Information," *Journal of Economic Theory* 27: 253-79.

Kyburg, Henry E. (1980), "Acts and Conditional Probabilities," *Theory and Decision* 12: 149-71.

Levi, Isaac (1975), "Newcomb's Many Problems," *Theory and Decision* 6: 161-75.

(1982), "A Note on Newcombmania," *Journal of Philosophy* 79: 337-82.

(1983), "The Wrong Box," *Journal of Philosophy* 80: 534-42.

Lewis, David (1969), *Convention: A Philosophical Study.* Cambridge, MA: Harvard University Press.

(1972), "Utilitarianism and Truthfulness," *Australasian Journal of Philosophy* 50: 17-19.

(1973), *Counterfactuals.* Cambridge, MA: Harvard University Press.

(1976), "Probabilities of Conditionals and Conditional Probabilities," *Philosophical Review* 85: 297-315.

(1979a), "Prisoners' Dilemma Is a Newcomb Problem," *Philosophy and Public Affairs* 8: 251-5. (Reprinted in Campbell and Sowden 1985, pp. 251-4.)

(1979b), "Counterfactual Dependence and Time's Arrow," *Noûs* 13: 455-76.

(1981a), "Causal Decision Theory," *Australasian Journal of Philosophy* 59: 5-30.

(1981b), "Why Ain'cha Rich?" *Noûs* 15: 377-80.

(1986), "Probabilities of Conditionals and Conditional Probabilities II," *Philosophical Review* 85: 297-315.

Lucas, J. R. (1966), *The Principles of Politics.* Oxford: Clarendon Press.

Luce, R. Duncan, and Howard Raiffa (1957), *Games and Decisions: Introduction and Critical Survey.* New York: Wiley.

McClennen, Edward F. (1972), "An Incompleteness Problem in Harsanyi's General Theory of Games and Certain Related Theories of Non-cooperative Games," *Theory and Decision* 2: 314-41.

(1978), "The Minimax Theory and Expected Utility Reasoning," in Hooker et al. (1978), pp. 337-67.

(1983), "Sure-Thing Doubts," in B. Stigum and F. Wenstop (eds.), *Foundations of Utility and Risk Theory with Applications.* Dordrecht: Reidel.

(1990), *Rationality and Dynamic Choice: Foundational Explorations.* Cambridge: Cambridge University Press.

MacIntosh, Duncan (1993), "Persons and the Satisfaction of Preferences: Problems in the Kinematics of Values," *Journal of Philosophy* 110: 163-80.

Mackie, J. L. (1973), "The Disutility of Act-Utilitarianism," *Philosophical Quarterly* 23: 289-300.

(1977), "Newcomb's Paradox and the Direction of Causation," *Canadian Journal of Philosophy* 7: 213-25.

Mele, Alfred R. (1989), "Intention, Belief, and Intentional Action," *American Philosophical Quarterly* 26: 19-30.

(1992), "Intending for Reasons," *Mind* 101: 327-33.

(1993a), "Justifying Intentions," *Mind* 102: 335-7.

(1993b), "History and Personal Autonomy," *Canadian Journal of Philosophy* 23: 271-80.

Milgrom, P., and J. Roberts (1982), "Predation, Reputation, and Entry Deterrence," *Journal of Economic Theory* 27: 280-312.

Montague, Richard, and David Kaplan (1974), "A Paradox Regained," in R. H. Thomason (ed.), *Formal Philosophy: Selected Papers of Richard Montague.* New Haven, CT: Yale University Press. (First published in *Notre Dame Journal of Formal Logic* 1, 1960, pp. 79-80.)

Moore, Omar K., and Alan R. Anderson (1962), "Some Puzzling Aspects of Social Interaction," *Review of Metaphysics* 15: 409-33.

Nietzsche, Friedrich (1956), *The Birth of Tragedy and The Genealogy of Morals,* trans. F. Golfing. New York: Doubleday.

Nozick, Robert (1963), "The Normative Theory of Individual Choice," doctoral dissertation, Princeton University.

(1969), "Newcomb's Problem and Two Principles of Choice," in N. Rescher et al. (eds.), *Essays in Honor of Carl G. Hempel.* Dordrecht: Reidel, pp. 114-46. (Reprinted in Campbell and Sowden 1985, pp. 107-33.)

Parfit, Derek (1984), *Reasons and Persons.* Oxford: Clarendon Press.

Pearce, David P. (1984), "Rationalizable Strategic Behavior and the Problem of Perfection," *Econometrica* 52: 1029-50.

Pettit, Philip, and Robert Sugden (1989), "The Backward Induction Paradox," *Journal of Philosophy* 86: 169-82.

Pink, T. L. M. (1991), "Purposive Intending," *Mind* 100: 343-59.

(1993), "Justification of the Will," *Mind* 102: 329-434.

Pollock, John L. (1983), "How Do You Maximize Expectation Value?" *Noûs* 17: 409-21.

Price, Huw (1986), "Against Causal Decision Theory," *Synthese* 67: 195-212.

Rabinowicz, Włodzimierz (1982), "Two Causal Decision Theories: Lewis vs Sobel," in T. Pauli (ed.), *Philosophical Essays Dedicated to Lennart Åqvist on His Fiftieth Birthday.* Uppsala, Sweden: Philosophical Society and Department of Philosophy.

(1985), "Ratificationism without Ratification: Jeffrey Meets Savage," *Theory and Decision* 19: 171-200.

(1986), "Non-cooperative Games for Expected Utility Maximizers," in P. Needham and J. Odelstad (eds.), *Changing Positions: Essays Dedicated to Lars Lindahl on the Occasion of His Fiftieth Birthday.* Uppsala, Sweden: Philosophical Society and Department of Philosophy, pp. 215-33.

(1992), "Tortuous Labyrinth: Noncooperative Normal-Form Games between Hyperrational Players," in C. Bicchieri and M. L. Dalla Chiara (eds.), *Knowledge, Belief, and Strategic Interaction.* Cambridge: Cambridge University Press.

Raiffa, Howard (1968), *Decision Analysis: Introductory Lectures on Choice under Uncertanity.* Reading, MA: Addison-Wesley.

Rapoport, Anatol (1966), *Two-Person Game Theory: The Essential Ideas*. Ann Arbor: University of Michigan Press.
Reny, Philip J. (1989), "Common Knowledge and Games with Perfect Information," in *PSA 1988*, vol. 2. East Lansing, MI: Philosophy of Science Association, pp. 363-9.
Roth, Alvin E. (1979), *Axiomatic Models of Bargaining*. Berlin: Springer.
Sartre, Jean-Paul (1956), *Being and Nothingness*, trans. H. E. Barnes. New York: Philosophical Library.
Savage, Leonard J. (1972), *The Foundations of Statistics*, 2nd rev. ed. New York: Dover. (First published in 1954.)
Seidenfeld, Teddy (1985), "Comments on Causal Decision Theory," in P. D. Asquith and P. Kitcher (eds.), *PSA 1984*, vol. 2. East Lansing, MI: Philosophy of Science Association.
Selten, Reinhard (1978), "The Chain-store Paradox," *Theory and Decision* 9: 127-59.
Selten, Reinhard, and Ulrike Leopold (1982), "Subjunctive Conditionals in Decision and Game Theory," in W. Stegmuller et al. (eds.), *Studies in Contemporary Economics*, vol. 2: Philosophy and Economics. Berlin: Springer, pp. 191-200.
Skyrms, Brian (1982), "Causal Decision Theory," *Journal of Philosophy* 79: 695-711.
———(1985), "Ultimate and Proximate Consequences in Causal Decision Theory," *Philosophy of Science* 52: 608-11.
———(1986), "Deliberational Equilibria," *Topoi* 5: 59-67.
———(1990), *The Dynamics of Rational Deliberation*. Cambridge, MA: Harvard University Press.
Sobel, Jordan Howard (1970), "Utilitarianisms: Simple and General," *Inquiry* 13: 394-449.
———(1971), "Values, Alternatives and Utilitarianism," *Noûs* 5: 373-84.
———(1972), "The Need for Coercion," in J. R. Pennock and J. W. Chapman (eds.), *Coercion: Nomos XIV*. Chicago and New York: Aldine & Atherton, pp. 148-77. [Chapter 13 of this volume]
———(1975), "Interaction Problems for Utility Maximizers," *Canadian Journal of Philosophy* 4: 677-88.
———(1976), "Utility Maximizers in Iterated Prisoner's Dilemmas," *Dialogue* 15: 38-53. (Reprinted in Campbell and Sowden 1985, pp. 306-19.) [Chapter 15 of this volume]
———(1978), "Probability, Chance and Choice: A Theory of Rational Agency," unpublished manuscript.
———(1980), "*The Emergence of Norms* by Edna Ullmann-Margalit," *Philosophical Books* 21: 124-7.
———(1982), "Utilitarian Principles for Imperfect Agents," *Theoria* 48: 113-26.
———(1983), "Expected Utilities and Rational Actions and Choices," *Theoria* 49: 159-83. [Chapter 10 of this volume]
———(1985a), "Circumstances and Dominance in Causal Decision Theory," *Synthese* 63: 167-202.
———(1985b), "Not Every Prisoner's Dilemma Is a Newcomb Problem," in Campbell and Sowden (1985), pp. 263-74. [Chapter 3 of this volume]
———(1985c), "Predicted Choices," *Dalhousie Review* 64: 600-7.
———(1985d), "Everyone's Conforming to a Rule," *Philosophical Studies* 48: 375-87.
———(1986a), "Notes on Decision Theory: Old Wine in New Bottles," *Australasian Journal of Philosophy* 64: 407-37. [Chapter 8 of this volume]
———(1986b), "Metatickles and Ratificationism," in A. Fine and P. Machamer (eds.), *PSA 1986*, vol. 1. East Lansing, MI: Philosophy of Science Association, pp. 342-51.
———(1987), "Self-doubts and Dutch Strategies," *Australasian Journal of Philosophy* 65: 56-81.
———(1988a), "Defenses and Conservative Revisions of Evidential Decision Theories: Metatickles and Ratificationism," *Synthese* 75: 107-31.

(1988b), "Infallible Predictors," *Philosophical Review* 97: 3-24. [Chapter 5 of this volume]
(1988c), "World Bayesianism: Comments on the Hammond/McClennen Debate," in B. R. Munier (ed.), *Risk, Decision and Rationality*. Dordrecht: Reidel, pp. 527-42.
(1988d), "Newcomb's Problem and Two Forms of Decision Theory" (appendix to "Metatickles, Ratificationism, and Newcomb-like Problems without Dominance"), in B. R. Munier (ed.), *Risk, Decision and Rationality*. Dordrecht: Reidel, pp. 438-501.
(1988e), "Maximizing, Optimizing, and Prospering," *Dialogue* 27: 233-62. [Part Two is Chapter 7 of this volume]
(1989a), "Partition-Theorems for Causal Decision Theories," *Philosophy of Science* 56: 71-93. [Chapter 9 of this volume]
(1989b), "Machina and Raiffa on the Independence Axiom," *Philosophical Studies* 56: 315-29.
(1989c), "One Mixing Event or Two?" *Journal of Behavioral Decision Making* 2: 197-8.
(1989d), "Utility Theory and the Bayesian Paradigm," *Theory and Decision* 26: 263-93. [Chapter 1 of this volume]
(1989e), "Kent Bach on Good Arguments," *Canadian Journal of Philosophy* 20: 447-53. [Chapter 6 of this volume]
(1990a), "Maximization, Stability of Decision, and Actions in Accordance with Reason," *Philosophy of Science* 57: 60-77. [Chapter 11 of this volume]
(1990b), "Newcomb Problems," in P. A. French et al. (eds.), *Midwest Studies in Philosophy*, vol. 15: Philosophy of the Human Sciences. Notre Dame, IN: University of Notre Dame Press, pp. 224-55. [Chapter 2 of this volume]
(1991a), "Non-dominance, Third Person and Non-action Newcomb Problems, and Metatickles," *Synthese* 86: 143-72.
(1991b), "Constrained Maximization," *Canadian Journal of Philosphy* 20: 25-51.
(1991c), "Some Versions of Newcomb's Problem Are Prisoners' Dilemmas," *Synthese* 86: 197-208. [Chapter 4 of this volume]
(1992), "Hyperrational Games: Concept and Resolutions," in C. Bicchieri and M. L. Dalla Chiara (eds.), *Knowledge, Belief, and Strategic Interaction*. Cambridge: Cambridge University Press, pp. 61-91. [Included in Chapter 14 of this volume]
(1993a), "Backward Induction Arguments: A Paradox Regained," *Philosophy of Science* 60: 114-33. [Chapter 16 of this volume]
(1993b), "Straight vs Constrained Maximization," *Canadian Journal of Philosophy* 23: 25-54.
(in press), *Puzzles for the Will*. Toronto Studies in Philosophy, University of Toronto Press.
Sorensen, Roy (1986), "Blindspotting and Choice Variations of the Prediction Paradox," *American Philosophical Quarterly* 23: 337-52.
Stalnaker, Robert (1981), "Letter to David Lewis: May 21, 1972," in W. L. Harper et al. (eds.), *Ifs: Conditionals, Belief, Decision, Chance, and Time*. Dordrecht: Reidel.
Swain, Corliss G. (1988), "Cutting a Gordian Knot: The Solution to Newcomb's Problem," *Philosophical Studies* 53: 391-409.
Talbott, William J. (1987), "Standard and Non-standard Newcomb Problems," *Synthese* 70: 415-58.
Taylor, Richard (1974), *Metaphysics*, 2nd ed. Englewood Cliffs, NJ: Prentice-Hall.
Tullock, Gordon (1967), "The Prisoner's Dilemma and Mutual Trust," *Ethics* 78: 229-30.
Ullmann-Margalit, Edna (1978), *The Emergence of Norms*. Oxford: Clarendon Press.
von Neumann, John, and Oskar Morgenstern (1953), *Theory of Games and Economic Behaviour*. Princeton, NJ: Princeton University Press.
Wakker, Peter (1988), "Nonexpected Utility as Aversion of Information," *Journal of Behavioral Decision Making* 1: 169-75.

Weirich, Paul (1980), "Conditional Utility and Its Place in Decision Theory," *Journal of Philosophy* 77: 702-15.
——— (1983), "A Decision Maker's Options," *Philosophical Studies* 44: 175-86.
——— (1985), "Decision Instability," *Australasian Journal of Philosophy* 63: 465-72.
——— (1986a), "Expected Utility and Risk," *British Journal for the Philosophy of Science* 37: 419-42.
——— (1986b), "Decisions in Dynamic Settings," in A. Fine and P. Machamer (eds.), *PSA 1986*, vol. 1. East Lansing, MI: Philosophy of Science Association, pp. 438-49.
——— (1988), "Hierarchical Maximization of Two Kinds of Expected Utility," *Philosophy of Science* 55: 560-82.
——— (1991), "The Hypothesis of Nash Equilibrium and Its Bayesian Justification," paper presented at the Ninth International Congress of Logic, Methodology and Philosophy of Science (Uppsala, Sweden).

Index of names

Abela, Paul, 119
Adams, Ernest W., 116
Allais, Maurice, 18
Anderson, Alan R., 304
Armendt, Brad, 165, 175, 187, 189-90
Arnauld, Antoine, 172

Bach, Kent, 119-20, 124-5
Bacharach, Michael, 250, 291, 304, 312, 328
Bennett, Jonathan, 253
Bernheim, B. Douglas, 291, 326, 329, 363
Bernoulli, Daniel, 18
Bernstein, Allen R., 281
Bicchieri, Cristina, 355-6
Binmore, Ken, 357, 361-3
Bolker, Ethan, 165, 186, 253
Bouyssou, Denis, 25
Brand, Myles, 239
Bratman, Michael E., 239, 247, 253
Braybrook, David, 324, 330-2, 339-40, 342

Campbell, Richmond, 43, 89-90, 92, 108, 119, 321, 349, 358
Cartwright, Richard, x
Castañeda, Hector-Neri, 239
Christman, John, 248

Eells, Ellery, ix, 31, 39, 41, 48-54, 56, 58, 74-5, 172, 173, 225, 227-9
Ellsberg, Daniel, 26

Falk, Arthur E., 116, 118, 131
Falk, W. D., x, 341
Farrell, Daniel M., 237-8, 241-4, 251
Festinger, L., 251
Fishburn, Peter, 18, 27-8, 165, 175, 185, 186, 187, 253
Freeman-Sobel, Willa, x, 150, 237, 283, 341, 345

Gauthier, David, x, 13, 87-8, 99, 122-30, 133-7, 244, 248, 250-1, 253, 276, 324
Gettier, Edmund, 117
Gibbard, Allan, 43, 59, 62, 77, 92, 108, 116, 151, 153, 163, 191, 198, 202, 242, 342, 345
Gilbert, Margaret, 320
Gillespie, Dizzy, 329
Ginet, Carl, 240

Hammond, Peter, 27
Harper, William L., 43, 59, 62, 77, 92, 108, 151, 153, 163, 172, 191, 198, 202, 224-7, 229, 233, 293, 328
Hart, H. L. A., 276
Herzberger, Hans, 316-17
Hobbes, Thomas, 341
Hodgson, D. H., 122, 281, 333, 342
Hollis, Martin, 253, 364
Horgan, Terrence, 108, 110, 115-18
Horwich, Paul, 49-51, 195
Hubin, Don, 116, 360
Hume, David, 124, 276, 280
Hurley, Susan, 96, 99

Jackson, Frank, 56, 346, 350, 359
Jacobi, N., 103
Jarrow, Robert, 5
Jeffrey, Richard C., ix, 5, 11, 13, 15, 18-22, 24, 27, 31-2, 36, 38, 41, 43, 54-5, 58-61, 69-71, 73-6, 80, 87-8, 90-2, 101-2, 114, 117, 120, 141, 143-5, 150, 154, 157, 160, 172, 175, 180, 193, 198, 202, 207, 216, 220, 228, 232, 233, 253, 264

Kahneman, David, 5, 8
Kaplan, David, 364
Kaplan, Mark, 328
Krantz, David, 12, 18
Kreps, D. M., 364-5
Kyburg, Henry E., 38-9

Leopold, Ulrike, 97
Levi, Isaac, 88, 164, 174
Lewis, David, 22, 35, 37, 43, 50, 54, 77, 86–7, 90, 102, 118, 154, 159, 172, 176, 189, 222, 241, 280, 308, 319–20, 326, 328, 343
Lucas, J. R., 282
Luce, R. Duncan, ix, 4, 9, 11–13, 18, 89, 280, 286, 333, 335, 337, 349, 364

McClennen, Edward F., 22, 248, 251, 319, 324
MacIntosh, Duncan, 245–6, 251, 254
Mackie, J. L., 42, 343
Mele, Alfred, 239, 242, 245, 248, 253
Montague, Richard, 364
Moore, Omar K., 304
Morgenstern, Oskar, 18, 294

Newcomb, William, 89
Nietzsche, Friedrich, 248
Nozick, Robert, 34, 60, 87, 89, 91, 112–15

Parfit, Derek, 138
Pargetter, Robert, 56
Pearce, David P., 296, 326, 329
Pettit, Philip, 345–7, 350, 352–3, 356, 359–63, 365
Pink, T. L. M., 243–5, 251
Plato, 135
Pollock, John L., 217
Price, Huw, 39–41, 63, 75

Rabinowicz, Włodzimierz, x, 8, 76, 154, 161, 172, 216, 227, 229–30, 232, 283, 288, 290, 293–4, 296, 298–9, 301, 310, 318–19, 322–3, 329, 345
Raiffa, Howard, ix, 4, 9, 11–13, 21, 89, 280, 286, 333, 335, 337, 349, 364
Ramsey, Frank P., 18
Rapoport, Anatol, 264

Regan, Donald, 99
Reny, Philip J., 356
Roach, Max, 329
Ross, Glenn, 116, 360
Roth, Alvin, 12–13

Salmon, Merrilee, 345
Sartre, Jean-Paul, 217, 250, 252
Savage, Leonard, ix, 22–3, 26, 253
Seidenfeld, Teddy, 103, 113, 115
Selten, Reinhard, 97, 250
Skyrms, Brian, 53, 62–3, 74, 87, 130–1, 163, 225, 323, 351, 364
Sorenson, Roy, 349, 355, 357
Sowden, Lanning, 43, 89, 90, 108, 321, 349, 358
Stalnaker, Robert, 74, 153, 176, 191
Sugden, Robert, 253, 345–7, 350, 352–3, 356, 359–65
Swain, Corliss G., 124

Talbott, William J., 42, 45, 47, 75, 97
Taylor, Richard, 117
Tucker, A. W., 89
Tullock, Gordon, 276
Tversky, Amos, 5, 8

Ullmann-Margalit, Edna, 324

van Frassen, Bas, 70–1, 73
Vansnick, Jean-Claude, 25
von Neumann, John, 18, 294

Wakker, Peter, 27
Wattenberg, Frank, 281
Weirich, Paul, 14, 22, 27, 142, 144, 190–1, 222–5, 225–7, 229, 231–3, 304
White, E. B., 125
Wilson, R., 364

Zeckhauser, Richard, 5